WHEN WE WERE HUMAN

Brain injury in an age of electronic memory

by

Sheila Newman
with
James Sinnamon

Countershock Press

BY THE SAME AUTHOR

Sheila Newman, *Demography, Territory and Law 2: Land Tenure and the Origins of Democracy in Britain, A new theory,* Print Edition, Countershock Press, Australia, January 2013.

Sheila Newman, *Demography, Territory and Law: Rules of Animal and Human Populations,* Print Edition, Countershock Press, Australia, January 2013.

Sheila Newman, *Demography, Territory and Law: Rules of Animal and Human Populations,* Kindle Edition, December 2012.

Sheila Newman, *The Urge to Disperse,* Candobetter Press, 2011
.

Sheila Newman, Ed., *The Final Energy Crisis,* 2nd Edition, Pluto Press, UK, 2008
.

Andrew McKillop with Sheila Newman, Eds., *The Final Energy Crisis,* Pluto
Press, UK, 2005.

Sheila Newman, *The Growth Lobby and its Absence in Australia and France,*
Environmental Sociology Research thesis, Swinburne University, 2002.

Author's website: https://candobetter.net/SheilaNewman
Email: countershockpress@gmail.com
sheilanewmancorrespondence@gmail.com

WHEN WE WERE HUMAN
Brain injury in an age of electronic memory

Cover design by Sheila Newman.

Print editions: 18 July 2016, 4 January 2017
ISBN 978-1-329-95962-0

Countershock Press
https://countershock.wix.com/countershockpress
P0 Box 5180 Frankston South,
VIC Australia 3199
countershockpress@gmail.com
sheilanewmancorrespondence@gmail.com
Tel: +61 0412319669

THANKS

James Sinnamon wishes to thank his father, Ian Sinnamon, for all the kindness and support he has shown, particularly since 2004. James thanks his mother for bringing him into the world and for her love and understanding. He is grateful to all his siblings and the wider extended family for their support and affection. Some are mentioned specifically in the book.

Sheila Newman and James thank Sheila's parents, Ruth and John, for their kindness towards him in his recovery and for taking care of the dogs while we were away. We would also like to thank Leanne Harris, Jill Quirk, and Sandra Garnier, Relationship Counsellor at Orwil Street Community Clinic, Frankston, for their interest and reading of drafts. We would especially like to thank Greg Wood, not only for his gift of enthusiasm, genius and laughter, but for being such a great support to us in Brisbane, because we somehow left out of the book his extraordinary input of days of hard labour on crucial home repairs for nothing whilst doing full-time shiftwork and looking after his family. We also cannot forget Ha Do's generosity in moving furniture, lending tools and transporting building materials. Rob Wilson, architect, has also, on James's behalf, done inspections, drawn and redrawn complex plans and liaised for many hours with other professionals for no financial reward. James's father has, of course, given much time to these matters as well. Thanks also to Ilan Goldman for helping James with computer building and with some legal ideas about bicycles.

Obviously many other people have been helpful to James's recovery. Most are acknowledged in the book, but some will have been left out inadvertently, due to editing down the length and more or less arbitrary decisions of what to include and what to leave out.

NOTES TO THE READER

Spelling is Australian style. Generally the names of institutions have been kept in this book but the names of most persons have been substituted for with an initial, a function, or fictionalized. In part this has been to avoid the need to contact every person mentioned and supply them with a copy of their part in the book. Where we contacted people and they did not reply, we have also changed their names, and this includes relatives. In some cases it has been due to legal circumstances beyond our control, but not associated with defamation laws. We have also removed or disguised names sometimes to avoid people being targeted, for instance, particular hospital employees. Some real names remain where we don't think that the person could be offended and believe they might appreciate having their kindness described. Then there were other names we would have liked to have included, for the kindnesses of their bearers, but did not, since we failed to conserve a record or never learned them.

Members of James's family were invited to contribute to this book or review passages mentioning them, but none did. Much more could be written about James's family's interaction with him before, during and after brain injury, but this would have made the book, already quite long, impossibly long, and there was always the danger of not doing justice to peoples' kindness or of misjudging their actions. In some cases this book attempts to explain events that subsequently gave rise to confusion and resentment due to their not having been explained at the time.

TABLE OF CONTENTS

Introduction:

Coming back from brain injury in an age of electronic memory is about trying to recover memory, function and personality after diffuse axonal brain injury in a political activist and history wiz with a computer science background. It concerns the irony of having years of emails and other electronic records preserving a past that links impossibly to the future like an Escher staircase. It is about his and my efforts to preserve the interactive political and environmental website we built together in order to try to change political trends we hated. It is about clawing a dear person's identity back from a potentially bottomless pit and building a new relationship over a period of almost six years.

This record of a partial recovery from a very serious brain injury is not the first of its kind, but it is different because it is written by a mental health professional and research sociologist in an era when the injured person has access to their former life, memories, plans, hopes and personality preserved in electronic detail as years of daily emails, articles, research, interviews and films. How does one recover lost memories and still find time to make new ones? Is it possible?

When my intimate friend and political associate, James Sinnamon, was severely injured in May 2010, I had a lot of questions. Although there were books and papers about brain injury, some important questions were left unanswered.

Those questions were:

What can delay a person's stay in a neurosurgical unit?

How do patients behave as diffuse axonal injury takes hold in the first couple of weeks?

What are the risks and uses of certain psychiatric medications in acute brain injury?

What brain imaging tests is it reasonable to demand from a hospital?

How can you help someone while they are still in hospital? What are their fears?

What kind of inner life and perception does a person in acute Post Traumatic Amnesia have?

How important is it to visit every day and what else might you need to reserve your energy, time and money for?

Ordinary neurorehabilitation does not deal with very high level technical skills in gifted persons; how do you test and restimulate them?

How do you work out who is at fault in arguments with a brain-injured person? Is there an independent factor?

Can changing the environment around a recently brain-injured person help them settle?

How does slower mental processing affect personal interaction?

Are brain-injured people really 'stupid' or is intelligence preserved but hard to access or demonstrate?

How I felt about changes in James. How he felt about them.

Is there some hope that higher functions, such as post graduate specialist technical and scientific skills, might take longer to return than two years?

In the first year or so, I read as many biographical books as I could then find on brain injury – totalling about six – for all of which I am immensely grateful. I only found one book, however, where I could compare the attempted return of a high-functioning brain-injured person to a specialised area of work. That was Cathy Crimmins's *Where is the Mango Princess?* [1] Crimmins's husband was a lawyer, but he returned to work with the help of a specialised neuropsychological work-coach. James's insurance did not provide for this because, although he was working several hours a day running a political website with a sophisticated Drupal content management system he had implemented when the accident happened, he was actually earning his living as a cleaner, due to termination of a research position in a university.

This book proposes, as well as a biography of a person's struggles with their brain injury, to look at how a brain injury impacted on a computer scientist's ability to design programs and write code and make professional decisions at various stages of recovery.

Whilst all brain injuries are different and so all recoveries are different, it gives the reader with hopes or an interest in this area a place to start. Since there is little or no literature available on the preservation of sophisticated information technology skills in a diffuse brain injury, it is a contribution to a new field. For this reason, in this book, I have preserved a certain amount of technical detail in the sections about computers. It is possible to read this book, however, without any technical knowledge of computers but you will still gain an awareness of how important computers and the internet may be for a brain-injured person.

For the information technology industry itself there is a vast potential area to develop that boils down to more than 'brain that changes itself' game-form basic brain exercises. Potentially open-source programs could be developed to test and re-teach mathematics, science and computer languages and application building skills, such as Drupal. Or even to build virtual and or physical robots to complement partially retained skills.

People who care about a brain injured person will also recognize how problematic the practice is of estimating a person's worth almost exclusively based on their earning capacity at the time of the injury. In an era of disemployment,[2] this is an inadequate test. Worse though is that our value is then almost entirely reduced to what we do to earn a living, potentially leaving all those other precious parts of us unrecognized and unrecovered. This must increase the widespread depression among brain-injured people who so need to be valued for themselves. Personal loss and suffering is actually a very small part of most compensation systems.

The book also records how important it is for independent adults with memory loss to be reunited with friends and to retrieve personal connections with people who know them as they were most recently. Parents and other relatives may be vital for survival but may in fact know very little about their children's former adult and independent lives.

Letter written in a semi-coma

19 June 2010 4.50 a.m.

Dear Friend,

I am now living in what appears to be a common household in which various individuals have authority over my life, but I don't understand the basis of that authority.

I also don't understand who pays for my food, who provides my food, and who pays for my other necessities of life.

Before I entered this communal project, perhaps five weeks ago, I thought I had a very good idea how I fitted into the world.

I thought I understood:

Who I was
What I did for a living
Where I lived
Who were my friends
Who were my trusted associates

Now, I am unable to confidently answer any of these questions.

Therefore I would like to talk with people I believe I know, who are associated with this project, who can give me some straightforward answers to these and other questions.

People I believe I know, who are associated with this project and who can answer my questions include:

1. My Friend Sheila Newman. Sheila has been contacted by (1) contacting her phone number (07) 33690819 which can be reached by dialing (0) 33690819 and then connecting her to my number 62899.

I need to talk to Sheila, Judith, [my sister] or perhaps others in order to understand

What I am doing here
Under whose authority I am here and go about my daily life

Why I cannot/shouldn't leave in order to resume my life elsewhere.
Thank you

[Signed] James Sinnamon

([Crossed out was:] Please arrange to have someone come and talk with me so that I can understand what I am doing here and why I should not just walk away as I would have thought any citizen could.

PART ONE

What the …?

"My name is James and I am in hospital ….

I am hospital because I was involved in a collision between my bicycle and a car on the 18th of May 2010.

I was riding my bicycle to work. I work at Royal Brisbane and Women's Hospital as a cleaner and I am now on Workers Compensation, receiving 85% of my weekly salary. So my income is guaranteed for months until I get well.

I also run a big internet site, candobetter.org and am a Linux computer language program writer…

This book has been designed by Sheila and James to help James remember."

(From ***James Sinnamon's Orientation Book for his hospital stay,*** by James Sinnamon & Sheila Newman, 19 June 2010).

So began the booklet which appeared mysteriously at James Sinnamon's hospital bedside one day in the neurosurgical unit and where he would repeatedly find some answers to the bizarre mystery that had engulfed him. He would consult it again and again, many times a day, for a long time, finding the same answers to the same questions and re-experiencing the same relief at finding them, always as if it were for the first time.

Intensive Care Unit 18 May 2010

ICU: "You are in hospital, you're being cared for, and everyone knows where you are..."

(Nurse's advice on what to say to the patient.)

A phone call - 18 May 2010

It seems to me that it was early evening when I answered the phone to hear James's sister, Barbara, who lives in Tasmania. "James is in hospital with a head injury. The hospital can't give us much information at the moment, but he is in the intensive care unit. He was in an accident when he was cycling to work." She gave me the hospital number to ring and then she said how members of James's family had been trying to get in touch with me for hours, and that she and her husband, unable to find me, had left a message on James's and my internet site and even on my you-tube channel to get in touch with them urgently.

James had been somehow knocked unconscious in a collision between his bicycle and a car shortly before 5p.m.

My immediate reaction was to suppose that the accident wasn't very serious until such time as I might find out otherwise. I rang up the hospital and, being a nurse, asked the right questions of the nurse looking after James in the intensive care unit (ICU). Although I am a psychiatric nurse, with a reasonable knowledge of the brain and behaviour, I had no experience of acute head trauma. That is, I had never dealt professionally with people who had just been seriously hurt. I was only acquainted with how some of them were when they entered the psychiatric system, sometime down the track, years later.

All I was able to find out then, was that James was in an 'induced coma' where the ICU team were attempting to reduce any damaging physiological and psychological reactions to his head trauma. I was instantly confused by the idea of an 'induced coma' when it seemed to me from what they were saying, that he had to be in a natural coma anyway, since he had been hit on the head and had arrived at Emergency in a state of unconsciousness.

I later realized that an induced coma is a way of maintaining a person in an artificially stabilized state, where, left to their own devices, they might be going in and out of various levels of unconsciousness, reacting to fear and confusion, struggling, and experiencing pain. Having the patient unconscious made painful and uncomfortable interventions, including oxygen intubation, the insertion and maintenance of intravenous drips, catheterisation, vital signs monitoring equipment, running CT scans and performing MRIs, possible. It was also a way of better regulating the more serious variations in consciousness, which epileptic seizures represent, where the person may experience lengthy deprivation of oxygen, tearing of muscles, and further injury during the convulsions. Indeed, James had had a full seizure at the scene of the accident and another in the Emergency Rooms.

On the telephone, the nurse told me that they were going to reduce the sedation and try to wake James up about four hours from then, around midnight, to see how oriented he was.

Orientation is an important test of consciousness and, to some extent, recovery from head trauma is about recovering continuous consciousness, not just about recovering intellect and physical skills.

Until that time they could not say much at all about James's immediate prognosis.

Day 2: Still in ICU – 19 May 2010

I rang again soon after midnight in the early hours of 19 May, but James had only responded to painful stimuli. The ICU staff were going to reduce sedation again and try to wake him up around 7 am.

It was now becoming clear to me that James was not going to simply wake up and walk out of there. At some stage I was told informally that he had two 'small' intracranial bleeds, one into the left lateral ventricle and one into the left 'quadragemina' – which I looked up to find it seemed to be an old-fashioned term for the sub cortical speech and sight area.

I asked the difficult question and the nurse I spoke to on the telephone tactfully conveyed that no-one could guarantee that James would live.

There were no flights in the small hours of the morning, but I booked a plane to fly down that evening, 19 May, as early as I could.

I woke up every few hours and phoned ICU to talk to the nurses. I looked at our website and saw that the last post that James had published was an affectionately kind comment on an article of mine. He must have done it just before he left for work, just before the accident.

Later I would discover that he had then left the house and hopped on his bike, only to find that he had a puncture. He had therefore rung the hospital where he worked, to tell them he had to repair the puncture and to expect him to arrive late.

He then repaired the puncture and set off on the series of bicycle paths that he always took to work, but at some stage, instead of crossing a small road and reentering the bicycle path, something happened whereby he finished up unconscious and fitting on the road next to the bicycle path. According to the policewoman investigating the accident, the driver of the four wheel drive who was coming from the tennis court car park, said that James was not riding on the bike path, but on the road itself of the cul de sac, albeit on the correct side, when they collided. If that is true, then she wanted to know the reason for the divergence from habit, but no-one seemed to know.

I was due to work on Friday night so I flew up to Queensland on Wednesday night, the night after the accident, and was booked to return Thursday night, ready for work the next day.

Day 3: Intensive Care – 20 May 2010

I arrived at the Intensive Care Unit half an hour after midnight, at the beginning of Day 3. James was in a 'pod' which is a kind of alcove full of machines with screens and tubes surrounding an uncomfortable-looking 'bed' – more of a plastic padded pallet, such as accommodated the space crew in *Alien*, in suspended animation, while their ship travelled between galaxies. James's face and neck were swollen; he was catheterised and wearing pyjama pants, spread-eagled in an uncomfortable-looking way, with eyes looking towards the ceiling and a very stressed expression on his face. The ICU nurse said, "Yes, go up and speak to him." I went up to him and held his hand and looked into his face and asked, "Do you know who I am?" He nodded with difficulty as though his swollen

16

throat and chin prevented much movement. I said, "What's my name?" He answered correctly, "Sheila," and added, slurring, "I knew you were coming." (I had told one of the nurses to tell him I would be arriving hours before.) I was delighted that he recognised me. Up until then he had not even known his father. I had dreaded that I might have completely disappeared from his memory, which would be my cue to exit completely from his real life and the probable end of our personal and creative collaboration.

Up until he recognised me I think that I was not really committed to staying with him through this. As soon as he recognised me, it seemed that I was committed. From that point on I constantly watched for improvements and the minute they came, I wanted more. I became very protective and for me he was like a patient as well as my intimate friend. By seeing him as a patient I was able to keep a distance from his pain and to look at his condition in a clinical fashion, which insulated me from every little up and down and uncertainty. By the same token, if something was not done that could be done, I did not stand meekly by.

As his friend and partner in so much, I was of course deeply anxious about the possibility of losing him. In addition I was extremely concerned at the possibility of our busy international interactive political website, https://candobetter.net, (or, as it was then, candobetter.org) which was so important for us both, sinking into obscurity for lack of management, or, worse still, with James completely incapacitated, being hacked and never getting up again. All I could really do was wait for him to get better and look for opportunities to help.

I caught a taxi back to James's house, for which I had my own key, and immediately went to the computer in the back room, vastly relieved to see that it was still humming. I climbed into his bed, where the electric blanket was still on, and fell into a deep and peaceful sleep, to be woken by one of his aunties far earlier in the morning than I would have wished.

"How was James?" she asked.

"He recognised me!" I replied.

She thought this was testimony to the 'power of love'. The year before, with several other relatives, she had contributed to send me and James snorkelling at Lady Musgrave Island on the Great Barrier Reef for James's 50th birthday. I remembered her for a singing voice so sweet that it actually brought tears to my eyes and for sculpting and lacquering a quirky looking full-size female figure in overalls out of old papers she would otherwise have thrown out. She was going to put it in the garden. There are a lot of architects and artists on James's father's side who often also cannot throw things out.

Apart from recognising me, James remained mostly at a coma scale of GCS 7 (Glasgow Coma Scale - explained further on) which means he was responding to verbal commands. As well as the 'minor bleeds' into parts of his brain, a CT scan had detected a fracture over his left cheek, but there was no deformity and it would never require any treatment.

There did not appear to be much other damage, except for nicks and cuts, but he had very clouded consciousness – as would be expected - and kept falling back into a sleep, then coming out of it in delirium, begging to be allowed up to go and pee. "Please, please, let me go and pee!" he would cry, even though he had a catheter in place. Or, craftily, "Just let me go for a walk, okay? I promise I'll come right back."

I stood watching him and noticed that his feet were arched and his toes were pointing downwards. His fingers were also pointing spastically[3] from time to time. "He's showing tonus. It looks like he's having a sub-clinical fit," I told the nurse. (I meant that it looked as though he was having a fit suppressed by medications, which in fact he was.) Possibly having no idea of what I was talking about, she shrugged noncommittally. "He's on sodium phenytoin," she said. Sodium phenytoin (Dilantin) is an anticonvulsant (an anti-epileptic fit drug).

I was not surprised that the nurse did not seem to recognise or care about the signs of seizure breaking through the IV drug therapy. On the one hand, I thought she should – since she was an intensive care nurse looking after a head injury – but on the other hand, I am used to nurses and doctors who seem to understand their specialties within the narrowest of spectrums. She probably wouldn't pay attention unless James had a full-blown *grand mal*[4] fit or total seizure.

I wondered if James's confusion was due to his sub-clinical fitting, and that if they could stop the fitting, he might become clearer. But he was getting an anticonvulsant and I didn't know how to stop the fitting I could see, so I stopped thinking about the matter. I hoped that the fact that he was receiving oxygen, plus the modification of the fit by medication, would protect his brain from any harm that a fit might otherwise cause.

The nurse thought that the catheter was perhaps irritating his urethra, but I wondered if there wasn't simply something - a lot - irritating his brain and giving him false messages as well as making him fit.

He became so agitated, struggling to get out of the pod and pulling on the various tubes attached to him, that he was given intramuscular sedation in the form of haloperidol. I understood that he had been given 1.5mg. At the time I didn't think of the fact that even a small dose of haloperidol lowers susceptibility to seizures, although it was something that I would have borne in mind with a psychiatric patient. I suppose I had to believe that they knew what they were doing. There were so many things that could go wrong.

It was only in 2014, when going over James's hospital notes (which were released to QComp – a public medico-legal assessment tribunal – which then released them to James) that I discovered that he had actually received a very large dose – especially for a major tranquilizer naïve person - 10mg in 5mg increments. This was followed by 0.5mg Risperidone, which has similar effects, even though marketed as a newer, better major tranquillizer. Notes for the afternoon, in a different ward, record an additional 4.5mg of haloperidol!

Neurosurgical Ward – afternoon, 19 May 2014

Medical/Nursing notes on James record that he was transferred to the Neurosurgical ward with Glasgow Coma Score (GCS) of 14 – which is only one less than fully responsive. The neurosurgery ward did not just do surgery, it monitored the recovery of patients with brain injuries, who might or might not ever have surgery for this.

James arrived at the neurosurgical ward at about 3p.m. Around 5p.m., his condition started deteriorating, and his GCS was recorded as dropping to 6.

I had one more hour before leaving for the airport, so you can imagine how anxious I was to leave him in a stable condition. While nurses tried more and more anxiously to work out what was happening, I could do nothing but watch fatalistically as his blood pressure went into freefall. James stopped responding to our voices and then even to painful stimuli, such as having his earlobe squeezed between two fingernails. I stood there by the bed, clutching his hand in mine, thinking, "Is this the last time I will hold your hand? Are you going to die in front of me? Oh, God I hope there is something they can do to save you."

Fortunately a neurosurgical resident arrived and, after a moment, pumped some more sodium phenytoin into James intravenously. Quite soon, James's colour returned along with his blood pressure. The registrar explained that James's level of sodium phenytoin had fallen below the safe threshold. James stopped struggling and relaxed. Suddenly he came out of the confusion enough to respond sensibly to "How are you?" with a very normal, "I'm fine." He held up his hand with his fingers formed in a circle.

Now he took my other hand, then he reached out to touch my face. Soon his Glascow Coma Scale went up to 10 or 11.

About this time I noticed that his eyes were quite bruised. I did not realise the significance of this. It probably meant that the frontal lobes of his brain had bounced hard against the bony areas around his eyes and nose and was a pretty good indicator of brain injury.

He could only open his bruised eyes with effort but looked at all of us a couple of times. I had kept him informed of the fact that I was leaving for the airport and would be back in six days to stay for nine. When I left, I said, "I'm leaving for the airport now, James," waving to attract his attention as he was surrounded by relatives. He focused on me and said "Bye" in a little voice, raising his hand in a weak wave. This was very good because it meant he knew that I was moving around and leaving, that he had some understanding of time and space and person. The bodies of his family, which had parted slightly, drew close around him again and I could no longer see him.

I had been told that it was quite likely that James would become completely better, or only with some minor deficit as long as he kept on slowly improving. Naturally I expected him to keep on improving, until I found out otherwise.

However, when I rang the next morning I was told, "No change, still at GSC 7." Since he had been at GSC 10 or 11, I supposed that the nurse didn't really know the patient well.

On reading the hospital notes in 2014 I discovered that he had stopped responding to speech even by opening his eyes during the 21st of May – hovering between GCS 6-8. On the 22nd he was assessed as GCS 13, however, with confused conversational responses. On the 23rd he began with a GCS of 13 overnight but became agitated again, pulling the naso-gastric tube out several times, and was given more risperidone in the early morning. After that he was found to be too comatose to assess and was incontinent of urine. On the 24th his coma score was 14-15 and his swallowing ability was assessed for the second time and he was put on a soft diet with close supervision. Although there is no record of sedation, there is a mention of 'analgesic' by a physiotherapist who would not have known the details, and he or she wrote at 16:45hrs that James was not responding to commands. An occupational therapist wrote that she could not commence Post Traumatic Amnesia assessment because James was unrousable. By 21.00 hrs he was rousable and 'agitated' and received another dose of Risperidone. He was also given Endone (a synthetic narcotic) for pain. He reportedly 'slept well' for most of the night.

25 May: Westmead Post Traumatic Amnesia Assessments begin

On the morning of the 25th he was assessed at GCS 14. The Occupational Therapist commenced him on the Westmead Post Traumatic Amnesia Scale. She found him sitting up in bed with a waist restraint to prevent him from climbing out of the bed. She described him as responding to her questions 'in complete sentences and correctly identifying the names of pictures in the picture cards.' James scored 4/7.

The Visitors and Information 'List'

Due to historic divorce-related rifts in the family, James had kept his address from some relatives. These people who did not know where he lived included his mother and his brother, although James retained some contact with his mother. Both now seemed to have been given authority over him in the eyes of the hospital.

His brother was nominated as James's next of kin and I was taken off 'The List'. This meant that I was not to be given any information. I feared that I would never see James again. The List was a reference for staff to say who might visit him in hospital, who could make decisions on James's behalf, and whom they might give information out to on the telephone or in person. I was devastated.

I discussed this with one of James's sisters and she said she was going in to try to sort things out that afternoon.

James and I were entwined – affectively, philosophically and in a mutual political and publishing activity, but we lived in different states. I had been commuting to Queensland on my days off every other month. The rest of the time we were in constant communication by email and on the internet in the editing and administration of the political environmental forum we had built together, *candobetter.net, 'a website for reform in democracy, environment, population, land use planning and energy policy.'* Maybe, if I had lived in the same state, we would have lived together, but we lived and worked in different states. I suddenly realised that, if you were not a blood relative, had no business for profit relationship, and were not married, you could lose all contact with your closest friend when they went to hospital.

The fact that we had for years shared the same bed when I flew up there on days off, would mean nothing if his mother and brother decided to pull rank and if no-one who was aware of our relationship did anything to stop that.

Professional and social importance of computers and internet for James before the accident

James's internet forum, which contained thousands of original articles by many authors, was the key to our relationship, the key to his personality and the device he was using to try to carve a platform for parliamentary candidature for himself and others. *Candobetter* had started in 2006, when James and I first began spending a lot of time together. By 2009 growth in visits to the site and reads were exponential as Australians and international site visitors found a voice and commonality unavailable elsewhere. In 2010 we had begun looking for a second, bigger site, to cope with the traffic and to secure our data against hacking.

James's political activities on the internet, however, were almost invisible to his closest relatives. Those on his father's side were rooted in the three dimensional world of arts and architecture and traditional political party membership.

All his relatives seemed to be internet naïve. James, however, was a web developer with a background in computer science research, with radical political ideas about real democracy, was actively engaged nationally and internationally via the internet, along with most of his friends and fellow activists.

On his mother's side there was also a focus on religion, whilst James himself was an atheist.

I imagined with horror James being taken home by his mother in a wheelchair, heavily medicated, to a house completely isolated from the internet, in a situation perhaps similar to the one at Bates Motel, in Psycho, but attended as well by his brother, Bill, whom James had told me he distrusted.

Indeed, I had met Bill for the first time in the hospital the Thursday afternoon before I had caught the plane back to Melbourne. At first I had mistaken him for a doctor, because he was standing amongst a bevy of them around James's bed when I came in during the afternoon visiting hours. Bill had a fox-like gleam in his eyes and sort of rolled back on his heels and stared down at me from a great height of 6 feet 4 inches to my 5 feet one inch and asked, "And who might you be?"

Suddenly I remembered James telling me one day that, as he got older, he was horrified to see, when he looked in the mirror, that he was getting more and more like his hated brother Bill (who was actually not bad looking).

"You're Bill," I guessed, correctly.

"And you are...?" he responded.

"I'm your brother's best friend," I replied, and pushed my way into the throng.

It turned out that Bill, aged 47, was in his second year of medicine, but had taken a year off. Somehow this qualified him to join the doctors in their rounds, to pick up James's chart from the foot of his bed and read it, and to arrive and depart outside of visiting hours.

Now I discovered that it had apparently also permitted him to assign himself and his mother as James's next of kin and to leave me off 'The List.'

It was never completely clarified, but the explanation we later got was that the ward staff became overwhelmed with the number of visitors and phone-calls for James and asked for a main point of contact to simplify things. Bill, being there at the time, nominated himself and his mother, with James's father remaining a more distant point of reference – possibly because he did not have ready access to email. I was probably left off 'The List' because I lived in Melbourne, so Bill and his mother, who had only just met me and rarely saw James, had apparently decided that James was unlikely to hear from me again and that I wouldn't need to hear about him. This was ironic and difficult because, until this moment, they had not even known James's address because that was the way he wanted it, but I had the key to his house and the administrative password to his internet site.

Complex family dynamics

When everyone met at James's bedside, after years without communication, it was often as though they were all still adolescents. James had told me that he felt that he had not done enough for his young brother Bill and thought that Bill would desire revenge. Their parent's separation and divorce had influenced the five young children and adolescents to take sides and Bill finished up on the opposite side from James.

Social workers would describe James's family dynamics as 'complex'. Indeed, on the morning of the third day that James was in hospital, I had met his mother and his ex-stepmother (his father was divorced from her as well) both for the first time and at the same time in a coincidence that seemed uncomfortable because they were not in the habit of meeting each other.

I added fuel to the confusion myself as for days I attempted to keep my role in James's life ambiguous. My simple reason was that James had kept my existence secret from his mother, as well as concealing his address. Who was I to question his pre-accident decision? I felt I could only try to maintain his original position. His mother, however, in a question and answer session in the lounge-area at the end of James's ward corridor while we waited for a procedure to be done, pinned my identity down via a weak link in the family chain – James's maintenance of relations with his sister in Tasmania.

After a few days I got back on 'the List' and the next time I saw James I was officially his next of kin, so-deemed by his father (who often visited James and me) and the social worker. James's brother was not seen about for weeks and his mother, although she was there almost continuously, did not have any formal say in his treatment.

I decided that it was likely that James would need someone to look after him for a while when he got out of hospital, so I decided to work my usual four night shifts in Melbourne per fortnight, flying up to Brisbane for most of my nights off, then perhaps to use my holidays and long-service leave to spend longer with him if he needed a carer in Brisbane for a while.

Day 4: Friday 21 May

On Friday 21 May, whilst members of the family who knew me worked to get me back on the list of people able to visit James and get information, I was back working in Victoria. I was, however, able to get an update from a nurse who said "I can't tell you anything but ..." and then told me a lot. That morning James's GCS had been around 14 or 15. 15 is the highest score. He had known where he was and the month. Knowing the name of day and date is the hardest thing for humans to keep up with if they are neurologically unwell or delirious.

Filled with hope, but knowing James would be out of action for weeks, I wrote to the main writers for *Candobetter* and asked them to keep writing articles to help me keep the site going in James's absence. Naturally they all responded enthusiastically, with many well-wishes for James.

Early signs of diffuse axonal decline 8-12 days post trauma

James's consciousness and behaviour fluctuated fairly slowly initially.

During the first few days, much of the time he rested and slept. When he was awake, he responded sleepily to those around him, or sometimes engaged quite intellectually, although five minutes later he would have no memory of this, and might reinitiate the topic anew, again and again.

He had near lucid periods, achieving close to full orientation, before he began a decline (which I did not then know probably coincided with the process of diffuse axonal degeneration which manifests some time after the initial injury). Looking back, the first indication of this decline was probably his 'agitation' (more about this later), which was first reported in the nursing notes on the 23rd, and which I first noticed on my return to Brisbane on the 26th of May. Perhaps it began during the five days I was in Victoria. Looking back, the periods of agitation and growing confusion were increasing right up to his birthday on the 29th of May, after which the decline really set in. So far, I know of only one other account of such a decline – in Lyrysa Smith, *A Normal Life: A Sister's Odyssey through Brain Injury*, 2014. [5]

That is not to say that it is a rare occurrence, but that it seems that the apparent recovery and then the sudden decline after a couple of weeks, are rarely documented. Diffuse axonal damage is still very poorly recognized as it is occurring. Nurses and doctors still do not seem to really be able to conceptualise it as it is happening, although they may after reading MRIs, be able to observe it in hindsight. I remember telling a doctor what it had seemed like and he interrupted with the common cliche - "Every brain injury is different". This was irrelevant. Diffuse axonal damage runs a course, which may be different in all individuals, but has in common the fact that patients drastically deteriorate after seeming to be recovering nicely. [6]

The 'System'

During James's semi-lucid periods, as he declined, he talked of his experience in terms of being treated within a system and observed to me several times his surprise to realize that there was a functional 'system' that could take care of him and save his life. "It reassures me to know that there is something that still works. So, this is the hospital system? And it has saved my life. I thought that all human social systems were breaking down and I never expected to be taken care of in this way. Isn't it wonderful, even though society is heading into trouble, that some of the systems still work, and that this one was there for me. I am so lucky. I could have been in some other place where the system doesn't work. What would have happened to me then?"

His brain seemed increasingly to lack specifics, but 'system' was a word he accurately used to describe so much of what he sensed around him. Although he did not know where he was or how he had got there, was not sure of his age or the year, he always remained aware of the existence of a political system. Specific subsystems and their units, like the hospital itself, were hard for him to associate. His situation in time seemed to fluctuate on a continuum which started around the age of 12 (which was, perhaps not coincidentally, the age when he had had a similar, but less devastating bicycle collision with a car.) This continuum did not reach up to the time of his employment with the hospital and he couldn't integrate the specific information that this was Brisbane Hospital. He identified the hospital as an institution (typically as the Australian National University or a National Park) although he sometimes lost that sense of an institution, and assumed more vaguely that he was dealing with a restaurant or department store or some kind of 'communal living system', as these terms came to his mind, stimulated perhaps by activities around him, like serving food, putting people to bed, and asking him if he wanted anything.

It wasn't until close to discharge from the Royal Brisbane that he even began to grasp the fact that he had been in a road accident, despite several discussions about this.

He did however realize early that something very strange and life-altering had happened and some kind of awareness accumulated of the uncertainty of his future and of the time he might be detained in hospital. He became very aware of his personal vulnerability. He realized that his only contact with the outside world and people who could tell him who he was, and who might care for him, was via his relatives and friends who visited. Highlighting this was the fact that he also began to voice concern for other people who were less fortunate than him and he talked of the need for a government policy to ensure that everyone had maximum opportunity to make friends and establish networks so that, if they were ever injured or in trouble, they would not have to rely on the institutions alone. Staff at the hospital who paid attention to what he was saying found this very unusual. The social worker said that she had never met any other patient in the neurosurgery ward so preoccupied by abstract concepts or the social problems of other people.

James later communicated to me again and again before he left the hospital that he did not know for sure whether he existed, nor whether his friends and relatives existed. He was not even sure whether they had ever visited him, let alone whether they might ever visit again, if they indeed had in the first place.

Coma or Post Traumatic Amnesia – what is the difference?

It was probably around this time that James's condition was treated as if he were moving from coma to Post Traumatic Amnesia (PTA). So the tests he received measured more and on a different scale from the Glasgow Coma Scale (GCS). What is the difference between a coma and Post Traumatic Amnesia? Psychologically, it isn't very cut and dried. Physically it may be signalled by a decreased reliance on complex life support, like drips and oxygen, and being able to sit, stand, or walk, drink and eat. Essentially James remained fluctuating around the top end of a coma – responding more or less conversationally, but not oriented – but was now treated as if he were in Post Traumatic Amnesia. With Post Traumatic Amnesia, there is an expectation of increasing responsiveness and engagement with one's surroundings and people – indicative of improving attention and consciousness, even though the person still doesn't lay down reliable memories. With coma the periods of near consciousness are few and far between, if present at all. With Post Traumatic Amnesia the person appears more awake than asleep, although their consciousness is still obviously impaired.

The two scales of measurement most used at the Royal Brisbane were: Glasgow Coma Scale and The Westmead Post Traumatic Amnesia (PTA) Scale.

The Glasgow Coma Scale

The Glasgow Coma Scale ranges between 3 and 15. Patients with scores of 3-8 are generally described as in a coma. The total score is arrived at by adding the scores for each of the three categories together. In head injuries, one convention has it that a score of less than 8 may indicate a severe head injury; a score of 9 – 12 may indicate a moderate Head Injury, and a score of 13-15 may indicate a 'mild head injury.' It should be kept in mind however that the coma scales can vary over short and longer times and that there are other criteria for assessing seriousness of the injury. No criteria can be relied on for detailed prediction of ultimate recovery. Where the injury occurs, for instance on the brainstem, may have impacts on consciousness but may ultimately resolve.

For adults the scores are as follows[7]:

Best Motor Response (6 grades)

1. No response to pain.
2. *Extensor posturing to pain:* The stimulus causes limb extension (adduction, internal rotation of shoulder, pronation of forearm) - "decerebrate posture"
3. *Abnormal flexor response to pain:* Pressure on the nail bed causes abnormal flexion of limbs - "decorticate posture"
4. *Withdraws to pain:* Pulls limb away from painful stimulus.
5. *Localizing response to pain:* Put pressure on the patient's finger nail bed with a pencil then try supraorbital and sternal pressure: purposeful movements towards changing painful stimuli is a 'localizing' response.
6. *Obeying command:* The patient does simple things you ask (beware of accepting a grasp reflex in this category).

Best Verbal Response (5 grades) Record best level of speech. If patient is intubated, a "derived verbal score" is calculated via a linear regression prediction.

1. *None.*
2. Incomprehensible speech: Moaning but no words.
3. *Inappropriate speech:* Random or exclamatory articulated speech, but no conversational exchange.
4. *Confused conversation:* Patient responds to questions in a conversational manner but some disorientation and confusion.
5. *Orientated:* Patient knows who he is, where he is and why, the year, season, and month.

Eye Opening (4 grades)

1. No eye opening;
2. Opening to response to pain to limbs as above
3. *Eye opening in response any speech* (or shout, not necessarily request to open eyes);
4. Spontaneous eye opening.

In the few minutes when James was on the road after the accident, his worst score was 2 and his best was 11 according to information probably recorded by ambulance officers, which reached James and me months later via a referral letter from a rehabilitation doctor.

Westmead PTA Scale

Below is a simplified version[8] of the Westmead PTA scale which can be performed by non-specific health staff when specialist testers are off-duty.

"Westmead PTA Scale

Ask the patient to answer the following questions once every 24 hours:

How old are you?
What is your date of birth?
What month are we in?
What time of day is it? (Morning, Afternoon or Night)
What day of the week is it?
What year are we in?
What is the name of this place?
Who do you have to remember? (Show set of 3 photos)
What is their name?
What were the 3 pictures that you had to remember?
Picture I
Picture II
Picture III

Record each answer and score 0 or 1 on the MR-120 Form. After each question tell the patient whether they were right or wrong and correct any incorrect answers. Sum the individual scores to give a total score out of 12. If the patient scores 12 out of 12, remember to show them a new set of picture cards to learn for tomorrow. Record these new cards in the appropriate area on the MR-120 Form. Finally, try not to discuss the questions with the patient until you test them again tomorrow."

The Westmead PTA test requires patients to be able to hear and speak. It is for daily use and requires the person asking the questions to do this in a standardized manner, initially asking the person the question in a straightforward manner, then, if there is no answer, asking the patient to choose between three suggested answers. Colloquialisms are not allowed; speech needs to be standardized. There are twelve questions, including the identification of picture cards. A regular person, often a speech therapist, does most of the tests and one of the main items is to get his or her name correct, indicating that the patient is able to remember a person in their environment. When the regular tester is away patients are supposed to be shown a picture of the regular therapist and asked to identify her by name. Successful emergence from Post Traumatic Amnesia, according to this scale, requires the patient to correctly answer the twelve questions three days in succession. In order to show that the patient has developed and maintained the ability to register new material, after the first time that he or she gets twelve out of twelve, the three picture cards are changed. When the patient gets those new picture cards right, three new ones are introduced. This happens until the patient is able to get all the questions right and memorise three different sets of cards successively over a three day period. The picture cards contain simple line-drawn images, such as bird, cup, flower, spoon. Before the therapist leaves the patient, the therapist is supposed to tell the patient what they got right and wrong, and to tell them the correct answers, then to brief them with the information that will be required the following day. There are some other conventions attached to delivering the test and more details which can be accessed on the internet.[9]

Brain injury in a hospital cleaner and computer scientist

James's preoccupation with systems and social causes seemed even more unusual for his treating team since James was not known to them as a person whose life was dedicated to developing a political paradigm on an internet site, but rather as a hospital cleaner. For some people, and for the law of workers' compensation in Queensland, James's job defined him, although he had come by it accidentally.

His work supervisor, Jan Dring, however, was not surprised. She knew what he was like and had encouraged him to apply for other jobs more suited to his interests and skills. James's low self-esteem after the loss of a doctoral opportunity and the diversion of his energy and time into his internet site and other political activities, running for political offices and writing about them as an amateur journalist, and encouraging and editing other peoples' writing, as well as exploring and publicizing social and environmental problems, meant that he found making time to do this very difficult. Prior to the accident, whilst employed at the hospital, he had gone for one clerical job interview at the hospital and he had designed one piece of software with commercial prospects.

Jan Dring took an interest in James before and after his injury. He was brought unconscious to the Emergency Department at the hospital, dressed in the cleaners' uniform he had put on to wear to work, so the staff there got in touch with her since they could find no-one else to identify him. By the time she arrived, his trousers and shirt had been cut off him, but she knew him immediately. She was very upset about his accident and used to come to see him every day. She also got his cleaning colleagues to visit him and, taking immense trouble, got permission from the ward and arranged for him to come down and have supper with the cleaning crew one night when he was out of what his father referred to as "the naughty boys' room", but still an inpatient. Apparently he seemed right at home at that supper. This was such a restorative experience that James actually remembered it when reminded, for a few days during his coma, however, when he came out of his coma and PTA, he had no recollection of it.

32

Naughty boys room: Haloperidol, Risperidone

On my next visit, James's father told me humorously, as he drove me from Brisbane Airport to the hospital, that James kept trying to climb out of bed and so had been moved to the "naughty boys' room." There were four beds there and a kind of nurses' assistant was on duty there 24 hours a day for the sole purpose of ensuring that no-one fell out of bed.

James, some weeks off 51 years old, was the youngest in the naughty boy's room. There were three others: two old men with crumpled faces who reminded me of the elderly gentlemen in the opera seats in the Muppet Show who exchanged dour remarks to each other about the antics of the on-stage Muppets. One sat taller and straighter than the other in his bed and was relatively calm. The second man had wild tufts of hair on the sides of his head and was permanently in distress. He kept saying, "I wish my wife would ring!" even when she had already rung that day. After falling off a ladder, his memory was so poor that he couldn't remember five minutes after she had rung. A nurse or a room-mate would remind him. Then he would say, "Well, I wish she would visit!" She was elderly and it was a very long drive, for they lived four hours away. One day he burst into tears and cried out, "Oh God! How I miss Audrey!" It turned out that Audrey was his cat.

In the bed to the left of James was a white-haired man whose wife came in daily and fed him lunch. Sometimes his son also came. This man was remarkably calm especially considering the fact that he had very little mobility. He had been a hospital patient for months and was in for the long haul. Gradually his bed was being tilted until he became used to taking weight on his feet. I never did find out what was wrong with him. Perhaps a broken back or a broken neck. His wife's eyes occasionally crossed mine, but she hardly acknowledged me. I got the impression of an exhausted family unit slowly making its way through the last leagues of a huge medical minefield, minimizing their energy expenditure and their focus.

Every post-comatose patient in the neurosurgical ward went through the Westmead PTA tests mentioned earlier in this book. These were conducted by speech therapists. The patients were asked to remember the therapist's name, the day, the month, the year, and three pictures. The goal was to get a perfect score three days in a row. This sublimely simple goal was well beyond the capability of most patients here, who hovered below the surface of human time in some realm of near-but-not-quite-consciousness. The first of the elderly Muppets once confided in me that he had written down the therapist's name, so he could look at it next time she came and finally get the answer right. He went home soon after.

Everyone in this unit but James seemed to realize at least that they were in hospital and that the staff were nurses and doctors. Sometimes he described them as if they were mere bystanders. Sometimes he referred to the nurses as waitresses. Sometimes he described the people around him as 'uniformed', but he often had little idea of the role or the presence of staff, as his conception of being in some strange kind of communal living arrangement [see prologue] indicates. He would answer the speech therapists' question about "What kind of place is this?" with a variety of institutional terms, including two favourites, as mentioned, "Australian National University" and "A national park." He also seemed to be convinced for about two months that the states of New South Wales and Queensland had merged. His understanding of distance was very odd.

Reading in a coma: Kennedy and Billy Pilgrim

In the 12 or so days before things went downhill, James spent a lot of his time reading, even though he was formally still rated on the coma scale. His father would read pages of the *Guardian Weekly* to him aloud and sometimes he would get James to read to him. James could read and apparently understand what he was reading, although obviously he couldn't take in an entire article. He read two books so repetitively however, that he actually took in quite a lot of their contents. The first book was David Talbot's *Brothers*[10], about John Fitzgerald Kennedy. He had been reading this book just prior to the accident. In fact, he had been on a JFK biography and assassination investigation jag for months prior to the accident. We had even watched Oliver Stone's long and engrossing movie reconstitution of the assassination and investigation the last time James had stayed with me in Victoria before the accident. James had reached a conclusion that John Kennedy had been the victim, not only of a CIA Conspiracy to assassinate him, but a later conspiracy to undo his good reputation. I must say that I also found the evidence rather convincing.

Within days of his regaining consciousness, albeit still in a coma, between 7 and 11, James somehow located his copy of *Brothers*, which had probably been in his work-bag, and would talk about what great men John Fitzgerald Kennedy and his brother Bobby had been. His main social-worker, Veronica M., was also a JFK fan and had seen the Oliver Stone movie, *JFK*.[11] She said to James, "One of the main reasons I became a social worker was because I couldn't stand injustice."

Because James was in a coma and had almost no recent memory or immediate recall, he would tell all his visitors and the staff again and again about this wonderful book about the wonderful Kennedy brothers. He might tell us all several times in ten minutes, possibly every two minutes. It was both comical and sometimes quite infuriating. Of course it would also have been tragic if James never recovered his memory and kept on talking about *Brothers* for years and years, always as if it was a completely new remark. But no-one was inclined to allow that thought for more than a moment, so we just treated the repetition as eccentric or funny. What was wearying was meeting James expectation of our responses. As fast as James renewed his old remark, we had to compose our faces into expressions appropriate to hearing the remark for the first time, and resist going, "Oh God, not that again!" Sometimes we did, however, and then James then tried to figure out how he could possibly be repeating himself. But, as soon as he had figured it out, he would forget the whole thing. There would be a pause and he would say, "Oh, I must tell you about this wonderful book, I've been reading. It's called *Brothers*. Here it is. You're welcome to borrow it when I've finished. I think that everyone should read it. It should be prescribed for schools. Until I read it, I was almost totally pessimistic about the political future of Australia and the world, but now I realize that there is evidence that people can reach high positions of leadership without being totally corrupted, and then work for good. ..."

At other times James would be reading a yellowing, tattered paperback version of Kurt Vonnegut Junior's *Slaughterhouse-Five*.[12] For weeks, it seemed, this little volume was constantly at hand. It bothered me, because my vague recollection was that it was a catastrophic volume and James began talking a lot about Dresden and disasters. To illustrate what I mean by his being able to take in what he was reading to some extent, he began to question whether the bombing of Dresden had really taken place. The hero of the Vonnegut tale was Billy Pilgrim and Pilgrim kept James company through many weeks, long after we all may have thought that he had gone away. Indeed, when I took James home on Day Leave, still in a state of traumatic amnesia (when he still didn't know what state he was in, what kind of institution, or what the day and the year were and still believed that NSW and Queensland had somehow amalgamated) he brought up the subject of Billy Pilgrim again. This was one of the occasions when I realised that Billy and I were, in James's mind, on the same level. For James we were both somewhat famous people who had taken an unusual and inexplicable – but very welcome – interest in him. For it turned out that Billy Pilgrim also visited James.

Coincidentally, I had, decades prior to knowing James, purchased a collection of Kurt Vonnegut Jr.'s novels. After reading one – which one I cannot remember – I became irritated by Vonnegut's stylistic device of frequently repeating, "And so it goes". This was the main reason that I had never read any more. More than a year after James's accident, I decided to read *Slaughterhouse-Five* and tolerate the "And so it goes's". I then saw the book in an entirely different light. It now seemed fantastically appropriate to James's amnesiac situation. For *Slaughterhouse-Five* is the story of a man – Billy Pilgrim - who lives his life with hindsight provided by an alien species, the Trafalmadorians. The Trafalmadorians know when they will die, but are not worried by this because they can relive any period they like. For them everything happens simultaneously. Thus the hero - rather the anti-hero - bounces around in his own lifetime, surfacing at odd moments filled with tragedy and confusion which he then contrasts with future outcomes, mostly quite counterintuitive. I wondered if James's father's decision to bring the book in to James's bedside had been a brilliant flash of insight. It turned out that James's experience of the world was very similar to Billy Pilgrim's. He kept waking up at different ages, in different places, with no control over his circumstances and little idea of what connected him to different memories, ages, people and experiences. For a long time after post-traumatic amnesia officially ended, James would be bothered by the idea that, just before he died, he would have to review every mistake he had ever made. Is this, I wonder, what happened as he lay anoxic and gasping on the service road while two boys wondered what to do and a nurse who happened to be a few hundred meters away at the time, and whose name we never learned, started running towards him?

Hospital noise-induced insomnia

James hardly slept at night due to the huge noise-level in the hospital. I learned from night-nurses that, in the neurosurgery unit, the night was alive with brain-injured patients wandering all over the place, desperately trying to find somewhere quiet to sleep. In fact hypersensitivity to noise is a common post traumatic brain injury symptom, but the ward was so bright and noisy at every hour of the day and night that normal people could not have slept.

As his mobility improved, James would try to sleep on two chairs in the patients' lounge at the end of the ward hallway. It was usually empty – because most patients were more or less confined to bed – but sometimes another patient would find it and turn on the television and sit up half the night watching night-TV, to James's dismay.

We first found the patients' lounge by accident, looking for a new route on one of our circuits through the ward corridors. D., James's step-brother joined us there and we had a discussion about whether the bombing of Dresden had ever happened and whether it should have happened, considering that World War II was close to the end, with no doubt that Germany had lost. And we also discussed the results of a speech therapy test where, given the adjectives, "small," "big," "huge" and "mega," James had picked "mega" as the odd one out. D. and I both felt it must be "small" of course, but months later, when I raised the matter again, James explained it as a number or a multiplier, meaning one million times something. I don't think this had occurred to the test-designers.

"Agitated" (First observed by me on 26 May)

The first I heard of James's akathisia (an infernal kind of involuntary restlessness)[13] was when a member of his family referred lightly to his attempts to stand on his head. If you asked him, you would find that he didn't really mean that James tried to stand on his head. He would say that James 'scrunched' his head into the pillow frequently. "It must be because he wants to do it for some reason." Other members of the family agreed and seemed to find James's head-scrunching activity amusing rather than worrying.

Akathisia is part of an involuntary complex of abnormal movements known as Extrapyramidal Side Effects (EPSE). Some of the other movement problems called EPSE are with groups of muscles, causing cramps and tics, which may cause distorted postures. Others cause slowing of movement, loss of ability to initiate movement, and generally a more or less subtle failure of muscles to coordinate at will, which can also affect balance and confidence. Several of these symptoms can occur in the same patient, either together or in succession.

The term, 'akathisia' comes from Ancient Greek and means 'inability to sit down'. People with akathisia are unable to sit still or remain motionless for as long as they might wish. They experience an inner restlessness and an unpleasant drive to constant movement. Part of this pattern may include a feeling of fatigue and the need to lie down and rest, but when the person lies down, they immediately get an impulse to sit up, then to stand up, even though they do not want to do this. If the symptoms are caused by a medication that is being readministered, the symptoms will not go away and may well become worse, so the person has no relief. People can become suicidally violent towards themselves and others because the symptoms are so distressing and inescapable.[14] They are a leading cause of medication refusal in psychiatric patients.

On Wednesday 26 May I observed this behaviour in James for myself, when I returned to Brisbane for the second time. (I had initially flown down on the 19th of May and returned on 20th May, worked four days and returned to Brisbane on 26th May.) In my diary I wrote, "Severe akathisia. James 1mg Risperidone one daily. Afebrile. I asked for them to do CK levels."

They told me they had not done the CK levels. James's neurosurgery registrar said, "Firstly, it's not a test for neuroleptic malignancy syndrome (NMS). [15] Secondly, he doesn't have akathisia." He may also have told me that CK levels rise when people are bruised from an accident and subsequent medical procedures. In fact, CK levels are a good test of NMS, when combined with other observations, and I maintain that James did have akathisia. I do agree that CK levels may rise in the absence of

NMS. I could have added that NMS was possible without akathisia – or rigidity - which is another common criteria. Essentially, the situation was very complicated and definitions of diagnostic categories and observations of signs and symptoms were not shared.

I became aware that headstands were part of a longer pattern, which his family hadn't really detected, even though they could describe it quite accurately, since they spent so many hours observing it. James would be lying down quietly, perhaps sleeping, when he would groan and sit up. Then he would get to his knees. At this point ward Staff would then run over to him or call to him, "Lie down!" At this point he would usually recline obediently. But it was as though the urge built up, and he would soon sit up, then get onto his knees, head facing towards the foot of the bed, then turn laboriously around so that his head was facing the pillow, and flop over on his side with his head on the pillow. He would do this again and again, despite appearing exhausted. The cycle had a habit of moving along itself and amplifying, so that, after he had sat up and got to his knees, he would often try to stand up. This caused so much panic, with people running towards him with the restraints, that he may have learned to modify it, so, getting to his knees, then swivelling on his bottom and swinging his feet over the side of the bed. He would sit there for a moment, and observers would hope he would stay in that position, but, inevitably, he would try to stand up, on the floor.

Family and staff reactions to James's strange behaviour

If his mother was there she would take his hands and guide him back into a sitting position. She also would sit opposite him in a chair and, when he swung his legs over the side of the bed, she would get him to put them on her lap and stretch them. He would do this, and then he would withdraw his legs, and put his feet on the ground. Then she would again ask him to put his legs on her lap, and he would do this. They did this for half an hour, an hour at a time, several times a minute. James's short memory assisted this displacement activity strategy and he would put up with the absurd compulsion-routine for a while. Sooner or later, however, his body would demand an escalation of activity and more space.

I understood from his mother that it seemed to her that James was trying out his limbs in a bid to self-rehabilitate and to exercise. She spent hours daily helping him by guiding his movements so that he wouldn't hurt himself. She would also help him to infuse some sense and purpose into what were automatic repetitive movements. For instance, sitting opposite him, when he went through a process of trying to stand, although harnessed to the bed she would mirror his movements and push back, but in a very gentle way, as you would prevent a baby from hurting itself. She talked to James through this very simply, almost as if he were a very small child, but not in a belittling way. She showed extraordinary patience dealing with these movements and seemed to assume they were benign, whereas the movements alarmed me.

The nurse-assistants had to deal with the movements 24/24. Their primary objective was to stop James from falling out of bed and doing himself further injury. (The risk of worsening things after a recent head injury with a further head injury are extremely high. There are two main reasons for this. People with head injuries often have poor balance and poor judgment so are more likely to fall over and a new head injury is still unstable, so easily inclined to bleed, for instance.) The secondary objective of the staff was to stop James's activities from disrupting other patients' rest and his own. A third objective was the reasonable need to reduce their own stress and work-load.

The first response of staff was to tell James to stay in bed and remain quiet. Then they would call out to him not to get out of bed. Like James's mother, they would try to guide James's movements safely. They would try to find out the purpose of the movements by asking him what he wanted, or suggesting what he might want. When they ran out of patience or had no time (he was not the only patient), they would apply restraints – what they called 'four point restraints' – which meant that they shackled his hands and feet to the bed.

James spent hours like that, still compelled to move repetitively, but unable to complete the movements. It must have been torture. He would groan and ask to be released. The staff would put off giving him extra medication for fear of running into limits early, the general aim being to queue up as much medication for the evening to promote good sleep, as far as possible. As he grew more disruptive, however, the nurse-assistants would get the nursing sisters to bring in some form of sedation. The sedation might work for a while, but then the sedation itself probably started the cycle up again. Pretty soon James's condition got to the point where his restlessness, his 'agitation' never stopped. Then other drugs were tried, with new results, none of them sustainable. There might be a short period of drugged stupor, but then the movements would break through and the patient's mental and emotional state would deteriorate.

The registered nurses described James as 'agitated', which was interpreted to mean aggressive and troublesome, overactive and trying to do things that he was told not to, intentionally. Most did not appear to spend enough time in the naughty boys' room to have a clear idea of what his 'agitation' consisted of. You saw them only slightly more often than the junior doctors (the 'residents'). They were general nurses, not specialised in psychiatry or neurology, and describing neurological symptoms and patterns clearly was not something they went in for. I wondered what they did spend their time doing. I assumed that they were writing reports and compiling checklists of particular procedures, including admitting and discharging patients and performing technical procedures using equipment. And, of course, running their parts of the ward – deploying their aids and accompanying doctors – and responding to life-threatening emergencies – where they excelled. They were also the staff responsible for the reports written about patients' progress on each shift. Their principal source of information was vital signs (blood pressure, pulse, respirations) and level of consciousness. These conditions were monitored with daily, four hourly, 2 hourly or even more frequent tests, when patients were unhooked from machines that performed these tasks constantly. Outside the daily, twice, thrice and four times daily routine measurements of blood pressure, temperature, and reports on fluids ingested and put out, most of the news they received about their patients had to come from the nurses' aides, who saw everything, but did not have the right to make written reports themselves. If any of them had attempted to describe James's cycle of movements in detail, they would have exceeded their status and challenged the knowledge of the registered nurses.

When you think of it, 'agitated' is such a general word that, without qualification, all it can mean for sure is some kind of movement. A washing machine agitates. A relative agitates a handkerchief to farewell a person on a receding passenger liner. Often there is an implication of some emotion or more than usual force.

The registrar (I never saw the consultant) assumed, as did the registered nurses, that James was acting dangerously, possibly aggressively. By this he assumed that there was a motive and that the behaviour was directed and chosen. However he didn't ever seek to understand or to find out what that motive was – why James might seek to get out of bed or struggle with nurses who tried to stop him from doing so. It was as though he assumed that James had motives but that they were not sensible or worth exploring because he had a head injury. Or, to put it another way, he assumed that the 'agitation' was driven by the head injury, that the head injury was non-negotiable, so neither was the behaviour.

Others might take a practical and completely different view, without necessarily even taking into consideration the possibility of akathisia. From Glen, of the Queensland Brain Injury Association (QBIA), for instance, I obtained the impression that figuring out the 'why' was the key to helping James. "It's probably something environmental," he said, because that is what the QBIA people had learned. "Take notes; look very carefully to see if there is some element in his environment that may explain why he is so repeatedly 'agitated'." 'Environmental' was a word I relearned in this context and a word I have taken back to psychiatry. It means the immediate environment and can include emotional situations created or fed into by other people as well as factors like noise, light etc.

I thought I knew, actually. I had seen it all before in psychiatric patients. It was a response to the psychotropic drugs he was being given. But when I raised this I received the arguments I have mentioned above, plus poorly backed assertions that I was supposed to swallow because they were delivered by an 'authority' that the doses were too small and that the particular kinds of psychotrophics were not known for those kinds of 'side-effects'. I had heard this before too. I knew though that I had to rule out other reasons, since it also seemed possible (as the staff at the neurosurgery unit assumed) that it was James's brain that was generating this 'agitation'. I would have had a bit more confidence in their explanation if they had appeared to be familiar with exactly what he was doing, and had noticed the repetition, but they had not – or at least no-one influential had. The social-worker intimated to me that she thought I might be right, but she couldn't step out of her role into the nursing and medical roles – even though she had once been a nurse. But that was how I learned that she had once been a nurse, because she said that she too had seen patients struggling with these kinds of side effects in the 'back wards' i.e. the chronic psychiatric and aged-care wards and ill-resourced communal houses where the severely brain-damaged finished their lives.

I tried the environmental approach given that a medical variation was ruled out in the short-term by disagreement on signs, symptoms and iatrogenic factors. I brought in earphones that James might diminish surrounding noise with and a CD player and a small electronic radio with which he might provide his own preferable and manageable noise. I also gave him sunglasses to minimize the bright lights and a piece of dark cloth to help him sleep at night when the lights were often only dimmed for short periods. James accepted all these things but was unable to keep any of them up because of his disorganization and the lack of systematic reinforcement by staff and other relatives. Pretty soon the sunglasses were lost. And it was probably around this time that he lost his prescription glasses, by putting them into the Sharps receptacle,[16] which he thought would be a safe place. Once they were there no-one would try to get them out. James now looked at most things through a blur. It turned out that the cd player which I had with difficulty found time to purchase, having sought assurance from the retailer that it could play into earphones, had no access point for earphones after all. And later I would find that for James listening to music was very difficult and often distressing for a whole array of brain injury related reasons. One thing in James's environment that we were able to change, however, was the length of the bed. It was pretty short. If James stretched out, his feet went right to the end of the bed. It turned out that things called 'bed extenders' existed and a nurse assistant was pleased to be able to help James by locating one. The bed-frame itself could be extended and the bed extender, which was like an oblong square cushion the width of the bed and about 18 inches long, could be inserted, with a sheet folded around it. My impression was that this did help the situation slightly in the short term, giving James more room to stretch and move, despite the akathisia. It also made James much more comfortable after his akathisia had subsided.

As long as the akathisia persisted, however, the basic problem remained. James spent a lot of the time, day and night, out of control, slowly or rapidly circuiting on his bed, constantly trying to widen the circuit to take him off the bed and beyond.

James's mother wasn't always there, and he didn't cooperate indefinitely with her either. Other relatives had far less success than she. So James would finish up in restraints, or bombed out on sedatives, with rails around his bed like a cage to stop him from falling out. The restraints were to stop him from climbing out.

When the rails were up, James had to modify his movement cycle, and that was probably when he stopped standing up and just did the circuit on his knees and flopped down on his side with his head on the pillow.

But as the accumulation of dosage and the combined effects of several drugs took effect, he was unable to restrain himself, so would stand up despite all demands to the contrary. And so he would be restrained and drugged again, and would lie there hour after hour, struggling to rise under his bonds, groaning and calling out, as though in pain.

He would sometimes be asked if he was in pain. And sometimes he would respond that he was. Over a short period he was given stronger and stronger painkillers, including morphine derivatives, which helped for a while. Whilst conventional pain was unmistakably present at times, notably in the form of a headache, observers were unsure at other times whether he was experiencing pain, discomfort or something else. Towards the end of the second week I had the impression that something new was going on, as well as the akathisia. One day he seemed suddenly to become very worried and I asked if he had a headache. He seemed not to be able to say exactly what the problem was. He looked at me and said, "Frightened." Then he pointed at his forehead and said, "Something here. Different." After this his condition deteriorated.

This brief conversation occurred perhaps in the second week of his time as a bedridden inpatient.

I also became frightened. Those words, 'frightened', 'something here [in my brain]' and 'different' translated in my mind to an irresistible darkness seeping across James's brain, removing light from his mind, energy from his body, and hope from his future.

James's orientation dropped several notches and his behaviour became more disturbed. The headaches stopped after several days, but James had lost weight and he spent long periods of time sleeping and his ability to recognise us was no longer reliable. The terrible movement cycle got worse and he spent days and nights tied to his bed and sedated. The hospital used pain-killers judiciously and reasonably, but despite the administration of laxatives, he became constipated from paracetamol and morphine derivatives, reduced muscle function due to paralysis, and the lack of exercise or decent food. He was probably also dehydrated because he was a big man who, under normal circumstances, drank large volumes of soda water mixed with small amounts of apple cider or copious amounts of tea. I told a nurse this and his fluid intake was increased.

Candobetter Internet site concerns

In the meantime, the outside world still existed and maintaining a place there for James was going to be difficult.

Back at his house, I walked into his computer room, which, from the elevated situation of a wooden house on stilts (called a 'Queenslander') looked into the middle of a huge Morton Bay fig tree, living apartment block for several native animals, magnificent guardian of remnant forest 20 minutes walk from Brisbane GPO. James's main computer, Tibrogargan, hummed in a carefree fashion, as though James had only just stepped out to put the washing on the line or get a cup of tea. At times I tried to pretend to myself that that was just what had happened. Around the graphics-animated screen were little sticky pieces of paper with fading clues to many different passwords in cramped, miniscule handwriting. Some I could guess, but I did not know what programs, lists and levels they had been contrived for, nor when. Many I could not actually read. Then there were James's notebooks. There were also a lot of these. For any new project he would start another. Often there were no dates, so I could not be sure what was current and what was old. These notebooks also contained clues to passwords of uncertain date.

There were two phones in the house. One of them, the Engine phone, from which I could make cheap calls to anywhere, also depended on Tibrogargan remaining alive and connected.

I dreaded any power failures that could shut James's computer down and require a restart because I feared I would then lose access to his email accounts and potential access to the Melbourne firm's virtual server, because you had to supply a password just to talk to them on the phone. If I lost that established electronic contact, I would have to negotiate with the virtual space providers over the phone and talk them into giving me information and access if James could not provide me with passwords. If the website itself got into trouble, I would need to pay someone from the server providers to fix it - if such a person could be found for such an idiosyncratic website. In all cases I would need an authority I could not easily provide for what was normally negotiated via passwords from the anonymity of a home computer. I owned our Canadian account for the other virtual server, which was supposed to become a mirror site for *Candobetter*, but I suspected that James had not got very far in moving candobetter.org (as it was called then before we changed it to candobetter.net) over there, so that it would not be much use if we lost our data in Australia. It was not until much later that I realized what a huge task moving such an enormous website would prove to be. I also vaguely recollected that James had thought that our Australian provider was scaling down or selling on. I hoped that I would still be able to negotiate with them if this were true and that they would not arbitrarily disconnect James's internet site before I could make contact with them.

Even if I got authority, I would still need to find someone who could work with Drupal[17] content management in a Linux[18] operating environment. [19] I used a Microsoft computer myself and knew very little about Linux and even less about Drupal. Although I had administrative privileges on *Candobetter*, I was only familiar with a tiny part of the PHP graphic interface with the deeper code-parts of the program. If PHP was like the dashboard of a car, Drupal was the way the engine was organized and Linux was what the engine parts were made of. The car ran in an environment known as an Apache web server platform. The passwords were the key. I didn't have a key. Right now the car engine was still running, but if the electricity stopped, or we were hacked, then the whole operation could potentially shut down forever.

I knew that our virtual edifice presented a potential innovation in the tools historically available in brain injury recovery and that it might provide significant assistance for James's reinsertion into life back on Earth. Such a trail of personal information had only become possible towards the end of the 20th century, twenty years before James's injury. From that time onward the river of information had flown ever faster and stronger. Brain-injured people notoriously lose access to long term and short term memories, skills and a sense and memory of who they are. Many of James's thoughts, however, about people, systems and things and his interactions with me and others on a daily and sometimes hourly basis, were preserved in email form in his computer and in mine, as well as partly with other correspondents in their own records. In addition his specialist contributions to software engineering were on an Australian National University site. The history of his activism in the local and wider community was also embedded in other websites that he ran or helped with. These web histories documented his movements, values and actions in real time and real space. He owned and ran the Citizens against selling Telstra site which recorded the political actions and responsibilities leading to increasing privatisation of Australian telecommunications. Whenever the debate about Telstra shares and CEOs resurfaced, journalists rang James, as the last spokesperson standing for the other side. He had created, donated and operated various interactive sites with a lot of citizens' political groups from Citizens against Tunnels, people wanting to save the Mary River and the surrounding farmland and town, from being turned into a dam, through to those of animal rights and wilderness preservation groups. He had also provided space for alternative political candidates and groups, and had documented his own interactions with mainstream politicians and mainstream media. A host of complex relationships were conserved in *Candobetter* and on various forums and lists all around the planet that James also contributed to. As well as being a political activist, James was a gifted political historian, but in this case, he was his own historian. If his brain failed to remember who he was, if he lost his 'soul', he might one day go to his site and email and look for them. Whether he would recognize or understand them was another problem we might have to face.

Our website, *Candobetter* was like a big newspaper written by activists, a newspaper that learned and evolved from its own history. Its importance depended in part on its popularity and reliability.

If *Candobetter* was 'off-line' for any longer than a few hours – a day at the most - the reliability that supported its popularity would be compromised. People easily lost confidence in unreliable sites and stopped using them. With repeated interruptions, it becomes very difficult to rebuild prior momentum.

The longer that James was out of action, the greater the chance of this happening to *Candobetter*. If that happened, a lot of what defined James and I individually and as a partnership, would dissolve into a few vague memories of what might have been instead of what was still a very clear and viable path forward in our lives and careers. Even though James was delirious and unable to take care of himself, the site was alive with his presence. If the site died, would he ever be able to restore it? If James's personality were irrevocably diminished, it would live on in the site like a photograph of a deceased. Moreover, superstitiously: if the site died, would James brain die? Yes, this was magical thinking, in the absence of any real power over James's fate.

29 May – James turned 51

It was James's 51st birthday. I went to the local Middle Eastern restaurant and takeaway, where James had been used to enjoying their little honey and nut cakes and zucchini omelettes and got a box of the cakes and a few omelettes and took them in to the hospital. Over the course of the morning, his father and half-brother, his elderly uncle and aunt, and one or two sisters and a husband arrived. James, who had lost a lot of weight and looked about 20, sat up on the edge of the bed with his legs swinging and chatted to us all, offering cakes and even smiling from time to time. He knew who his father was, and his uncle, and me, of course. I think the social worker made an appearance and there was a nice nurse on that day who joined in. Towards the end of the party, his friend Chris also turned up. James recognised him, although he didn't recognise his half-brother at first.[20]

I think all his visitors were cheered up by his responsiveness and I went home feeling quite happy. During this time I noticed, not for the first time, that James would often angle his ears towards the person he was talking with, rather than look them in the eyes or look at their face. I got the impression that he heard sound in a different way. I later found was that it was probably an adaptation to free more brainpower by reducing visual attention, for the visual cortex uses more energy than other parts of the brain.

30 May – Unable to stand or walk

The next day James's condition worsened. He responded feebly and he didn't always know who I was, calling me by the name of a woman he had lived with ten years prior. Once he peered at me closely and gave a start, as though he had thought I was someone else.

Later, in notes I wrote to a solicitor we engaged, about this period over the weekend of 29-30 May 2010, I described James's orientation and judgment as improved - although I also said they remained wildly fluctuant - to the extent that he had become overtired and suffered from severe headache and restlessness and required sedation and 'four point restraints' on 30 May. ('Four point' means that he was tied to his bed by both feet and both hands.) He had a coma scale fluctuating to 14 on that day. I recorded that I had put my laptop on his knees the same day (in a more peaceful moment) and that he had been able to scroll immediately and had said that he recognised his net site (although I would need him to say more to convince me). He later developed a headache and had to lie down and sleep. He seemed not to be able to smile anymore.

When he was stood up by the physiotherapists, they had to shoulder him by the armpits and his feet dragged behind him, although he would try to stand on his toes when they instructed him to try to stand up, and then clumsily drag one foot on its toes in front of the other. All his long muscles were slack. His arms hung by his sides and he didn't hold his head up. It wobbled slightly from side to side, propped on his chin against his chest. His facial muscles had become paralysed. His whole face drooped, like a very old person's. Worse, in fact, he looked like a corpse.

I remember when, with one of his sisters, I encountered the two male physiotherapists hauling him along the corridor. James's sister got my attention and wiped her hand down her own face from forehead to chin and allowed her own muscles to slacken, in imitation of James's. She looked like a mime artist going from 'normal face' to 'sad and grotesque'.

James was still gripped by frequent long bouts of the infernal restlessness as well. Gradually he regained some strength and supported walking seemed like a sensible option, but the sides of the bed stayed up and the restraints continued to play their part.

Others who visited him, including, I think, his sister Judith, expressed concern about the fact that he kept asking to get out of bed. "Why don't the nurses help him to walk around, instead of tying him up?" was the question repeated to me. "What's the point in giving him more drugs to stop him from getting out of bed? What good does it do? Surely it is reasonable for him to want to walk?"

I mumbled once again that side-effects from the drugs were probably making him inordinately restless and driving him to want to get out of bed and walk continually, even though they also made him feel totally exhausted. I felt bad every time I said this because I realized that a lot of the doctors and nurses would contradict me and that there wasn't much that James's other relatives could do about the side-effects – so why conflict them with this knowledge? There seems to be low interest in describing movement patterns over time. To many they seem incidental. Some nurses and doctors spot unusual movements quickly; others never. Psychiatric and medical textbooks on "extrapyramidal" and "Parkinsonian" side effects mostly rely on studies involving small numbers of people, wildly varying definitions, scatty indicators, and unreliable collection methods, giving results that cannot be compared from one study to another.[21] With regard to neuroleptic-generated movement problems, nurses and doctors suffer from the blind clinicians describing the elephant syndrome. Few can agree on what the elephant looks like and whether it is really an elephant or possibly several different animals.

The nurses said that they had to remain in the four bed room (the 'naughty boy's room') to watch the other patients. They didn't have the authority or time to walk James up and down the corridor. Sometimes they added, "The physiotherapists do that."

But the physiotherapists only came round once, in the morning, Monday to Friday. Then, depending on how James presented, they might walk him up and down only 20 feet, with one of them on each side, holding him up by the armpit. Sometimes they walked him right round the circuit formed by the corridors running down 8AS and the adjacent ward and the connecting corridors in between. That might take 10 minutes. For a normal person walking, it took perhaps one minute. The rest of the time, James was supposed to spend lying on his bed, or sitting up on a chair next to it.

At this time three antipsychotics – haloperidol (Serenace), risperidone, and olanzapine (Zyprexa) and one benzodiazepine – diazepam (Valium) were being used on James in different doses, at different times and in combination – with the aim of reducing James's 'agitation.' In fact, the anti-psychotics were almost certainly exacerbating it. When there is really a lot of literature around that flags the many problems associated with neuroleptic medications and brain injury, why did the doctors here seem to have such slight awareness of this problem? Neurologists are aware of the problem and they have the language to describe the symptoms, but non-specialists generally take their instruction from the companies that market the drugs – which diminish the importance of the 'side-effects' – or generalist reference books that summarise a lot of different areas relatively superficially. The neurosurgical unit where James was did not have neurologists, and they sought the opinion of a psychiatrist rather than a neurologist when I complained about James's situation, although there was a neurology department at the hospital as well. Finally, the antidote for the side-effects of neuroleptic medications, benztropine and similar anti-parkinsonian drugs, can increase delirium. For this reason it is often avoided. But the neuroleptics themselves can cause delirium, so why not avoid them? I think that the answer is to use antipsychotics very sparingly in situations such as intensive care, when sedation may be absolutely necessary for brief periods of time.

Around this time, Ian (James's father) and I were sitting on a bench outside the four bed room James was lying in, probably tied up again. I was telling Ian about James's injury, as I understood it. He said, suddenly, "God! I thought this would be all over by now. How long is he going to be here?" I asked him how long he had expected. He said, "About two weeks." I wondered if that was how long Ian had been hospitalised when he had had a stroke.

And he had been told that he would never recover his balance enough to get up on roofs, but he had been up on roofs since that (well over seventy years old). He was still expecting James to snap out of it any day now.

Well, we all were. That is all we could hope for.

But James's injury was more serious than Ian's stroke.

31 May - Delirium with akathisia and perseveration: "Alright, alright, alright! ..."

On the 31st of May I wrote in my diary, "Alright, alright, alright…"

Overnight James began to avoid eye contact and he also stopped responding with speech. Naturally this only made his visitors and the staff doing the tests try harder to get responses. Then the only response we got was an infuriated, tight, "Alright." And James would purse his lips and look away. But he was unable to suppress a terrible urge to perseverate, and would follow on with multiple, 'alrights' in higher pitch and each more emphatic, his eyes squeezed tight with effort to control the involuntary outburst: Alright, alright, alright, alright, alright, alright, alright, alright, alright, alright, alright, alright, alright, alright…! Until he seemed to suppress them by sheer force of his lips and by burying his head in the pillow and refusing to look at anyone.

Staff who were aware of this seemed mystified by these symptoms. "Brain damage" they would say obviously, yet mystically, when relatives asked. It made the brain-injured seem like aliens, with an inexplicable culture which earth-people might never fathom, so there was no point in trying. The brain-injured were inscrutable, as the Chinese were once thought to be. I remember one nurse (the only cruel one I met there) who said, loudly and contemptuously to his colleague, in the naughty boys' room, apropos of nothing, "Isn't it fascinating how the brain-injured can just stare into space for hours?"

James's repeated vocalisation just looked like a neurological sign to me and I wondered what was causing it. Because it was apparently involuntary and it involved movement and perseveration, I suspected the usual culprits, the neuroleptic drugs that James was getting. I did also wonder if part of it was that James was getting impatient with hearing some visitors interpret and comment on his behaviour to the room as if he were a small boy or even an infant.

The involuntary movement cycle also continued, with the addition of these involuntary vocalisations – "Alright alright alright!"

That evening probably also ended up badly with a big dose of neuroleptics to knock James out, ironically with the result that he didn't sleep much.

The next morning I rang Judith to see how James had been. I had had something to do that morning and would be going in the evening.

Judith said, "Dad wheeled him about in a wheelchair this morning and he seemed to find that okay, however, the wheelchair belonged to someone else and we had to give it back and we couldn't find another.

A wheelchair seemed like a good idea.

Wheelchair (2 June 2010 to 16 June 2010)

I hired a wheelchair on 2 June 2010

I rang up to ask if James could be given a wheelchair, but was told that only patients who were completely unable to walk were allocated wheelchairs. The practice was to scavenge them in corridors anywhere they had been abandoned. I wanted, however, to ensure that James had a wheelchair so that visitors could reliably wheel him around. I worked myself up into a state of some indignation then had the idea of telephoning the commercial pharmacy in the hospital food and shopping court to ask if they hired out wheelchairs – and they did – for $25.00 per week.

I got to the hospital at half past four and quickly hired a wheelchair and dashed up to the 8th floor in the lifts, impatient to try my experiment.

I wasn't surprised to hear that James had had a bad night and had been thrashing around in his bed, where he was attached, once again by restraints. I asked if I could untie him and take him for a walk. If I thought I could handle it, was the response.

James was a bit bemused as I labelled his chair (as advised), "Property of James Sinnamon. Please push him in it if he requests." When I asked him if he would like to sit in the chair he nodded, avoiding my eyes and keeping his lips closed muscularly.

As soon as I began to move the chair forward, he stopped leaning forward and relaxed although he still didn't say anything. When I asked him if he wanted to keep going, he nodded. When I stopped, he said, "Keep going.... Alright, alright, alright, alright, alright, alright ...oh, dammit, alright alright alright ALRIGHT ..." until the alrights wore themselves out. By that time he seemed very angry, radiating irritation, so that I could not help feeling that he was angry with me.

"Are you angry? Are you angry with me?" I asked him. He shook his head, still appearing furious, and avoided my eyes. I told myself not to take this personally and to seek other cues from James as to what was motivating the irritation. Apparently James was having some kind of problem which got worse if I spoke or if I stopped pushing the wheelchair.

After about 20 minutes I was still pushing him around. Although he had lost weight and the floor was level, at 6'2" he still weighed quite a lot. Fortunately his 'evil' little brother turned up and helped me push him.

We walked round and round the ward and up and down corridors. If we stopped for a few seconds, James's akathisic inner propulsion would cause him to stand up and start to walk forward, but he wasn't steady enough on his feet to do this. If we stopped and held him by the shoulders, he would make signs to make us keep going, and every so often he would explode into a volley of "alrights!"

I was beginning to get the idea. "Are you avoiding peoples' eyes because eye-contact prompts you to speak and then you can't stop saying, "Alright"?

"Yes!" said James.

"How about we try to avoid the alrights by getting you to answer yes or no?" Do you feel as if there is an inner tension in your stomach and that you are compelled to move forwards?"

"Yes!" James said, emphatically.

"Am I a genius, James?"

"Yes!" said James.

"You do think that I'm a genius?" I repeated.

"You are!" said James, and unleashed a volley of 'alrights' then buried his head exhaustedly in his hands for a moment.

His brother the second year medical student was impressed and appalled. "Do you mean to say that the drugs are causing this and no-one is recognising it?"

"Yes," I said. "At least it seems the most likely explanation to me."

James's brother stopped at the nurses' station to enquire about something and I kept pushing James. We went to the end of the corridor and turned to the right and I wheeled him to the window at the end. There was a small couch and a couple of upright chairs. When I stopped the wheelchair, James automatically stood up and moved forward, then sat down on the couch. There was room for me to sit down as well. I sat down next to him and he turned towards me, put his arms around me, almost looked me in the eyes, and kissed me passionately on the lips. Then we rubbed noses and he leaned his forehead exhaustedly on mine.

It felt wonderful. I was enchanted. He wasn't angry with me.

But there was no possibility of remaining in one spot. With my help, James climbed back into the wheelchair and we continued our endless journey, picking up his brother as we passed the nurses' station.

His brother and I decided to take James down to the food court via the lift. Did James want to go? "Yes!" he said, sounding irritable.

The food court was closed. Stars shone through the glass atrium. We wheeled James all over the place, finishing up at a bus-station on the mezzanine. Perhaps James thought we were going to take him out of the hospital. I would have thought he would have welcomed this, but he felt cold and had an instinct for survival, and asked to go back to the ward.

We took him back and put him to bed. He fell asleep almost immediately. Apparently, however, he soon woke up, with the same restless need to get out of his bed. It was another terrible night, bound to his bed and sedated, but still 'agitated'.

Olanzapine and sweeties

The second night with the wheelchair was scary. James was completely different. He wanted to be in the wheelchair and he still couldn't stop moving, but his personality had been transformed by the drug Olanzapine (Zyprexa).

When I offered him the wheelchair, he got onto it slowly, silently, without eye contact. I placed my handbag in his lap and wheeled him down the ward corridor, into the vast hallway that connected all the wards on the eighth floor of that wing of the hospital. Identical corridors lay beneath us, all the way to the ground.

I became aware that James was ignoring me totally, but groping through my handbag, which I had placed, as usual, in his lap. With some surprise, because he had never behaved in this rude and invasive way before, I asked him what he was doing. "Sweeties," he mumbled. "I'm looking for sweeties. Where are they?" I had placed chocolates from his bedside in my handbag for our tour of the hospital, but I had not expected James to want more than one or two. To my surprise, when he found the chocolates he began eating them, one after another, like a caterpillar munching its way through leaves.

I asked him how old he was, and, after some hesitation, he mumbled, "99, I think."

Sitting down opposite his wheelchair at the end of the long corridor, I hoped he would engage a little more. He appeared unaware of the city lights twinkling in the distance through the window. He barely seemed aware of himself or me. He gave me no eye contact, appearing preoccupied and slightly stuperosed.

"I'm cold," he said.

"There's a blanket on your knees," I said. He lifted the blanket from his knees, looked at it with some puzzlement, then like a confused geriatric patient, he placed it over his head.

"Take that off your head, James," I said, "You look demented."

"If you say so," he said, uncomprehending.

Alarmed, I said to him, "Do you know who I am?"

There was a long pause, during which I could tell that he had no idea. He said finally, "My mother?"

"No, I'm not your mother."

"Oh."

I removed the blanket and wheeled him back along the corridor towards the ward. "Where are you? What is this place, do you think?" I asked him, in my continuously testing fashion.

"I don't know! I don't know. Oh God, I feel like such a fool!" James bleated.

I told him where he was and who he was and who I was.

Suddenly he said, "Of course I know who you are. You're Sheila. I'm so sorry. You're my girlfriend. Oh, what is happening to me?"

I took him back to his bed early, and he climbed in quite gratefully. I could tell that he just wanted to be alone. His main interest was still the chocolates. They were obviously playing with his mind, so I left them within his reach.

I found out from the RN on duty that James had been given five milligrams of olanzapine[22] earlier that day. "Just 5mg," said the nurse.

"Well, it's had a big effect on him. It's turned him into a human caterpillar. Please don't give him any more," I pleaded.

Olanzapine makes people incredibly fat, frequently gives them diabetes and, obviously in James's case, blunts their personalities. It was hard for me to sleep that night, at the thought that the doctors might give James more of that stuff. I knew the callousness with which olanzapine is prescribed in psychiatry, having seen lithe young men and women transformed into blimps and then treated by those who had made them this way as objects of contempt. I would fight against this happening to James.

The episode with the olanzapine convinced me to make a formal complaint.

Complaint re concerns about slow recovery

This complaint, sent on Saturday 5 June, received a reply on 7 June 2010, thanking me for my email, stating that it had been 'forwarded to the appropriate area,' and inviting me to contact the hospital directly by phone if the matter was urgent. Since the matter was urgent, I did so, as soon as I received the email, but I was told that the department had been contacted and it was up to them to respond.

Here is what I wrote in the complaint:

"Sir/Madam,

James Sinnamon is in Bed 21 8ASouth. He was admitted with a head injury to RBWH on 18th May, subsequent to being knocked off his bicycle by a car. He is employed by your hospital and is on Workers Compensation."

The next paragraph anticipated a repetition of the responses I had already received verbally from staff:

'Firstly, I want you to realise that I do understand that head injuries cause confusion and various forms of agitation, indeed many different behaviours, that every head injury recovery is different etc. However I have read widely and asked questions of doctors and nurses with knowledge of head injuries in Melbourne in the last two days, who agreed with me and encouraged me to voice my strong concerns."

Then I stated that I was concerned that staff had failed to diagnose that he had developed akathisia, which was quite probably caused or exacerbated by the neuroleptics (antipsychotic medications) he was receiving, as a 'side-effect'.

"I am concerned about 8ASouth response to James Sinnamon's slow progress and ongoing confusion, the fact that he has not had an MRI for this, only a repeat CT scan, and the fact that the neuroleptics he is receiving from time to time may be exacerbating the situation in that they could be a cause of what the doctors describe in a loose manner as 'agitation' but in fact involves a series of repetitive actions which I would call akathisia, and which my colleagues would also call akathisia, which from decades of psych nursing I know well as side effects of the drugs he has been on and which he continues to get from time to time: haloperidol, risperidone and olanzapine. Despite older advertising around the 'atypical' antipsychotics they do - each of them - cause akathisia. Not always in everyone but sometimes in some people. It does not matter how small the dose in a sensitive person and brain injury makes a person much more sensitive to these drugs."

(In 2014 I would read through the hospital notes and discover that a radiological consultant had recommended an MRI on the 21st of May, only four days after James's accident, but I was not aware of this when I wrote my complaint.)

I described my efforts to have the matter dealt with to my satisfaction and alerted the complaints department to a verbal description of James's behaviour by the night staff:

"Today I rang the ward at 3.30 am and asked about James's behaviour that night. I was trying to verify what exactly was justifying the ongoing claims that James is 'agitated'. Apparently he had been restless earlier that night. I asked for a description of the restless behaviour.

Here is what the nurse assistant said: "He gets up and out of bed, asks for a drink, goes to the toilet, then goes to bed, then gets up again to the chair, then gets up and goes to the toilet or back to bed - all night long until he goes to sleep. Also, if we don't let him get out of bed, he repeatedly sits up, gets on all fours, kneels, then flops on one side and lies down again, only to sit up, get on all fours, kneel, flop down on one side, lie down ..." (almost endlessly.) That is, he tries to increase the area and activities in his circuit to make them more meaningful, but if confined, he continues to cycle in a circumscribed fashion like an animal in a cage because he cannot stop this.

This is exactly what I have observed and what other visitors have observed and what we have recorded. I have not seen any literature describing anything like it in acute brain injury under the heading of 'agitation', except where antipsychotics were involved. I have, however, seen this kind of behaviour many times in back-wards and aged care and Acute Management Units and it disappeared when the neuroleptics were removed, reduced, or sometimes when benztropine was administered. (Yes, I know there is controversy about whether benztropine works in akathisia, but the literature I have read on this issue says that the definitions and stats are not reliable enough to say whether it works or not in such cases)."

At the end of the paragraphs above, I anticipated objections to chemical antidotes – such as benztropine for this problem of akathisia – and suggested that another solution could be to simply withdraw the neuroleptics.

As I concluded:

"In the end, these neuroleptics don't seem to be working, anyway. Why don't we try stopping them?"

I then went into some other reasons why neuroleptics were undesirable in James's case, being careful to clarify my initial willingness as his next of kin to allow them to be used in small doses. For instance, they were associated with slower recovery from confusion (PTA):

Neuroleptics associated with slower recovery from confusion

"Neuroleptics associated with slower recovery from confusion. Let me say at the outset that I initially had no objection to James receiving small doses of anti-psychotics. I only objected when I saw how he appeared to be responding to them. My first query about them was to the night duty charge on Wednesday 26thMay. [I had rung the nurse in charge at 0300 hrs to say that I was concerned at James's akathisia. "Aka-what?" had been her response.]

Importantly, some research has also associated these medications with slower recovery from confusion. Here is one very clear example (among many) which showed that "the neuroleptics affected cognitive recovery with almost 7 more days required to clear PTA in the neuroleptic treated group." i.e. confusion lasted 7 days longer on average than in patient groups where neuroleptics were used."

I then cited the abstract from Mysiw, WJ et al, 2006, "The impact of acute care medications on rehabilitation outcome after traumatic brain injury."[23] The endnote contains more details, but for the purpose of this book, the Main Outcome and Results were as follows:

"…The narcotics, benzodiazepines and neuroleptics were the most common categories of CNS active medications (92%, 67% and 43%, respectively). The three categories of medications appeared to have no significant outcome on the FIM [Function Independence Measure] outcome variables. The neuroleptics affected cognitive recovery with almost 7 more days required to clear PTA [Post Traumatic Amnesia] in the neuroleptic treated group. The presence of benzodiazepines did tend to obscure the impact of neuroleptics on PTA duration but the negative impact of neuroleptics on PTA duration remained significant. Conclusions: [My emphasis] The results suggest that the use of neuroleptics during the acute care stage of recovery has a negative impact on recovery of cognitive function at discharge from inpatient rehabilitation. Due to the paucity of subjects with hemiplegia in this cohort, conclusions could not be drawn as to the impact of acute care medications on motor recovery."

I continued:

"James's restlessness can be stopped by walking him or by wheeling him in a wheelchair. Because he doesn't have much strength yet, a wheelchair is necessary. The hospital would not provide one, so I hired one from the Chemist in the hospital court on the first floor. It is due back there on Wednesday.

Obviously this repetitive behaviour waxes and wanes, depending on the amount of neuroleptics and their breakdown in his system at any time, and possibly in response to environmental factors.

I asked the nurse or nurse assistant to record what she had said at 3.30am. She said that she could ask the nurse to do so and that she would. At 0800 I called O-, who is in charge today, and asked her if the observations had been recorded. They had not been recorded. What had been recorded was a 'settled night' - so the period of restlessness had not even been mentioned.

I have spoken to one registrar and one resident. I have asked to speak to the Consultant, Mr C-., but got the impression that that simply doesn't happen. I have spoken to B-A., the Nurse Unit Manager, who said that a registrar would call back (several days ago). No-one called. I talked with the new resident in 8ASouth, who repeated what the registrar had told him, telling me that I was not seeing what I and others are seeing, albeit in a nice way. He also said that a single room was being considered. Today I asked to speak to the consultant again but he was not on duty because it was a Saturday. I asked for an email address to write my concerns to him, but I was told there was no email address. It was suggested to me that I wait for a family meeting. I said we have one scheduled on Thursday, but I consider the matter too urgent to wait for that long.

I asked O- to tell the doctors how concerned I am and to insist on an MRI. She said she would carry my message but could not guarantee that anything would happen. She read from some notes that an MRI was 'planned' but there was no date. I asked that the MRI be scheduled for today. I don't care if it is Saturday. I consider the matter urgent. I consider it negligent that no MRI has been done already when James is not getting better and no-one knows why. An MRI and a few days trial without the neuroleptics seems the least that the ward should do."

There was more:

Second CT Scan and Progress Report

"Second CT scan and progress report:
See at the end of this letter the most recent CT scan report, which I have transcribed.[24]
This is the third CT scan that James Sinnamon has had,

for the purpose of comparing progress. What it says is that two small bleeds have been resolving over time, which is good news, and that there are no new bleeds. The scan could find nothing remarkable at all. But it cannot show what is happening at a microscopic level in the brainstem or anywhere else.

What James's new resident (i.e. Intern doing 3 month rotation on the ward, who started on Thursday) told me (after he spoke to the registrar) was that they still cannot rule out axonal damage in the region of the brain stem due to a possible contusion (twisting) of it in the accident, although they don't actually expect to find axonal damage. Axonal damage is bad news because it is permanent damage. They would expect to be able to see such axonal damage with an MRI scan, but he said that they don't see the point in doing one in the short term since it wouldn't change their treatment of James's symptoms and it would be hard to get him to be still.

Discussions I have had with doctors who have trained in head injury at the hospital where I work have said that they would expect an MRI to have been done well before now. "If you don't do one, how can you know there is nothing else wrong?" "How can you say that you wouldn't change your treatment after an MRI if you haven't done one?" are two remarks that doctors where I work have made.

The Nurse Unit Manager also indicated to me one evening verbally that there was involvement of the Reticular Activating System (RAS) near the brain stem in the accident and this part of the brain controls consciousness, therefore the slow resolution of James's delirium could be explained by this. Until the delirium resolves, James will only have tiny windows of clarity in which to lay down new memories. When the delirium improves or resolves, his ability to remember should improve."

I anticipated a reiteration of an excuse that had already been made to me several times.

"But the CT scan tells us nothing in particular about this; without an MRI it is all conjecture. Perhaps an MRI would not tell us much but without one, how can we know? An excuse has been made that James is too restless for an MRI but I am reliably informed that he could be sedated for one and that this is done frequently, just as was done with the CT scan.

Whilst there is hope that he would improve back to near normal, that is not a given and the chances deteriorate statistically with the length that recovery from confusion takes. As the CT Scan request reason shows, we don't know what is causing the confusion. As well as the confusion, there is the ongoing trauma. I know that James is scared and isolated, in pain and crippled, and that he has insight into his memory loss. I consider that his human rights have been neglected."

I described James's physical condition prior to the accident and his physical presentation on 5 June 2010 in Royal Brisbane and Women's Hospital, noting that the akathisia and probably associated confusion, were exacerbating his physical deterioration because he could not take adequate advantage of physiotherapy:

"James, who until the accident, was a tall, strong, fit 50 year old man who engaged in rock climbing, walking, cycling and various kinds of manual work, now has some left sided motor weakness, most noticeable in his inability to walk unassisted, to stand alone, and his tendency to drag his left leg.

His continuing confusion is preventing him from using the gym and having physiotherapy there. He is growing weaker every day."

I was writing the email after having done a ten hour nightshift and I got a bit repetitive at times, trying to remain clear. It was better to be repetitive than to leave anything important out. (In fact, I did forget to say that James had begun to 'drool', which is another sign of problems with neuroleptics.) I had meant to make clear the usefulness of the wheelchair movement for stemming akathisic compulsion to move forward, but I failed to do this. As well as seeking better treatment for James, I hoped to make the treating team think a bit more about the nature of akathisia and additional non-chemical responses that might be helpful. However, these last paragraphs summed up the problem:

Restlessness, medications, wheelchair

"*Restlessness, medications, wheelchair:*

Various visitors and other observers have noticed how James restlessness has recurring features, such as the tendency to bury his head in his pillow, then flop down on one side, turn over on his back, groan and sit up, then get on all fours, then kneel, then sometimes to try to stand up, then, from the kneeling position, flop down on one side and repeat this sequence continually, at different rates. Sometimes he does the circuit rapidly; other times it is so slow that you are unlikely to pick the pattern. Most recently he has also been making involuntary vocalisations, notably "Alright, alright", as he sits up. When he goes through this routine rapidly and noisily, the staff describe it as 'agitation' and he gets neuroleptic tranquillisers.

Recently the problem got worse and James avoided eye contact or direct interaction because it set off these involuntary vocalisations and some new ones including echolalia which were part of the evolving sequence of the repetitive cycle. He told me this when he was more relaxed as we wheeled him around and around the corridors of the hospital a few nights ago.

Perhaps because they do not spend long periods watching him, the doctors have not spotted this repetitive phenomenon and think that I am making it up when I report it. A

?registrar came to see James when he was not 'agitated', then claimed that the phenomenon described did not occur. Perhaps I have also confused a doctor by describing the behaviour as 'akathisia' when it doesn't fit a narrow definition of this term, such as dancing on one spot.

I have told them that I am concerned that the antipsychotic medications- called 'neuroleptics' - that James is on cause this behaviour as a side-effect. I cannot get anywhere with this because the repetitive nature of the behaviour is not acknowledged by the doctors. This really bothers me because, if I am right, the behaviour will be made worse by the medications. It is indeed possible that I am wrong and that this pattern is part of his injury behaviour, but in that case why aren't the doctors investigating with MRIs?

Olanzapine causes a very high rate of diabetes and morbid obesity. James has become very sedated and food focused since he has received doses of this medication and his orientation has dropped even further, to the extent that he has got his own identity confused at times and has become emotionally flat. This is an overall deterioration. He is also on Risperidone, which has similar side effects, although not such a high rate. James might spend several more weeks in RBWH and if he remains on these meds, especially olanzapine, I don't like to imagine the consequences."

I also included an electronic transcript of the second CT report, which disclosed no reason for James's slow recovery. In fact it showed that the bleeds into the ventricles (the fluid-filled spaces in James's brain) had got smaller, which was a good thing.

Plasticine and a strange map of Australia

One night I came to push James around in his wheelchair and found one of our mutual friends and his mother sitting with James on the bench in the corridor outside James's room. James's mother was talking in some detail about shoes made by a Norwegian shoemaker. What she had to say about shoes was actually quite interesting and made me want to know the name of the shoemaker, but the conversation was unhelpful for James when his grip on the most basic realities was so slippery. I asked James where he thought we were, and he replied, "It's a shoe factory, isn't it?"

Soon after his mother left and a friend and a sibling arrived. I had brought some plasticine and we sat down and got James to do a map of Australia to see what that might reveal of his perception. As I mentioned before, he had for weeks an unshakable conviction that the state of New South Wales had merged with the state of Queensland.

His plasticine map was recognizable as Australia, except all the towns were squeezed in a south-westerly direction. Darwin was very close to Alice Springs, near the centre of Australia. Sydney, Brisbane and Canberra were almost merged. We talked to James about this and made suggestions, which he tried to take on board.

It was quite a lovely evening because James was the centre of attention but we were all in tune with him and relaxed. His akathisia seemed to be hardly a problem anymore and he was walking better and better. And he was so happy to be in a small social gathering.

We would shortly dispense with the hired wheelchair. And soon he would be moved from bed 21 to bed 8, out of the naughty boys' room and into a room where he was next to the window and had a view of trees and a garden, a bridge between two buildings and some old architecture.

Family meeting – response to my complaints

By the 10th of June, when the family meeting took place, James's medications had been reduced and he was improving.

At this meeting were the resident, the registrar (who I had not seen before), the social worker, a representative from the complaints department, James's father, his mother, a sister, a step-sibling), and his social worker, Veronica M.

I probably shocked them by handing them about sixty pages of references and notes on cases of akathisia in neurological patients. Nearly all the articles I cited said that the main culprit in akathisia was neuroleptic medications. I had only been able to find two cases where the akathisia seemed to be caused by the brain injury. In both these cases, diffuse axonal damage in the frontal lobes was involved. The treatment was anti parkinsonian drugs, the ones used as an antidote to neuroleptic side effects. Again and again, researchers and textbooks warned of the high risk of akathisia and other extrapyramidal side effects where neuroleptics were used to treat 'agitation' in brain injury. A few had also noted the loose use of the term, 'agitation', although I don't remember if I included these. I talked about the completely inadequate definition of the term and the irresponsible and inaccurate manner of observing and noting patterns of behaviour in the neurosurgery unit.

A registrar, typically dressed in his operating garb, tried to counter what he supposed were my objections with explanations – such as, "But agitation occurs in brain injury." And, "The novel antipsychotics don't cause akathisia. And, "but he was on very small doses of these neuroleptics." And, "His movements were purposeful, not repetitious."

I had already anticipated and reviewed these in my written complaint. I explained this and gave him more details, but he just responded as if he had heard nothing. It is a good tactic, when dealing with such people, to ask them to repeat back what they have understood, but I forgot to do so. About half-way through the meeting the registrar excused himself for a surgical appointment and left none the wiser.

The resident was the only medical staff member present who seemed to have read and understood my complaint and seemed to be able to refer to specifics. He admitted that James had received up to three different antipsychotics, volunteering that he had received three within six hours of each other. He raised the point I had made about definitions of akathisia, that different definitions made for misunderstanding.

"How would you know whether or not he had akathisia," I asked. "You've only ever seen him for at most a few minutes at a time and your nurses don't know how to describe this kind of phenomenon."

The resident said, "We did have the consultant psychiatrist come down to examine him, and he said that the regime he was on wasn't likely to cause akathisia."

I asked if the psychiatrist had observed the full cycle of James's movements and learned that he had just had a quick look at him and had asked him a few questions.

"So he didn't observe him for any longer than any of you. And asking James questions, when he has no capacity to learn or retain, isn't going to elicit anything useful.

"Well, no."

I finished off what I had to say by quoting a nurse (who I did not name, of course, for his job would probably have been at risk) who had said, "Nearly every patient here is on Risperidone. No-one gets benztrophine unless they are absolutely *rigid*."

I explained to those present that this 'rigidity' means suffering from extreme stiffness that pulls muscles tight and caused severe pain and could go on for days and weeks and months and years if the cause was not withdrawn. I could have added, but I didn't, that the same rigidity often causes chronic constipation, difficulty in swallowing and constant drooling.

"That is cruel. You have to stop doing this in this hospital. Your nurses have to be educated and your doctors have to pay more attention to the impact of the medications they prescribe."

The Complaints person reacted with a kind of shudder.

After I finished the social worker asked the others assembled if they had anything to say. No-one else had much to say. James's mother gave a short speech to say that she was very impressed by the care that James was receiving there. Everyone she knew had spoken highly of the Royal Brisbane and she could only thank the doctors and nurses for their superb work. Obviously nothing I had said had made the slightest impression on her, or, if it had, she was at pains to distance herself from my opinions.

The resident (who turned out to be a quite helpful and kind doctor), said that they were going to be tapering James down from his antipsychotics anyway, because he wasn't psychotic.

I then remembered to ask for an MRI.

The otherwise somewhat kindly resident had obviously been primed to deal with this expensive request and sank rather low. He said that we should all be aware that, as with any medical procedure, things could go wrong with MRIs.

"What could go wrong?" asked James's mother.

"Well," said the resident, "He could be allergic to the dye they have to inject."

I couldn't believe these scare-tactics and rolled my eyes. "And of course, if he should have that rare reaction of anaphylactic shock, the Royal Brisbane and Women's Hospital would just let him die, rather than perform the usual treatment for anaphylactic shock."

"No, of course we would perform the usual emergency treatment," said the resident, trapped now.

"Well," said his mother, I wouldn't be in favour of him having that test, with those risks."

"Well, as his next of kin, I am in favour," I said. "There are no big risks with MRIs, except financial ones. They cost something like $700 each, so I can understand why the hospital is avoiding doing one. However I want him to have an MRI because we don't know what is holding up his recovery and an MRI will show much more than a CT scan can.

"Alright," said the resident. "We will do an MRI since you want to have one."

"It's funny," I said later to the social worker "I don't recollect dye being injected as a matter of course with MRIs."

"That's because they don't use dye in MRIs," said the social worker. "I can't believe he said that."

I also asked if James might come home for Day Leave. They said they would consider it.

Day Leave! 13 June

We kept the wheelchair for two weeks, from 2nd June to 16 June. By the second week, James akathisia had diminished and his walking had improved.

The 13th of June was his first day out of the hospital whilst still an inpatient. It seems, in my mind, to follow on quite suddenly from James being tied to his bed to being able to leave the hospital, probably due to my having spent some of that time working in Victoria. Looking back, a significant improvement coincided with a rapid reduction of those psychotropic medications.

I was later told that James was the first brain injury person ever to go on Day Leave from the Neurosurgery ward. Perhaps it had not occurred to anyone else to request it. The fact that I was a psychiatric nurse helped the decision.

By coincidence the 13th of June was also the day I was to pick up our friend Jill at midday from the airport train at Roma Street station. I had asked her to come down to Brisbane and keep me company on what had been a lonely time visiting a very ill James. Before I was due to pick her up, I went in to visit James at the usual opening of hospital visiting hours at 10 a.m. There I discovered that James had been granted Day Leave. Obviously he wanted to come home with me. It was going to be difficult looking after him and meeting Jill.

James and I wheel-chaired it to the hospital lobby. It was about five minutes walk up the hill to where I had parked the car and James had to sit down on a low brick fence and rest, before we had got more than half way. He didn't recognise the outside of the hospital where he had worked for the past few years and he guessed that we were in New South Wales. He received the truth politely but it made little impression on him because information that he worked as a cleaner and that this was Brisbane really didn't make much sense in the scheme of things that his mind currently retained.

He also didn't recognise his car, but he was kind of interested to see what kind of car he drove anyway.

Things only started to look a little familiar as we drove down his own street.

We had just time to fix coffee and for me to study a map in order to know how to approach the station in the busy built-up Brisbane CBD. I usually walked to and within the CBD, so I didn't know how to navigate the streets and tunnels in a car. I couldn't make head or tail of the approach because, due to the incessant rebuilding of infrastructure in Brisbane, the map and even Google-Earth, had not kept up with the changes.

We were going to be late and Jill did not know Brisbane. James, however, had known this area like the back of his hand. He had participated in protests and had content-managed political internet sites dedicated to fight the specific infrastructure changes involved prior to the accident.

Upon his return to his home on this first Day Leave, he had recognised his own street and had immediately found his way around his own house, which he had lived in as a child and then returned to in the past few years.

"Do you think you might remember how to get to the station, James?"

"I think I might," said James.

And he did. Although he would not subsequently remember any of this day, he correctly navigated all the way. When we got there he expressed disgust at the cost of parking, when we left the car under the station to go and meet Jill. Jill was, of course, surprised to see James out of hospital. James recognised her and said how pleased he was to be on Day Leave.

On the surface, it may have looked as if James was aware of what was going on and where he had come from, but it was more as though he had limited access to the physical world and common realities, via his visual and spatial memory (despite his belief that New South Wales had amalgamated with Queensland).

While he seemed to have some grasp of space, he still had huge problems using telephones and spent part of his few hours home dialing the phone in his bedroom from the phone in his study (which was on a different line). He would often mis-dial and this gave him much disquiet. Had this system failed too? The telephone seemed to symbolise one of the ways out of his mysterious captivation, with his inability to do what he wanted or to keep track of things or to be the free person he thought he had been.

When it was time to go back to the hospital he was very worried. Worried that he wouldn't sleep. Worried that he wouldn't see me again. Worried that the telephones would not work.

12 June MRI performed

I arrived late the morning of James's MRI. James's father was already there with James, who looked foggy, having received 5mg of Valium.[25] To demonstrate to the MRI technician that James might need extra attention, I asked James where he thought he was. We were standing in a small anteroom decorated with medical posters and exit signs, near a window where on could see the MRI technician was preparing his computer to test James and, through another window, a large MRI machine with the plastic bed that feeds the patient through it. James looked around briefly then said that he thought we were in a national park.

"Well, you're right in one way, James. It is a kind of institution, but I'm pretty sure that we are actually in a hospital, not a national park," I said.

James took this correction good naturedly, as though he wasn't constantly interrogated and corrected. In fact, of course, he had little or no memory of those interrogations. He didn't have any concept of how far off the mark he had been. I think he saw my correction as being some kind of specialist precision thing, which he couldn't really be expected to know – a little like the difference between two different kinds of classical Greek architecture. Okay, if I said it was a hospital, I probably knew what I was talking about. The MRI technician raised his eyebrows infinitessimally at James's father, who shook his head wryly.

15 June Interview with Stacey B., policewoman investigating accident

This interview was conducted at the end of a hospital corridor, where we sat at the window on a bench. The deficits to James ability to reconstruct who he was and what might have happened were evident in the important part confabulation played in his responses. These responses were not lies, they were simply the naïve rendering of his injured brain, which was doing the best it could to fill in blanks. It was a bit like going, 'la, la, la' when you don't know the words to the music. Prompted, James agreed that he had been on his way to work. When asked what kind of work he did, rather than say that he did not remember what he did, he came up with a mishmash that I cannot remember in detail, but it seemed to have him tutoring and running computer services at the hospital in his capacity as a patient support officer (the hospital's technical name for a cleaner). After this interview James became very worried that any investigation might lead to finding him 'guilty' because he could not remember what had happened. He only took in a small portion of what the policewoman had to say - I now realise this was due to his slowed processing skills which were only explained to me later. He retained anxiety (for a few days) that he had been unable to account for his actions to a member of the police.

The policewoman drew a sketch of how the accident might have played out. The sketch above is much more detailed than hers. The road narrows at the bend where the 'x' is. The byway was too small for two cars to pass each other without one going up on the bicycle path, so the fact that the passenger described the car as over the middle of the road as it was proceeding around a blind corner seemed damning.

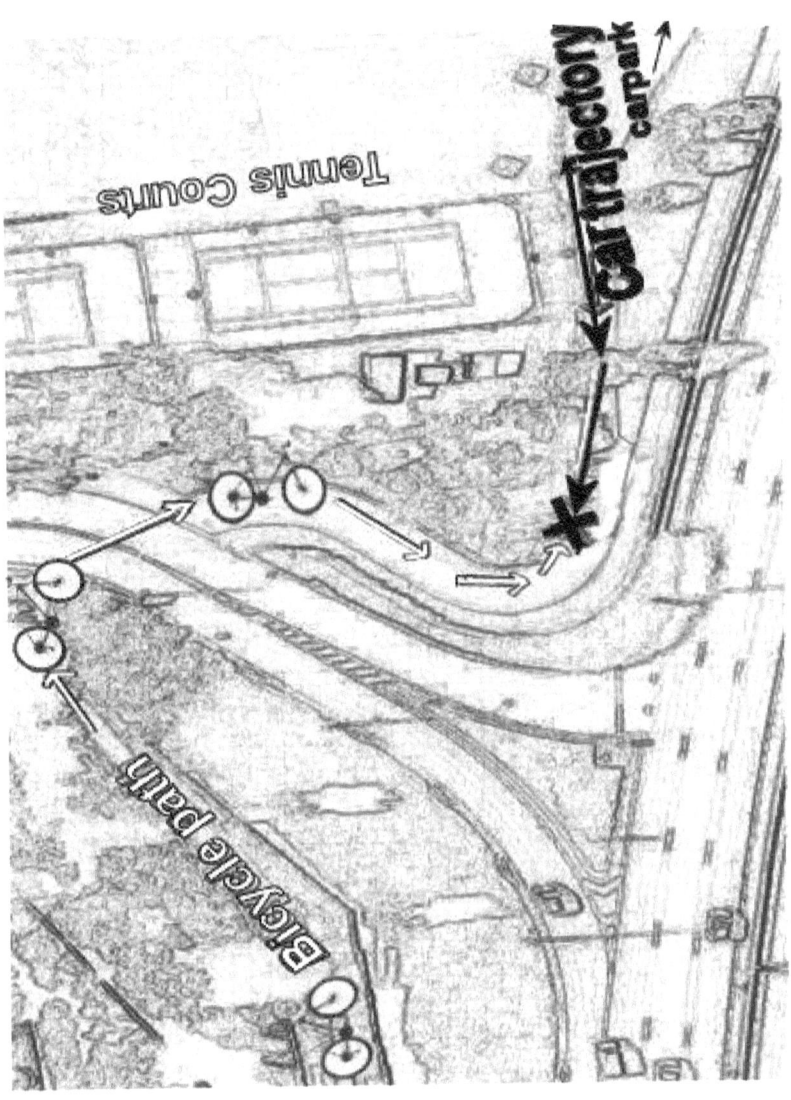

Fig 1. Diagram of location of accident with assumed trajectories and 'x' at presumed impact.

The corner would not have been totally blind if a vine had not grown over the chain-mail fence, which would otherwise have permitted a view round the corner. The police kept asking why

James had not been on the bicycle path. The last thing James ever remembered was a vague impression of people occupying the bike-path. It seemed obvious to me that there must have been an obstacle on it, such as pedestrians or cyclists coming the other way, preventing James from immediately accessing the path. Since it was legal for him to ride on the road, I could not see the relevance of this question anyway. I could not recognize the names of the streets or the area, in part because I had misunderstood which direction the hospital faced, when the policewoman described where James had been coming from. I did not therefore realize that I was walking that precise route every day to the hospital to visit him.

Revisiting the scene of the accident - Attempts to find out what James might remember

After the policewoman's visit, James's solicitor managed to impress on James the importance of understanding as much as possible about the accident. James was able to force himself to focus on the task. On one Day Leave, on his solicitor's advice, we walked the area in an attempt to jog James's memory about the accident and took photographs. James could remember details of going back home to fix a puncture and ringing work to say he would be late because of this. He could remember parts of riding the route to the hospital. He could remember right up to crossing the road to access the bike path on the access road and finding that it was blocked by other people... and then his memory stops.

Each time I was obliged to revisit that place, in person, when examining police photographs with James, or via Google-Earth especially, the whole mystery of fate and the unchangeability of the past would compel my mind. And this would be caught up with the desire to know the truth of what actually happened. Modern technology meant that I even went to Google Earth wondering if that near-fatal moment had been preserved with the car that had hit James caught cutting into the wrong side of the road as it approached that blind corner. But the space satellites had been filming elsewhere at the time.

Two years later I still had daydreams of going back in time and avoiding the accident, meeting James on the bicycle path and stopping him from going ahead at just that moment. I still imagined visiting him in Brisbane, as if he continued to exist there in his pre-accident person. Although I knew this was not reality, it was refuge for that part of me that used to relate to the parts of James that had changed. When we actually returned to Brisbane in later years I would keep a stealthy eye out for the old James in the new James.

15 June James wrote a letter to his mother

On the night of the 14th June James and I had walked around the hospital atrium, before I left him to sleep. I remember that James stopped for a moment outside the hospital chapel. A propos of nothing, it seemed to me, he said, with a rather strange, bland look on his face, "Personally, I would describe myself as non-denominational." I said I was surprised, having known him to be an atheist.

The next morning when I was to take James on Day Leave, he asked me if I could help him with his mother. A problem had arisen, he told me, the evening before. I was pleased to hear him able to remember that something had happened the evening before. His mother, he told me, had upset him by telling him that he needed to pray to God for forgiveness in order to be healed. Although that seemed reasonable for her to say, given her religious point of view, James was so bothered by it that he had retained the memory.

"How can I tell her to stop doing this?" he asked. "I don't think I am strong enough to tell her to stop visiting, but this is really upsetting.

It might seem petty to take something like this so seriously, but it was clear that her behaviour fed into dark childhood conflicts and James was truly afraid of losing the few fragile threads of his adult identity that he had so far retrieved.

I suggested that he write a letter asking her to respect his beliefs and their relationship as it had been prior to his accident, where they only met by mutual agreement, and that they meet, furthermore, in a neutral place, not her home and not his home.

James thought that would be a very good idea. Jill, who was visiting Brisbane still and I, helped him to compose it, but we did not interfere with its message.

We gave it to the social worker to deliver it to James's mother. The social worker said that it came at a good time because James's mother had been telling people on the staff that James only trusted her and something to the effect that she should have control over who visited him. Pretty soon it got to the point where his mother spent so much time on the telephone to the nursing staff on this issue as she perceived it, that all her calls were diverted to the social worker for special handling.

More than two years later James's mother thanked me emotionally for taking care of James. James and I went to meet her in a church where she was singing in the choir. I tried to tell her about the circumstances of the letter, but she didn't want to hear. It turned out that she had been afraid that James's accident had been an attempted murder. Perhaps she was trying to protect him. These fears all went back to conflicts in his early childhood. There was no evidence of foul play and we were able to reassure her then, although I don't think she will ever be permanently reassured.

As it turned out, James began to stay out later and later with me on Day Leave, only returning around 9 p.m. after visiting hours, so he saw almost no more visitors, including his mother.

Thinking back on this period, it is possible that some of James's relatives logically construed his sudden disappearance from the ward on Day Leave in my company as a threat by a strange woman to take over James's life. Why would they trust me? If James made the decision to accompany me, was he in his right mind? I am seven years older than James. What was he doing with me in the first place? Was his mother concerned that I was kidnapping him? Then again, his father and closest sister had recommended me as Next of Kin, and knew me, and thought James should be with me.

The Day Leave permission came out of the blue. One day a doctor said that James could leave with me. My being a psych nurse influenced this permission. No-one knew of anyone else ever having had Day Leave or even having asked for it. I simply grabbed a wheelchair and took James out as quickly as possible, in case the opportunity disappeared, but also because I had to meet Jill's train from the airport. I never thought to ring his relatives before I left the hospital and I didn't ring most of them after that, because looking after James took all my energy. I failed to consider how his visitors would take it when they came to see him, as usual, that day. I left it to the hospital staff to explain what had happened and most of them probably didn't know either. Ultimately James's father probably told them what had happened, but he had to ring James's home phone number to find out where he was after he had gone to visit him and found an empty bed. So I probably inadvertently caused a lot of confusion.

Slaughterhouse-Five: Billy Pilgrim and me

As I have mentioned, I found out later that James considered me on a par with Billy Pilgrim in *Slaughterhouse-Five*. This meant that I was not quite human in his eyes. He treated me like a saint or a famous person. At the time his manner and expressions seemed quaint and rather stilted. He referred to me as "Miss Sheila Newman" when talking to the staff about phoning me or waiting for me to visit. It made me feel a bit like Mary Poppins, Holly Golightly, Miss Hargreaves, or some other eccentric literary character. I was not too phased by the strangeness of it all because I knew that you can expect all kinds of changes in vocabulary, delivery and personal quirks as a person comes out of a coma and I knew James had a long way to go.

Not only did James speak of me with reverence, he treated me as if I could do no wrong at all. There was nothing he would not do to please me, if I would only stay with him. And, when he was on Day Leave, he kept me in sight continually, the same way a small child or an insecure dog will.

I got rather used to this and came down with a bit of a bump when, weeks later, after a few hours with his own room at the Mater Private secondary rehab hospital, he realized that I was an ordinary person whom he knew through ordinary events, even if I was his special friend. It wasn't that he treated me rudely after that, but he realized I was not perfect and therefore would argue the point where he had not done so before. Since he quite often got the wrong end of the stick, this was the beginning of a period with a lot of arguing and indignation. Being perfect was definitely more restful.

Sex on medication and off medication

Whilst James was still an inpatient at the infernally noisy and rigidly institutional Royal Brisbane Hospital, but had Day Leaves, he relished lying in the quiet of his own bedroom. The windows were shaded by crafted metal hoods, and the muted tropical sun cast silhouettes of shifting bamboo stalks, dancing butterflies and birds on long drifting curtains. Coincidentally, I had purchased these sumptuous two meter cream curtains from the Brain Injury Foundation opportunity shop in Paddington a couple of years previously. I remember thinking each time I went there of how lucky I was and how unlucky they were, however I would later ring up that very foundation for supportive listening and helpful advice.

I knew that brain injuries could affect sexual performance and sexuality itself, so, when I first found myself in bed with James after the accident, I wasn't counting on everything proceeding as it had before. I had read somewhere that couples after brain injuries had reported *better* foreplay than before - possibly attributed to loss of inhibitions associated with frontal injury – but I also knew that there was a risk of erectile dysfunction and of weird new behaviours. In Cathy Crimmins's *Where is the Mango Princess?*[26] Cathy's husband's frontal lobe damage caused marked loss of inhibition and he would, for a time, masturbate in front of visitors at the hospital. Fortunately James didn't do anything like that. Although James had not given me any cause to expect new weirdnesses, he had made one or two vague remarks in the hospital that made me think he might experience problems getting hard or ejaculating. He had not tried to masturbate in the hospital so really had little idea, but I got the impression that he did not experience the same rate of erections in the morning that he was used to. When we were alone for the first time, James was very gentle, very affectionate and very attentive. Although he could get an erection, he had some difficulty maintaining it and attaining sufficient physiological arousal to ejaculate. I was so fortunate to be a psychiatric nurse and therefore to realise that the drugs that James was on - Risperidone 1mg nocte[27] and sodium phenytoin (I cannot remember what dose) made it hard to get erections and ejaculate. I mentioned this problem to one of the resident doctors at the Brain Surgery Unit and he ceased both medications. With both withdrawn, erections and ejaculation were not a problem.

I was actually a little worried when the doctor so quickly withdrew the phenytoin sodium (Dilantin) because it could have been the only thing preventing James from having fits. The nice young doctor explained to me that James hadn't had any fits, so there wasn't any problem ceasing the drug. Getting things so drastically wrong was typical of the hospital, so I gently informed him of James's *grand mals* on the scene of the accident and in

Accident and Emergency. The doctor then said, "Yes, but he hasn't had any fits since then." I said that he had had a series of complex partial seizures which had only stopped when his intravenous sodium phenytoin dosage had been raised. "Oh, well, he hasn't shown any signs of fits since then and he's missed his regular dose for several days now, so just be aware and look out for the signs," countered the young doctor.

In *Where is the Mango Princess?* the brain-damaged solicitor's masturbating disinhibition could have been related to his concern about impotence caused by his Dilantin dose – although we don't actually know. If that had been the case, however, it just might have helped if the problems he may have been experiencing could have been repeatedly explained as temporary and due to necessary medication until he understood this. Quite a lot of dysfunctional behaviour in brain-injured patients can be traced back to confusing experiences in early treatment and recovery. Early treatment and recovery in brain injury contains similar elements and risks to early development and emotional learning of children. The application of this knowledge is very important in constructive neurorehabilitation.

Whilst in hospital, on sodium phenytoin, James had spoken without the occasional stutter which had afflicted him before the accident and his visitors would remark on the clarity of his communications, albeit marred by lack of memory and his tendency to repeat himself. In fact as the weeks went by after his sodium phenytoin was ceased, James's thinking and speaking deteriorated ever so slightly. I asked his rehab physician about 8 months after the accident if this might be a reason to recommence some form of milder anti-epileptic medication, but she strongly advised against it. Of course the side-effects of all those medications are not limited to sexual dysfunction, which is actually one of the lesser problems in the long term. I do still wonder sometimes if James's concentration might be improved with some anti-epileptic medication but knowledge of the costly side-effects stop me from encouraging such an experiment. These side-effects include sedation, which is, of course, not helpful for concentration.[28]

The importance of psychotropic drug knowledge for anticipating and responding to these problems like drug-related sexual dysfunction, which might have been misconstrued by patient and correspondent as long-term effects of brain injury and never spoken of again, is obvious. Yet it must be very hard for many people ever to mention these issues among so many new challenging behaviours and handicaps they may have to come to terms with. I have, as a psychiatric nurse, also come across men who were suicidal and whose marriages had broken up due to sexual dysfunction related to the use of antidepressants. These men's doctors had simply never bothered to tell them of this side effect. In my experience, the public would be unwise to think that most doctors and nurses in the 21st century have overcome the general shyness that surrounds sexuality. In fact, the protective barrier of modesty that protects patients and medical and nursing staff from inappropriately exploiting the various forced intimacies of the therapeutic relationship (i.e. lack of clothing, lack of privacy, intimate revelations) also often prevents them from asking vital questions about sexuality.

James's Orientation Book

Although James was coming home on Day Leave and could walk, talk and follow simple instructions, this did not mean that he could function independently at all. He was still an inpatient in the neurosurgical unit, suffering from Post Traumatic Amnesia and not fit for discharge. He did not know the day, month or year and, although he recognised his own home, he did not recognise his situation. He was only allowed home with me because I was a psychiatric nurse and could manage him.

A gauge of his continued confusion was the need to invent *James Sinnamon's Orientation Book for his Hospital Stay.*

This was a 'book' written soon after James commenced Day Leave. It is dated 19 June 2010. I remember the day we began it together. James could not bear to have me out of his view for a moment and would even wait outside the bathroom when I went to the toilet. So it was easy to get him to collaborate on a book to help him to remember the important points of his basic circumstances, such as why he was in hospital and how and when he got there, and what he was doing before the accident. The book was designed to reassure and to inform. It contained important instructions about how to telephone friends and relatives from the hospital, with detailed explanations of area codes and phoning procedures for various kinds of telephones. Due to persistent difficulty, these were eventually supplemented with pictures.

Telephone testing

Perhaps symbolic of the blocked and confused communication system in his own brain, his confinement to the hospital, separation from friends, family and work, and the unaccountable interruption to his normal life, James became obsessed with the Queensland and Australian telephone system, which he found very unreliable and difficult to use.

His memory problems made it difficult and near impossible to retain new telephone numbers and he was rarely able to remember old ones. Complicating these memory problems was his lack of dexterity, which meant that he could easily make fingering mistakes selecting numbers on a keypad. At the hospital, patients could either request a portable phone from nursing staff or they could purchase a pre-paid telephone voucher. Getting the portable phone from the nursing staff was already quite difficult for James. Until quite late in his time at the hospital, James remained disoriented about his immediate environment and the roles of the people who inhabited it, so he was not able to differentiate pyjamas from uniforms or other clothing and thus to spot nurses. Nurses, being nurses, were always busy doing something else and usually seemed to flash by anyway. Bringing the telephone, rather than a drip, medication or a chart, to a patient's bed was usually well down their list of priorities, and so days could go by without the fulfilment of promises to bring a telephone soon. Then the nurses could unpredictably bring two different kinds of telephone. One was a portable phone that dialled out through the hospital switchboard. Another was a phone which could dial-out direct. Several times, I gathered, when James was given a telephone, he tried to dial whatever number he had in front of him, but did not get anywhere, either because he omitted the '0' to dial out, or he omitted an area code.

His *Orientation Book* gathered in importance because I now included in it specific instructions about using telephones. I had to have several goes at this because of James's faulty perception of the telephone system and its various access points. It is a challenge to illustrate the telephone system, in all its modern variations, for viewpoint of a person hardwired to something general and with faulty software. I had several tries, beginning with words and culminating in pictures.

Using the telephone from the hospital

"Using the telephone from the Hospital
I can ring Sheila and my friends and family. There are two ways I can do this from the hospital.

First way:
I can use the white phone near my bed if I have paid for phone credit.
On 23-6-2010 I purchased $50 worth of credit from the Hospital Newsagent in the form of "Telstra PhoneAway."*
To use PhoneAway on the white phone near my bed:
Dial '0'.
Dial 18933 and follow the prompts.
(Be careful to punch in the digits properly.)
Your account number is 8089 1795 4352
You will be asked to include an area code with any number you dial. For instance, 07 is the Brisbane area code that you use for my numbers in Brisbane. 03 is the area code for my numbers in Victoria.
(If I have problems, dial 1800 616 606)

Second Way:
If I haven't any credit for the phone near my bed I have to ask the nurses to bring me the ward portable phone. Then I have to dial "0" to get a line out. Then I dial the normal number without the area code, if it is a Brisbane number."

Using the telephone(s) at home

At home there was another complication in the form of a second phone, an "Engin" phone, which made its connection via the internet. This phone always required the area code to be specified, even when dialing within Brisbane.

James would spend large parts of his day-leaves home dialing from one phone to the other to ensure that the connections were functioning. Due to his continuous confusion about area codes and the dial-out requirements of phones at the hospital and at home, he made numerous mistakes. Clumsiness and poor memory caused similar suspicions about his computer.

From its inception in 2006, maintaining control over access to content and contributors' identities on *candobetter.org* (later *candobetter.net*) had been a priority. This was why James used Linux instead of Microsoft or Apple and why he didn't use commercial webpage builders. He used Apache webserver architecture (common throughout the web), PHP language and a Drupal content management system.[29] The 'content' to be managed consisted of articles, comments, pictures etc.

James's concern over controlling access to the site was based on knowledge acquired from historical studies of the tendency for people to be harassed if they criticized the established political regime and our website, *Candobetter*, was a political website. He was correct in feeling that it was entirely possible and likely for governments to go after the identities of contributors and to then harass them – even by such apparently innocuous methods as targeting them for parking fines, speeding fines and other financial and administrative inconveniences.

This awareness had not been lost in the accident and James, who had only the barest, fluctuating insight into his condition, naturally became suspicious that someone was tampering with the phone-lines at his home and at the hospital. In fact, that there might be something gravely wrong with the whole state of Brisbane and NSW telephone system, if not Australia's and the world's. He muttered darkly and walked from one end of his house to the other, dialing from the phone in the bedroom, then going to listen to hear if the computer phone rang in his study. Quite often, of course, he would misdial a digit and the phone would not ring. Sometimes he would pick up one phone and dial its own number. Then there would be no connection at all, which of course, he found alarming. As he went from one phone to another, he carried *James Sinnamon's Orientation Book* in an easy to spot pink shopping bag. Often he would stop to check whether the telephone extension cord connections were properly sealed. He would put his pink bag down on the floor beside him while he checked the phone extensions and forget to take it with him when he proceeded to one or other of the telephones to redial its opposite. Arriving at his telephone destination, in the bedroom or the computer, he would cry out in horror as he perceived that the pink bag was missing. Still with poor balance and walking, he would totter through the house, scanning his surroundings for the pink bag. When he saw me, he would demand that I stopped whatever I was doing in order to tell me the "extremely worrying news" that he had lost his *Orientation Book*. "It's in a pink bag. I had it with me. I'm sure I didn't put it down. I have no idea where it has gone. It's terribly important to find it. It tells me how to check the phones and it has everyone's phone numbers in it and it says who I am and what has been happening. I'm terribly worried! Please help me find it!"

It was never very far away.

At other times I would again be interrupted. "You must listen. Something terrible has happened. I cannot dial out. The phone system isn't working. You may think that I am exaggerating, but I truly think this may be a sign of something sinister. ... No, no. I assure you I haven't made a mistake in dialing. This is the number I dialed from this phone...."

We would find that he had failed to dial the area code from Engine, or that he had dialed the wrong number or that he had dialed the number of the phone he was dialing from.

Sometimes he was relieved to find this out, but sometimes he refused to believe my explanation. Always, when we found the *Orientation Book*, he would start to make notes to ensure he did not make the same mistake again, by underscoring the relevant area codes and making me watch while he practiced dialing. The *Orientation Book* soon became a palimpsest of extra handwritten notes, underscores, diagrams and remarks to self, an archive of James's interactions.

I came up with the following pictures to illustrate what seemed to become so jumbled with words. He couldn't relate the picture of Australia to the concept of area codes, but he had more success relating the pictures of telephones to their locations in the hospital and in his home, and then to the instructions about which dial-out digits and area codes were applicable.

Phone by James's bed.
Dial '0'
Then dial 13933 and follow the prompts.
You will be asked for a voucher number.
You will need to dial area codes in front of any number you call.

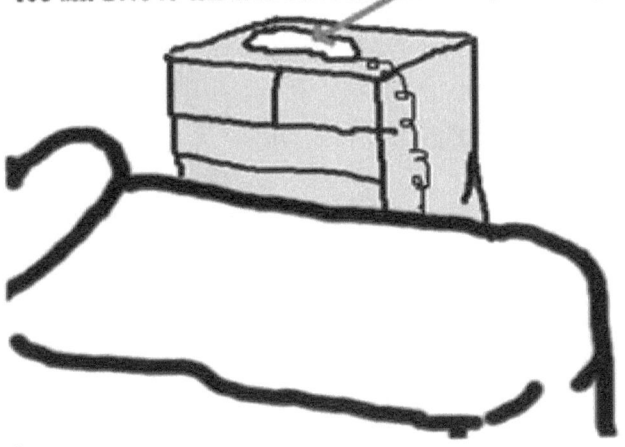

Phones at home
Don't dial '0'
If you are using PhoneAway, just dial 1800 150 117
then follow the prompts.

"Area Codes and lines out"

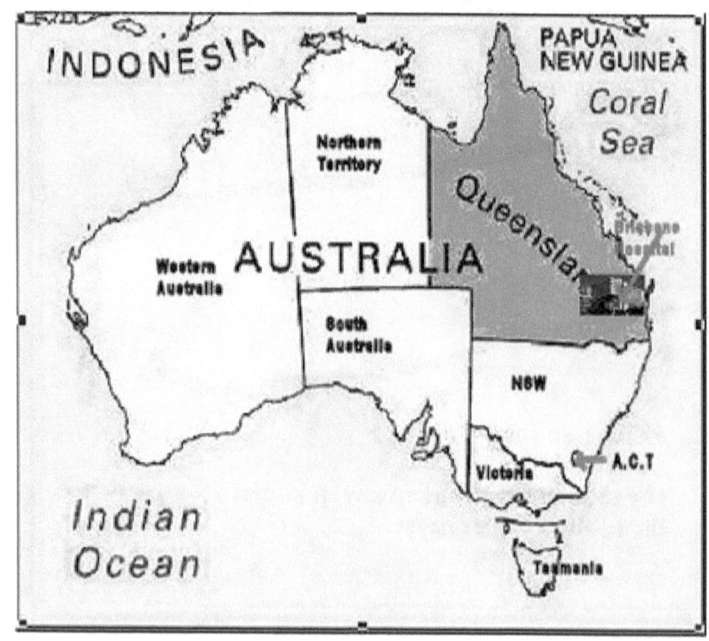

"The hospital has its own network. Most organizations have their own phone network, which you have to dial out of to get a line out.

Queensland has its own network.

Australia has its own network.

If you are using a remote telephone system, like the PhoneAway one, then you need to specify area codes, no matter where you are located.

If you are using a home, private telephone, then you only need to use an area code if you are phoning a number outside the state you are in."

I soon learned that James could take things in better from diagrams and pictures than from writing. Although he was writing all the time, his memory still didn't function enough for him to take in any rich information conveyed in writing. Illustrations convey their meaning instantly, so no great memory factor is required to understand them. And we eventually received formal confirmation that James's visual-spatial memory was more intact than his verbal memory.

Computers and brain injury: lost emails and lost passwords

When James came home from hospital for the first time he went almost immediately to his computer. This was very cheering, like bringing a dog home from a serious operation at the vet clinic and finding that it headed straight for its food bowl and barked at the postman.

He searched for emails from me first of all, in an attempt to find records of lost memories. Unfortunately, he could not find them, although there were thousands, and he immediately formed similar suspicions to his telephone system ones about interference with his email. He later found a lot, placed somewhere unexpected – under "Population" or something like that. He then lost them all again.

Fortunately the password for collecting email was automated, so he could get new emails, despite not remembering what the password was. Unfortunately, however, he could not properly remember his complicated system for automatically filing them, so he would catch a tantalizing glimpse of an email heading as it arrived, and then the message would disappear, heightening his suspicions about malevolent cyber-interference. Unfamiliar with the several open source browsers he used interchangeably to manage his email files, I could not find them either.

I thought it would be awful if James were to lose all his records, and especially email records of people he knew and had interacted with over years. Email records now represented a modern possibility of artificially recovering memories. No-one knew James's potential for recovering memories locked up or isolated by scar tissue in his brain, nor whether it was possible to stimulate his memory by providing old cues, but we certainly would have plenty of material to do this with if it were possible to do – as long as the computer didn't die or lock up on us. Fortunately the computer didn't stop working. For weeks and months there were no power outages or his computer was able to reboot without the need for a password.

Although James felt he knew his way around inside the computer, he didn't really. I knew though that there would be months and even years of improvement ahead, so the mere fact that he recognized his office and knew what a computer was, meant that he was on the right track. Looking back, I can see that he operated with the memory of knowing what he was doing, but filling in a host of blanks with the executive function equivalent of "tra la la la." He had a kind of generic, theoretical idea of what was in his computer and how it was organized, but he didn't have access to the actual detailed memories. Whether those memories still existed in his brain or had been completely destroyed or damaged, only time would tell.

James memory performance just out of hospital reminds me of the difference between a Windows Movie Maker Project file (MSWMM) and a Windows Movie Video file (WMV). The first kind of file is the one you work with as you edit film, and you can save it, however, without the Movie Maker program to view it with, it only contains a facet of the much larger work it represents. If you try to move a Windows Movie Maker Project file from one computer to another, you won't be able to open it on the new computer. That is because it only contains visual links to much larger files, which remain on the first computer. With James it was as though he still had the Windows Movie Maker project files, but he had lost the power to operate the Movie Maker program, which uses those project files to open up the real data. So the data itself was lying around his brain in piles that had no connection to each other until some other program assembled them together. More generally speaking, the accident had knocked out a huge amount of James's brain's indexing system. How permanent this was would be the big question.

As I have mentioned already, early fears of mine had involved the idea of James's computer shutting down and of our never finding the password to turn it on again, and thus losing, not only programming access to candobetter.org and some other websites., but also access to backups plus James's records of earlier professional work, articles in progress and all his email.

James did once turn Tibrogargan off, in the certainty that he knew the password. Jill and I were both present and, when we realized what he had done, we stared at him in horror and berated him hopelessly. Naturally James found us unreasonable – until he realized that he didn't remember the password after all.

Jill and I frantically tried to decipher the faint scrawls on the post-it notes around the computer, while James stabbed combinations of letters impotently and inaccurately on his keyboard.

Tibrogargan's screen was now dark, the computer silent, and we could not use the cheap Engin telephone. *Candobetter* still functioned on the remote computer, but if anything went wrong, how could we get to it?

Then, suddenly, a day and a half later, James called out, "I'm in!" And he was. Tibrogargan was humming again. James reminded us that he was in the habit of using parts of popular song titles for passwords and that there were some he used more than others. One of them had worked. That meant that his own brain indexing system also still functioned to some degree. I also noticed that James could often remember the numerical internet protocol (I.P.) addresses (invisible and uninteresting to most people) of his computer sites and of his computers, even when he got confused about their names.

When James confidently turned his Apple computer off, however, he found he had no idea what the password was. And so that computer remained off for many weeks, until finally James found a way to start it up without the password by searching on Google for ways to get around the problem.

Finally James gave in to us and adopted a general password that Jill thought up and we all agreed on. He would frequently need to be reminded of this over the next few months and even that password grew complicated by different spellings and hyphens, capitalizations and numbers substituted for letters.

Although James could get into his computer and access all the programs, he was not able to remember his 'root' access password, which was necessary to make fundamental changes to programming, content and record storage in Tibrogargan and on the remote host. I was relieved about this in the short term because he was easily capable of destroying the function and content of his computer in his current state; by trying to reorganize what he incorrectly perceived did not work, getting distracted and starting all over again in mid-process, furthermore.

Of course at some stage I hoped James would recover enough to begin running *Candobetter*. For this he would need his root passwords for Tibrogargan and Reykjavik. Reykjavik was the name of *Candobetter*'s remote virtual computer host. Tibrogargan was the name of the 'local' computer 'hosting' *Candobetter* in James's home.

James still felt that he would eventually come up with the root passwords. Perhaps he did come up with them, but tried them out in such a disorderly and restless fashion, with his fingers prone to press the wrong keys or two keys at once anyhow, that we could not be sure of anything.

The way James handled his computers and his passwords are clues to the way he had organized and prioritised information in his brain prior to the accident. It seemed that passwords, I.P. addresses and concepts associated with dealing with his computer and websites remained prominent and uppermost in his mind On the other hand, if I had not taken him back home and supported his reconnection with Tibrogargan, these important facets of his pre-injury life might easily have been lost to other externally determined priorities. Hospitals, rehabs and internet-naïve parents are likely to underestimate the importance of a brain-injured person's virtual life. The only time that the retrieval of software engineering and programming skills may be supported is when they are integral to the person's immediate prior occupational function. James's professional software engineering career had been interrupted, and he was actually employed as a Paatient Support Officer to clean offices in the hospital where he was now an inpatient. Furthermore, even when there was a later attempt to help James access computers, the general naïve bias towards Microsoft would inevitably overlook James's considerable and sophisticated engagement with Linux. It was up to me to retrieve and preserve these skills until (if ever) James could use them reliably again.

An Enthusiasm for stationary

James's office and home showed that he loved RJ45 ethernet cables and new stationary. The cables snaked out of a huge old router and festooned the walls and ceilings, one of them 50 meters long winding all the way to the main bedroom, where it hung from a hook originally designed to anchor mosquito netting over the bed. There were many containers for pens and pencils full to bursting and no surface was absolutely free of these small rolling cylinders. Desks and shelves held many lined A4 and foolscap notebooks with records of James's computer, political, and personal activities over the years. Most of these notebooks were discontinued abruptly as the identification of some new exciting phase in his life gave James an excuse to invest in a new notebook. Ah, the pleasure in turning back the designer cover and holding one's pen poised over a new first page! I know it too.

There is, however, such a thing as having too many notebooks going at any one time.

One of the well-known features of brain injury is damage to the frontal lobes whereby emotions become less well-controlled and enthusiasms less restrained. Just the general inflammation of a brain post injury, even where this will not be permanent, can cause transient personality and behavioural changes. James was no exception. Fortunately, as I have already mentioned, he did not suffer embarrassing disinhibitions like Cathy Crimmin's husband in *Where is the Mango Princess?* He did, however, display some new behaviour, for instance, a hypertrophy of his enthusiasm for notebooks and complex record keeping.

One day, whilst still on Day Leaves from the Royal Brisbane, he announced that he really needed some new notebooks to get some order into his life. This meant that the highlight of our day was a trip to Officeworks.

James came back with twelve bright red school notebooks with lined pages. He intended to have the first book as an index to the rest and then to have a table of contents for each separate book.

Each volume had to be numbered.

I can still see James on the couch in his computer room poring over these little red notebooks, numbering each page. I wondered how many he would succeed in numbering before his energy ran out. In fact, he ran into a different problem, caused by his poor concentration. He couldn't number the books consecutively. He realized this because, after numbering three, he found that they each seemed to have a different total number of pages. He considered the idea that some big notebook manufacturer was short-changing the public by surreptitiously supplying fewer pages in some units of a standard notebook series, but also reluctantly considered the alternative, which was that his injury was preventing him from counting reliably. To disprove one or other explanation, he asked me to number the pages. He conceded that each volume did indeed contain 96 pages, and was momentarily appalled at the extent of his disability, but this insight quickly receded and disappeared until the next confronting experience. If he could not concentrate long enough to number 96 pages at a time, he also could not concentrate long enough to feel sorry for himself or frightened by the consequences of his injury.

Perhaps harking back to earlier school-life, James used artistic balloon lettering, with unusual facility, to write the headings in his red notebook series. The writing underneath the headings, however, was just barely legible, with b's indistinguishable from p's.

I don't think the balloon lettering lasted much longer than a month, when it disappeared as if it had never happened. A universal b and p confusion persisted for perhaps a year and a half. It was a very occasional feature of his pre-injury writing.

James completely numbered only three of the twelve books, with my help. He ultimately used four, one of which where he only numbered the pages he used, up to 20. A fifth book has his name on it, but no volume number or title. The numbered and titled volumes are 1, 2, 3 and 12. Each of these four is titled, "JAMES SINNAMON – IMPORTANT NOTES, Post Accident 2010, Vol x of 12."

The notebooks show that James was trying to keep a systematic cross-indexed diary of his activities. Although he often protested that everyone was exaggerating his memory problems, these diaries show that he was trying to deal with them.

Inside the title page of Volume 1 is written, "Contents." Underneath are a series of headings, of which the intent is not always obvious. They read like this:

"4. Webserver Notes
4. From Content Management Notes
4. Hosts
6. Who.
[Drawn line.]
Related vol.3 [?]4 Copying large files with rsync

Vol 3, 10 Mysql administration
7. Getting there. [Another line drawn.]
8. TO DO 6 Sep 2010
9. TO DO 9 Sep 2010
Help Desk at Big Pond
12. Internet Connectivity lost 15 Sep 2010.
[Another drawn line.]
NOTES IN DATE ORDER
p. 17. 2 September: Phone calls regarding routers, modems, hubs etc
p. 19. 22 Sep, 25 Sep Installation of Apache, Mysql.
p.22. 12 Oct 2010 TO DO
p.24. 13 Oct 2010 Phone Commonwealth Rehabilitation Services.
25. 13 Oct 2010."

The last entry was a record of a discussion with his psychologist about his difficulty with the noises that my dog in Melbourne, Nubi, a Staffordshire, made when he was cleaning himself or drinking from his bowl.

The Table of Contents inside Volume 2 was:

"Page 1. Names
Page 4. Phone numbers
Page 6. Counselling
Page 8. To do lists

8. Candobetter
9. House (C- St.)
19.   Meeting

(intended 3 Aug

11. Legal Affairs"

The Table of Contents of Volume 12:

"Page 1. Sheila Newman and James Sinnamon
Page 2: Move to Mate-Rehab Unit (Tues 29 June 2010)
Page 3: Transfer between RBWH and Mater Rehab

hospital Tues 29, between 10.00am and 12.30 bm
Page 4. Sheila Newman's visit to Victoria (1 July
– 8 July)
  Page 6. Sheila Newman is going away for 9
days (1 July 2010-8 July 2010.
  Page 8. Sheila's flights to and from Victoria (1
July 2010-8 July 2010) See also page 18.
  *Page 9. Instructions for **PHONE USE**.*
(leaflets inserted for now)
  Page 10- Article by M.P. on Oil Spill etc (with
A4 insert for now – 1.35 bm. 5 July 2010.
  Page 12. Unusual messages
  Page 13 – Electrical problems of C- St and
going there instead of on picnic
  Page 14 – Get weekend leave for Sheila for
weekend 10-11 July she comes back and Thursday
  Table of Contents (continued)

At the beginning of this volume 12, under the heading, SHEILA NEWMAN AND JAMES SINNAMON, is written:

> *"This note is to make clear that*
> *James Sinnamon deeply loves Sheila Newman*
> *and hopes that Sheila also loves James as*
> *deeply."*
> *[I wrote] "Yes, of course I do, darling James –*
> *[signed] Sheila"*
> *[James continued:]*
> *"Sheila Newman will have to spend time in*
> *Victoria for up to approximately two weeks after I*
> *am put in the Mater rehab unit as inpatient.*
> *Obviously I will miss her and await her return*
> *and visits with great anticipation."*

James's last entry in the red notebooks was October 19, 2010. They preserve a useful record of his preoccupations and function over the period from about the 20th of June 2010 whilst still an inpatient at the Royal Brisbane. In these volumes he anticipates and plans for the move to the secondary rehabilitation hospital, Mater Private, and his preparations for my returning to Melbourne (where I worked) for two weeks. He also records the names of staff at the Mater Private, and his first day there.

A sudden spike in memory and concentration

One day I was just amazed. James wanted to have a sleep, but didn't want to be alone. I wanted to work on my computer in his study, so James chose to have a nap on the couch there. He said he would like to read for five minutes, first, and floored me by asking for my pride and joy, my theory on Incest avoidance and the Westermarck Effect in population spacing. I happened to have a copy printed out, so I passed this to him. Turning back to my computer, I fully expected him to put it down and fall asleep within 30 seconds, or simply forget it was there and start re-reading the ubiquitous *Orientation Book*.

Remarkably he read for about three hours, very closely, making notes and highlighting passages. Even though he probably never got past page 10 (of 60 pages), he took those ten pages in in detail. I know this because he asked me many questions and expressed pleasure that I had been able to show him that overpopulation was not, after all, something naturally built into all species. His questions and observations were intelligent and appropriate. He even wrote me an email to say how he felt, and so I have the following record with its lovingly formed typos preserved:

On 24 June 2010, James Sinnamon wrote:

> *"Dear Sheila,*
>
> *This is more a demonstration of my intellectual interest in incest avoidance than an attempt to add to your knowlege. I am impressed that incest avoidance*
> *1. has its intellectual elegance*
> *2. appears to be based on soild observation an therisation.*
>
> *With love*
>
> *James*
> *XXXXX"*

21 June verbal communication results of MRI 2010

I had for some time suspected that James's injury was far more profound than tests had to then indicated, because of his deterioration and the delay in his recovery. Although these delays could be caused by the drugs he was given, I kept asking staff at the RBWH what the results of the MRI were, but got nowhere.

On the 21st of June, when I brought James back from Day Leave, I managed to get some sense out of one of the residents, who probably regretted it. I never apologised directly to him, although I did ask one of the other residents to pass on an apology. The resident who told me of the results did so, perhaps naively, in a background where other staff were far cagier – possibly because they thought I might give them trouble by demanding more tests or attributing the results of the test to the medications I had objected to and the delay in reviewing them. What the resident did was, however, good. We needed proper information to make decisions about James's immediate future.

In an electronic note to myself I recorded from G., a medical resident at the RBWH on 21-6-2010:

> *"In response to my query re results of MRI which I requested for James, Re Why slow progress in recovery: G. told me a bunch of stuff that the CT scan had not revealed.*
>
> *No new injuries, original injury was quite severe.*
> *Frontal areas white ... Diffuse axonal damage.*
>
> *Neurology dept.*
>
> *He needs to be here for review.*
> *?Ventrical bleed into the*
> *Left frontal sub cortical*
> *Right corpus calossum*
>
> *No residual hemorrhage."*

So, what G. said was that James had severe frontal lobe injury in the nature of diffuse axonal damage. Yes, the haemorrhages had been reabsorbed and cleared away within the brain, but a whole bunch of tiny message conductors had died forever. Diffuse axonal damage is permanent and far reaching. Damage to the frontal lobes affects personality and memory. It changes a person in unpredictable ways. It is the stuff of horror movies.

I remembered James saying to me early on in his deterioration, "I guess what has happened to me is the worst thing that could happen to a person."

I cannot say when exactly he made this insightful remark. It seems to me that, except for when he was almost continuously delirious that, even though he did not know where he was or his age at times, he sometimes had a deeper sense of what was happening. I do believe that that deeper sense remains there and that it is the original James.

Despite this awful news, it was too early to know how James would deal with these injuries. As long as I could help by my presence to get James the best options, my effectiveness staved off that sense of hopelessness that surrounds so many brain injury cases.

Printed Report on MRI 2010

The scan was done on 12 June 2010. Some weeks later we received a copy of the report dated as printed on the 27/7/2010 as well as CDs containing all James's scans and X-Rays. This is how the report read:

"HISTORY:
Pushbike accident - traumatic brain injury, slow recovery.

TECHNIQUE:
Sagittal T@; Axial T1, T2 diffusion weighted imaging, FLAIR, SWI sequences.
Movement artifact was noted in a number of sequences.

FINDINGS:
Comparison is made with the series of Head CTs

performed dating to 18/05/2010.

Numerous foci of low intensity demonstrated on the SWI images within both cerebral hemispheres. This is most marked in the left frontal lobe posteriorly at the subcortical white matter and within the right cerebral hemisphere adjacent to the body of the corpus callosum. This has areas of corresponding high signal on the phase images and are consistent with haemorrhagic shear injury.

The cerebellum and brainstem appear normal (there may be a small haemorrhagic focus in the left midbrain, but movement artifact obscures the detail). Vascular flow voids are preserved.

Midline sagittal structures appear unremarkable. The craniocervical junction and visualized spinal cord is normal. No restricted diffusion detected.

Mastoid air cells and paranasal sinuses are clear. Normal bone signal.

COMMENT:
Diffuse axonal injury localized to the left frontal subcortical white matter, the right corpus callosum, and possibly in the left midbrain – suggesting a severe injury."

Another scan was done in mid-2012, with a much more detailed report, which is actually far more devastating to read than the one above. You can find it in the Appendices at the end of this book.

Interviews for Rehabilitation Hospitals

22 June Dr D. from BIRU comes to Royal Brisbane to interview James

James eagerly returned home to C. Street every day from the first Day Leave. Getting him back to the hospital each night became an increasing battle. He would think up really lame excuses that he would propose I offer to the hospital as reasons for him to remain home overnight.

He was really not well enough for me to look after alone 24/24, but he no longer needed the emergency support of a surgical ward and was well enough to go into rehabilitation on an inpatient basis. The major reason he had to return to the RBWH in these last few days was that the rules dictated that he could only access inpatient rehabilitation at the Brain Injury Rehabilitation Unit (BIRU) if he was still an inpatient. BIRU was purportedly the only reputable option for brain injury rehabilitation in Brisbane. Without rehabilitation and subsequent support for reentry into the community (we hoped) James's outlook would be greatly compromised. I knew we had to play by these rules, however silly they seemed.

One day we were rung at C. Street by a member of the assessment team at BIRU. This person, it turned out, after some confusion, had been misinformed by a member of the RBWH staff that James had been discharged. BIRU were proposing that James apply for an outpatient rehabilitation course of very low intensity with a waiting list several months long. What she was proposing was clearly inadequate. James would deteriorate significantly if left without supervised physical and mental rehabilitation. He was not safe to be by himself. He had not yet even emerged from Post Traumatic Amnesia. The person on the other end of the phone took some convincing, but ultimately conceded that there may have been a mistake.

We returned well before 5p.m. to the RBWH and asked the medical residents what was going on. The medical residents were just as surprised as us. They contacted BIRU and told them that James was still an inpatient. BIRU's representative doctor arrived the next morning and gave James an assessment.

What she told us was that, because Workcover Queensland would be funding James's rehabilitation, the chances of his getting a place in BIRU, which was the only free neuro-rehab unit in Brisbane, were next to none in light of the number of other, unfunded, brain injured patients competing for the same scarce BIRU beds. When I pointed out what a waste of hospital resources it was to keep James for weeks waiting for a bed at BIRU, the doctor said, "There are people who have been waiting and wandering these wards for months. They have nowhere else to go."

Unfair as it was, she said, the reality was that the system was quite capable of making James wait here for months too. She added that even if he got into BIRU, there was no guarantee that he would ever see a neuropsychologist. Furthermore, we were probably overestimating the extent of the programs available there. "Most days there is nothing to do in the afternoons and the programs in the mornings are rudimentary and don't last long."

"But BIRU is not the only place that does neurorehabilitation," she added. We were surprised, since we had been told that it was, by doctors, nurses and a physiotherapist family member, who had worked at BIRU for years prior to retirement.

"Because you have insurance funding, you can go to a private facility," said the doctor. She told us that she had worked in both sectors and felt that the private sector had a lot to offer and often provided much better programs and service, in much nicer environments. "Of course the private rehabilitation hospitals don't have the MRI and CT scan and emergency equipment that the public hospital related facilities have," she said, "But James doesn't need those now. If he should suddenly take a serious turn for the worse, they will simply transfer him by ambulance, anyway."

She gave us the names of a couple of facilities fairly close by and a bit of a run-down on their pros and cons. I said that we had still not given up on BIRU, but I would look into the matter.

I spent the next day telephoning the private rehabs, asking them what neuro-rehab experience they had and what they could provide. I found two. I selected one – the Mater Private Hospital - because they said they could provide access to computers for James.

We then visited them just before closing. The coordinator showed us around. The rehab area was located in a four story building with a pleasant wood-featured interior on human scale, lacking the sterile institutional imprint. It was likely that James could have a room on his own. Admission would depend on an interview with the rehabilitation doctor, which could be scheduled for the next day. Because the hospital was not a locked unit, the staff would need to be convinced that James would not wander out onto the extremely busy road in a state of confusion or throw terrible tantrums.

The human scale, the quiet and the prospect of a room of his own and therefore a good night's sleep made James desperate for admission.

Friday 25 June interview with Mater Rehab

I tried to impress upon James the importance of his showing the staff at the Mater Private that he would not wander. He needed to look relatively on the ball. It would not be good if he told them he was in a National Park in Sydney and lecturing classes in a university when the rest of us knew he still occupied a bed in the Neurosurgery unit.

I only wanted to get his attention enough for him to take in where he was going and not to leave unaccompanied, but I must have overdone it. The prospect of waiting indefinitely for BIRU in the Neurosurgery ward and my insistence on his need to reassure the Mater Private staff unnerved James so that he over-prepared for the interview by the Rehabilitation doctor. The day before he began writing notes and rehearsing what he would say. His preparation went well beyond basic orientation and promises not to wander. It began to resemble the preparations for a high level job interview. It seemed that job-interview preparation was the closest he could get to the idea of meeting the hospital admission requirements. He wanted me to ensure he was dressed impeccably, that we arrived early, and he tested a variety of attitudes, from intellectual and studious, to keen to work and interested in medical matters. He went over jobs he had held and work published.

The rehabilitation doctor had been expecting something much lower key and was slightly startled by James impassioned plea to be admitted and his promise to be an outstanding patient, but he took it in good humour and reassured James. It was Friday. There would not be a bed before Tuesday when, hopefully, James would be out of Post Traumatic Amnesia.

Last days at RBWH

When we went back to the hospital and told James's doctor and social worker at the RBWH, they were happy to refer him to the Mater Private. They seemed fascinated to hear our report on the place, since most of them were also only familiar with the reputation of BIRU.

James wanted to spend time with me in the hospital atrium and grounds as usual before I went home. I was pretty tired and so was he, but he was also very revved up by then about his interview with the rehabilitation doctor at Mater Private. As I wheeled him about in his wheelchair, for he had done enough walking that day, he babbled on about how he would not let the Mater Private down; he would make them proud; he would be the best patient they had ever had. He would work unstintingly and surpass their expectations... It sounded like he had now exited the job application concept and entered Olympic competition or something. He became so strident and manic and pathetic that I turned a little savage and begged him to shut up. He then became hurt and offended and I felt like a brute. But then he began to perseverate indignantly about my inability to appreciate his spirit and so we escalated into a ridiculous argument out there in the warm night air with James in his pyjamas in a wheelchair, raving on about performance and reliability, endurance and excellence, with a pyjamaed audience of patients wheeling drips and holding cigarettes.

The only hitch about getting discharged was that James had to be declared out of Post Traumatic Amnesia (PTA) in order to be discharged to a lower acuity hospital. Fortunately he had, for the first time, scored 100 per cent on his PTA test that morning. For him to qualify as out of Post Traumatic Amnesia, however, James had to get full marks three days in a row.

As I mentioned, the interview was on a Friday. Over the weekend no-one assessed James. The reason seemed to be that there was no-one in the entire neurosurgical ward prepared to do it over the week-end. The occupational therapist didn't work on weekends and a nurse qualified to do the test was too busy or something, though I would have thought that any nurse should have been capable of doing this. This was, if you will excuse the cliché, typical of bureaucracy. We suspected that bloody-mindedness on the part of some nurses might also have been involved, since one nurse seemed particularly resentful of James going out on Day Leave.

On Monday we were in a state of dread about the PTA score. We found someone responsible though, who tested James that morning. He got one question wrong but the hospital simply overlooked its own criteria. James was discharged on Tuesday, as though he were out of PTA. He was admitted to Mater Rehab with the idea that, even though he wasn't really out of PTA, he was safe to treat there as long as he gave his consent not to leave the hospital by himself until he was truly clear of PTA.

Discharged to Mater Private Rehab

Tues 29 June discharged to Mater Rehab from Royal Brisbane after stay of 42 days

From the 18th of May to the 29th of June made a total of 42 days. When James was discharged, he was still in Post Traumatic Amnesia, both technically and really. In fact Mater Private determined that James did not come out of PTA until the end of his second week of inpatient rehab there.

James's Diary: Mater Rehab —Private Hospital

James was literally counting the sleeps before he got to Mater Private Rehabilitation Hospital.

In balloon letters in his red notebook series, volume 12, he wrote:

> *"MOVE TO MATER REHAB UNIT Tue 29 JUNE 2010"*
>
> *"Am moving between 10.00 a.m. and 11.00 a.m. with help of Sheila Newman. Sheila has packed a large case in which I should be able to pack most of what I need for day to day use in Mater Rehab."*

He also gave a detailed account of his arrival:

> *"TRANSFER BETWEEN RBWH and Mater Rehab*
>
> *Tuesday 29 Jun (between 10.00 a.m. and 12.30 p.m.)*
>
> *1. Packing and formalities of exit from RBWH took longer than expected.*

2. Road system in South Brisbane near Mater Hospital is very confusing causing us at least once to lose our way.

3. Streets close to Mater Hospital don't have much parking and parking has to be paid for in parking meters which require coins or credit cards.

As a consequence, we got there at 12.30 p.m.. None of the staff was cross as a result of our lateness. They were used to new arrivals having difficulty with the local roads, it seems. They were very friendly, welcoming and everything that was needed to settle me in. Sheila was also very helpful in the process (as usual). Without her help, things would be harder for me everywhere, even at the Mater Rehab Centre, where the staff have a much better attitude to patients."

Admission to Mater Rehab

James relaxed by an order of magnitude (to use an expression of which he is fond) as soon as he saw his room on the Tuesday he was admitted to Mater Rehab. It was a quiet single bedroom with a telephone, television, and a nice view of a busy urban village down below, with Queenslander-style wooden houses pitched on a hill behind the shops. James even recognized the street at the top of the hill where his sister Judith had lived before moving to Maleny in the country.

There was a pamphlet explaining exactly how to dial out from the phone, so he didn't need me to do new drawings and explanations for his *Orientation Book*, although he still made sure it was at hand at all times.

I helped him to unpack and hang his clothes and put his notepads and books away.

A succession of staff came in to introduce themselves, including the occupational therapist and two physiotherapists. The kitchen staff let James know when meals were, took his dinner order and offered us both afternoon tea. He was told that the doctor would arrive soon to do a neurological examination.

Before the rehab doctor and his understudy arrived, one of James's sisters rang up and asked to come and see him. It did not occur to me then, but, because I took James home for Day Leave every day during the last ten days or so, his usual visitors had found his hospital bed empty. I did belatedly email everyone, but as I have noted previously, relatives for whom I had been a stranger prior to James's dreadful accident may have had an understandable concern that I was gaining an unhealthy influence over James. Furthermore, James had written that letter to his mother. The sister who rang up was still quite close to James's mother. James said that she could come up and visit for about 15 minutes but indicated he was expecting the doctor. He meant her to understand that she should leave when the doctor arrived, but he did not make this clear.

One of his sisters was still there when the doctor came and took it upon herself to talk to the doctor about James's prospects. She also queried my ability to take care of him and seemed to want to discuss other arrangements.

James was more on the ball than I had imagined. He took offense on my behalf, whereas I was disinclined to get on the defensive. I could see that his sister meant well. Afterwards he told me that he had been very upset by his sister remaining there during the doctor's examination without his permission and was furious that she presumed to speak on his behalf and gave the impression that she did not trust me.

Whilst all the above is understandable, including his sister's attitude, since she did not know much about my role in his life and may have been fearful that I was taking advantage of him in his weakened state, James's response was much harder for everyone to understand. He did not have the mental means to deal with the situation appropriately. As the doctor asked him questions and stood him up to assess some aspect of his posture, James suddenly said to me, "Sheila, we have to go down to the car."

Disconcerted, I said, "Not in the middle of an examination, maybe later."

"No," he said, "Now."

"Why?" I asked.

"Because we left my brown trousers down in the car," replied James.

This was sounding and looking quite odd. I feared that the Mater Private might think that James was going into a dangerous

107

panic and was about to flee the hospital and that they might then judge him unsuitable for admission and send him back to the RBWH.

"Please stop making a fuss," I said. "All your trousers are hanging there in the cupboard. There is no reason to go down to the car."

"We must go down to the car. Now," James insisted loudly.

I walked up to him and put my face very close to his and whispered, "What is going on?"

"Get her out," said James, indicating his sister with an oblique movement of his eyes. "I want her to leave. I can't stand her being here."

"Alright, but calm down or the staff will think you are going to be a problem, James."

I then went over to James's sister who was sitting in a chair looking very settled and told her that James wanted her to wait outside while he was examined.

"Why can't he tell me?" was her response, and she remained there.

I went back to James, at the other end of the room: "Would you tell her yourself?"

"I can't," hissed James, beside himself. "You have to tell her." I was caught like the meat in a sandwich. My priority was to try to keep James from looking wild and unstable. The hospital staff members were observing the situation closely. To them it probably looked like an 'outburst' from a new patient who was an unknown quantity. James's sister also did not understand what was going on.

James was getting more and more upset, so I went back to his sister and said, "Look, you really do need to go. He's not feeling right and there are too many people here. He wants you to leave. I'm sorry."

All James's sisters are tall, willowy and beautiful in different ways. This particular sister, at over 50 years old, still had long wavy blond hair, fantastic posture, and looked like a Valkyrie. She fixed me with an electric blue glare and said, "It is you who wants me to leave, not James."

I denied this of course. She got up after a minute and left.

The doctors completed their examination. The coordinator was present during the examination and had seen James's odd performance and his sister's exit. The coordinator's phone rang twice after James's sister had left and I thought it must be his sister because the coordinator said, "Yes, I promised you I would meet with you after the examination is over. It isn't over yet."

When we were alone I asked James what on earth had got into him. He said he realized now that he hadn't handled the thing well, but had felt compelled to assert himself. His feelings harked back to a very difficult childhood, with sibling rivalry and conflict between his father and mother. .

We discussed the matter some more and then I convinced him to write the coordinator a letter to reassure her that this was an unusual outburst and that he understood that his behaviour had lacked judgment. He wrote four detailed pages about his motives and perceptions, apologizing and swearing he would never run away. For someone who could barely hold it together when the unexpected happened, James did surprisingly well expressing himself in writing on issues he had time to consider. He rewrote the letter about four times, exhausting himself, but the coordinator was reassured.

"He put his little heart and soul into it, didn't he?" she remarked, when I asked her. And he had. His heart and soul were bare for all to see at this stage.

James remembers who I really am

James had many visitors on his first day at Mater Rehab. After the doctors' examined him, I left him for the day, but I returned in the evening. Perhaps it was the nice treatment and therapeutic interactions with new staff in pleasant, human-sized surroundings with a view of his sister Judith's old neighborhood, which he recognized, but James had taken a few big strides closer to reality. When I returned that evening I sensed that his attitude to me had subtly changed. In response to my questions, James said that he now remembered how we had known each other in real life, and that details of our relationship were beginning to come back. He explained that he had thought of me as a kind of famous person who had taken this welcome but inexplicable interest in him. Now, however, he could see more mundane connections. He could remember how we had met and times we had spent together, although they seemed very vague and lacked detail. So I was an ordinary person now, not a star. That was the difference that I sensed.

Barbara comes to look after James's house

I went back to Victoria for nine days to work and James's amazing, generous and delightful sister from Tasmania, Barbara, came to look after the house. At nearly 50 she was tall and willowy like his other sisters, and reminded me of the actress Julie Christie in *Dr Zhivago*. Barbara was also there to visit and attend to anything James needed in hospital and in case James got overnight leave. She took time off work and travelled to Brisbane with her foot in a caste, due to an injury. She then proceeded to hire a cleaning company to tidy the house, which owned by his architect father until recently, had been a repository for timber, old furniture and plumbing fixtures. It had also been used by James's extended family as a storage depot for decades. In Brisbane's tropical climate, it was an environment shared by growing populations of rats, mice and cockroaches, as well as the less numerous and less disconcerting geckos, possums and bats. In an act of generosity that would make such a difference to James's life after the brain injury, his father had given it to James a few months previously. They had been building a flat together underneath it. James had only just begun to put some order into it when the accident happened.

For several years prior to owning it, James had rented the house from his father and had not felt entitled to change much. Neither did I. Now, Barbara not only hired cleaners, she hired people to take the rubbish away and then she limped around with soda bicarbonate and bleach until the place looked starkly gracious. She also negotiated with James's father to vacate the furniture, house parts, mattresses, packing cases, old books, rocking horse, white upright piano, boxes of photographs, and piles of newspapers, that various members of the family had stored in the front room over more than half a century. She then got the rubbish removers to help her move furniture around using to advantage the well-designed spaces of the 19th century timber Queenslander house.

As James still says, it is a sign of real deterioration in quality of life that ordinary working people for whom these timber suburban houses with their ornate wooden facades were built in the late 19th century, can no longer afford to live in them. The suburbs are now gentrified, as population pressure has inflated their prices, whilst forest depletion has made wood unobtainable at affordable prices.

Sadly, these wooden houses on stilts in Queensland towns are all that remains of a primeval forest of giant trees that once stretched like Eden from coast to hinterland.

Marooned in State Park at night

You have to go up into the Mount Coot-tha hills 7 km from the centre of Brisbane and then drive a winding road for 40 minutes to find any semblance of that once great forest, but you won't find trees the size of the original giants, even there. (The biggest trees remaining in Australia are in the Atherton Tablelands forests.)

When I came back from Victoria, James was still an inpatient at Mater Private Rehab and very keen to get out for the afternoon. For old time's sake, I decided to take him to a state park in the foothills of Mount Coot-ha to do some low-key short distance bushwalking in the drylands forest surrounding the reservoir at the base of the more luxuriant highlands and hopefully remind him of past expeditions. It was supposed to be a pleasant little adventure, but I was really testing his limits. I did not however intend for us to get stranded in the dark on the outskirts of the forest.

James had still not scored the correct answers three days in a row that would put him officially out of Post Traumatic Amnesia and his sense of humour was non-existent. He had Day Leave on the week-end. I picked him up at the hospital and drove him out to the foothills, where we parked in the car park near the tourist shop and restaurant, and walked down to the reservoir. James's balance and strength were still quite poor, but we walked for about an hour on a trail where we had often been, although James could not remember it, and I thought he did well. This was in a landscape where, a couple of years ago, we had had a furious argument about the shortest walking route to a landmark. It had turned out that I was the confused one then.

The sun was descending quite rapidly in the sky, so we began to hike back alongside the water. Finding our way on the track to the car park was difficult in the fading light, and when we arrived there it was nearly dark. It took me a couple of minutes circling around to find the way out and, to my great surprise, when we got to the park gates they were shut.

I had cut it fine before and no-one had closed them. Didn't they look to see if any cars were left in the car park? I got out and went to the gates just to make sure. Yes, they were shut. James leapt out of the car and began dancing around in the dark with his poor balance, making noises as if he thought we were lost in rangeland in the vast desert interior of the Australia, instead of adjacent to an outer-suburban reservoir perhaps 15 km from Brisbane CBD. In fact, that is exactly how he perceived the situation, I later gleaned from him. He thought we were lost in the bush. Unsurprisingly, in view of his altered orientation, he actually feared we might die 'out there'. Under normal circumstances, James would have had a mobile phone with him, but it had been destroyed in the accident. With him still protesting, we went stumbling in the dark to look for a phone. There were none. We could find no numbers to call park management anyway. We eventually felt our way through the undergrowth to a light source, which turned out to be a small house near the park, where a young man was practicing guitar inside a brightly lit lounge room with uncurtained windows. It took some banging and shouting to get his attention. He then came to the door but had no telephone. He indicated an invisible track to the nearest public telephone, half a kilometer down the road. I supposed we would have to call a taxi from it.

We trudged off in the dark, separated from traffic by a ditch, tall trees waving above our heads in the dark sky. James continued to expostulate about my monstrous, unforgivable crime of getting the car locked in the park. Although his diminutive Mazda was close to 30 years old, James held grave fears that it would not be there when we returned, if we ever returned.

I could just see a telephone up ahead, when suddenly James began running in a dangerously unbalanced way, waving his arms and yelling for attention. The ground was uneven and I feared that he would fall and further injure his brain, so I ran after him, yelling for him to stop. He paid me no heed. His adoration for me was definitely a thing of the past. How the worm had turned. I saw that he was running down towards a bus. I caught up with him at a little bus depot. He was at the door of a bus, asking the driver for help. I wondered what crazy explanation he had given the driver for his circumstances and winced at the idea that he might have told him he was brain-damaged, since I thought this could put the driver off, along with James's agitation, and cause a call to the police.

The driver, however, seemed immediately to grasp the situation. He was extremely kind and helpful and drove us to another bus connection. He was so helpful that James insisted the next day that we ring up and let Brisbane City Council buses know how helpful the driver had been.

I, however, had not been helpful. I had caused the whole thing, as James pointed out repetitively. We walked from James's house back to Mater Private. It took nearly an hour. We arrived there an hour late, but no-one cared.

In the morning I took a bus out to the same terminal, walked back to the car and retrieved James's car. Despite his very faulty memory, James remembered that I had got us locked inside the park and that he had had to run to get a bus. He retained the impression that we had been in great danger and that finding a bus had been a stroke of luck, without which we might not have made it through the night. He has never forgotten. Emotion is the stuff that burns memories into a person's hard drive.

This was one of several such examples where James retained a new memory early in recovery. Most or possibly all these new memories had in common that James was an active and central participant. Furthermore, I think that this was characteristic of the way that James had formed many of his autobiographical memories prior to his injury, going by what he retained after the accident. Perhaps we only remember things that we actively interact with, so that I might remember a landscape, but James, who might have been thinking about our relationship rather than the trees, might not have.

James rang his sister, Judith, and grumbled vividly about my appalling judgment and our near escape from some grisly end. His sister apparently said that these things could happen to everyone and he should try to see the funny side. James said that there was no funny side and that he resented her suggesting that there might be.

We went through quite a few months where the funny side was pretty much off the radar.

Emergency Counselling Sessions

James's sister, Judith, is a special education teacher as well as an artist. She had the inspiration to suggest to James that he attend relationship counselors with me. I was amazed that he took so readily to the idea, since prior to this, although our relationship and work together were blighted for months by severe quarrelling over different perceptions of an event, he would not spare the expense or the time, unable to believe that they could help.

James now began to drag me off to relationship counselors at the drop of a hat. Not that I was against seeing them because we needed any help we could get to get through this patch, but I could not help being irritated and stressed at the way he interpreted small events as crises. This apparently exaggerated kind of response to the unexpected is common in brain injury when emotions run too high and events go too fast for a slowed brain to deal with.

One morning I was driving him to Mater Private when James was an outpatient there. From a side street, I was trying to make a right hand turn into a four lane highway that led to a five-way intersection. Cars were streaming down in the direction I wanted to turn into, but the lanes I needed to cross were empty. I could see traffic bearing down from the changed light in the distance, so I crossed the empty lanes and stopped with the car's nose ready to insert into the stream going in the direction I wanted to turn. The car's rear end was pointing out in the path of the oncoming traffic but I had figured that that stream of traffic would just have to go round me while I waited for the stream I wanted to enter to stop when the lights turned. As they turned, I nosed the car into a gap.

It all felt a little dramatic. I commented, "Well, I don't think I'll try this again," but James didn't hear me. There was too much going on for him to take in what I said.

"How could you put us both in such danger?" he cried, and continued to berate me all the way to Mater Private, causing me to lose my temper and argue in a similarly heated fashion. When we got to the hospital he refused to go in until I promised to take him to a counselor that lunch time. So I promised.

It took the unfamiliar counselor (who charged us about $150) most of the session to work out that James was only recently out of the neurosurgery ward and that the incident he described as having happened the day before had only happened that morning and, finally, what had actually happened. She suggested that he was going a little hard at me and James suddenly stopped in his finger-stabbing tirade and burst into tears. He apologized for being 'an awful person' and begged my forgiveness. I said that 95% of the time he was great to get along with – which was true, even then.

The psychologist repeated this message to him and he wiped his tears away and thanked me.

Considerably washed out we walked down the stairs and into the street in searing heat to drive back to James's home.

Where to after Mater Rehab?

At the end of three weeks as an inpatient at the Mater Private it became urgent to decide where James would continue his treatment as an outpatient. A family meeting was scheduled. His rehabilitation doctor, Dr G. came in prior to the meeting and said to James, "Where are you going to live? Brisbane or Victoria?" James said that he couldn't decide. He said that he thought his father wanted him to stay here.

"Why don't you go to live with Sheila in Victoria, James?" asked Dr G.. "You're always telling us how much you love her. You never stop talking about her. The rehabilitation is much better there, anyway."

These remarks went over James's head, only stimulating him to relaunch into how special I was, causing Dr G. to roll his eyes at his colleague and say, "Too much information, James."

I was mildly embarrassed, but still interested that Dr G. was himself proposing for James to come back with me to Victoria, something I had felt until then to be a very remote option. I asked Dr G. what he meant about Victoria being better. The doctor said that there were more neurological rehabilitation options, including several 'neuro'-wards in major hospitals, and that the support network post discharge far outstripped Brisbane's.

It would have helped if Dr G. had stayed for the family meeting, which most of James's other therapists attended, along with his father and his Queensland Workcover contact. The other therapists knew nothing of Victoria and felt that it would be better for James to live in the most familiar surroundings ready to go back to work at the hospital.

The Workcover person added that it wasn't possible for James to have any alternative occupational rehabilitation if he went to Victoria. Queensland Workcover just wasn't liable for occupational rehabilitation anywhere but in Queensland, although they would fund treatment for the injury in any state. She was flagging a legal peculiarity of the Queensland Workcover Act.

I repeated what Dr G. had said, although it sounded rather deprecatory to the Mater Private team, who I thought looked somewhat shocked, hurt and unconvinced.

Thinking back on their reactions, I felt I could have put things differently. The Mater Private staff did not deserve to feel unappreciated. Mater Private made such a difference to James. They provided a caring and safe environment, small enough scaled not to dwarf a patient diminished by his serious injury, where all the staff knew their patients and the patients knew the staff. There was a positive atmosphere of respect and safety. Mater Private Rehab was an outstanding hospital in Queensland, but Queensland laws and services made it a poor cousin of Victoria for the brain-injured.

At the meeting we were given a few more days to deliberate.

I felt worried about James's lack of choice in staying in Queensland. I felt that he would do better with me to help him but that his current rehab team underestimated this – apart from Dr G., who had not turned up to the family meeting. I also thought that his return to being more fully James depended to some degree on what I could promote in him of his old life, such as his writing and website design, which no-one else considered important if they even knew of it. James was only about 10 per cent present in his current manifestation, and I was hanging on to the idea that more of him would come back. I missed the old James but his recovery was so rapid I kept feeling he was just around the corner. If I left him, who would be there to remind him of what he cared about, to help him with computers? Furthermore, what would happen to our website if James were not supported to keep it up?

James, who needed help to select clothing to put on each day, wasn't really up to making a sound decision about his welfare. I knew that in his somewhat childlike, dependent state, loyalty to his father was a big thing influencing him to remain in Queensland.

Looking back on this period from nearly 18 months down the track, I tried to understand my own feelings. My relationship with James at the time was peculiarly ideal. We saw each other daily and his conception of me was heroic or goddess-like. Except for those occasions when his eyes were temporarily shocked open to the true depths of my inadequacy, such as when I got the car locked in the park, I was still a semi-idealized figure who could do very little

116

wrong. Not only did he depend utterly on me outside the hospital, he seemed to know on some level that I could complete many of his most important memories, keep him in contact with his contemporary friends and maintain his internet and political interface to some extent. These were things that no-one else could do. It was as if he understood, through me, what was now missing, and sought to keep a communication there in the hope of reviving, retaining and restoring those missing parts of the whole. I could not help but share his intuition and feelings on this. Although he had been so seriously injured, there was hope now. I was on leave from work, so my ability to dedicate my time to James, on top of this hope, made this period somewhat holiday-like, away from generally unpleasant work politics, in another state, responding to a unique challenge for which I knew I had unusual skills.

There was, however, no way that I could remain in Queensland indefinitely because I was employed in Victoria, had a house there, a dog, and two 86 year old parents who occupied one half of the house and depended on me to drive and attend to emergencies and repairs.

I had acquired the dog, Nubi, a two and a half year old, perhaps four months prior to James's accident, from the RSPCA. My then 85 year old father had emailed me recently that Nubi was getting increasingly out of hand. Nubi was a muscular black English Staffordshire, named after Anubis, the Egyptian jackal-headed god who weighed the souls of the dead, in my typical romantic dog-naming style. My father had taken him down to the local park and Nubi had jumped on my father, flattened him, removed his hat and taken off with it in between his teeth. My father and the dog had made their way home separately with Nubi leaving a trail of upset dog owners on his way, due to his violent idea of play.

In an effort to help James come to grips with what his options were, I sat down at his computer and we wrote out a letter together to Dr G., outlining our concerns.

Workcover and Where to do outpatient rehab? Queensland or Victoria?

"Dear Dr G.,

We were sorry that you were not at the conference with

117

Workcover about James Sinnamon because you had earlier raised certain issues with James and Sheila which we felt needed to be discussed at that conference, but which we were not able to deal with adequately without you there.

These are the questions which we are unable to answer without some more information which we hope you may be able to provide:

1. What did you mean by saying that services for brain injured people are better in Victoria than in Queensland?

2. Were you talking about out-patient rehab or about later reinsertion into society?

3. If you were talking about better neurological treatment or rehab options, we would be grateful if you would please give some specific examples of treatments or options not available in Queensland so that we can evaluate the positives of Victoria against the positives of Queensland options? (We feel we must bear in mind any possibility of better medical/rehab options in our decision and we would appreciate your frank opinion.)

4. Could you organise/recommend/support action by Workcover to find ways for James to be given assistance to get a job in Victoria (e.g. via Commonwealth Rehab), rather than in   Queensland?

5. Could you organise/recommend/support action by Workcover and RBWH Personnel [James's employers], in the event James managed to get a job in Victoria post rehab, to give him leave of absence for 6 months to a year to do such a job in Victoria, thereby leaving the option to return if things did not work out in Victoria - either work-wise or relationship-wise?

Social benefits/disbenefits discussion:

Both Sheila and James share many virtual friends from their political activities on the internet and these friendships and the related social activities (writing, editing, intellectual discussion), which are among those most important to Sheila and James are not affected by where they live.

The positives of Victoria, apart from any you may be able to suggest, include James having a mature relationship based on established mutual affection and intellectual commonality with Sheila in a situation where statistically his ability to make new friends may be compromised by his illness in the medium to long-term. This is a very important consideration.

Queensland benefits of rehab and social reinsertion here are mostly that James has most familiarity with his house here and has his father and a favorite sister for moral support and a couple of friends and political acquaintances.

Statement of Problem

James and Sheila are worried that, if James does all his rehab in Queensland, then returns to work in the Royal Brisbane and Women's Hospital, it will then become virtually impossible for him to choose to go and live with Sheila Newman in Victoria. Sheila feels that his future will be set in concrete and located more or less irrevocably in Brisbane. This would be because of the lack of choice of employment for injured workers which would mean that he would probably have to remain in his RBWH employment indefinitely, even if he found a different position there. (This is putting the worst complexion on the recovery outlook, of course.) On the positive side, employment in Brisbane would be more secure, we think - although James is not sure.

Sheila also finds it difficult to put her case for James coming to Victoria because she knows that, on the face of it, the

familiarity of Brisbane and his home will stand James in good
stead in early rehab. She wonders, however, if the poor formal
injured worker social reinsertion options available in Brisbane
compared to Victoria might make Brisbane a lonelier option
in the long term."

Letting go: James decides to remain in Queensland

After we drafted this letter, James went back to spend the night
at the hospital. In the morning his father visited, and James then
changed his mind and decided that he should remain in
Queensland in order to occupy the house his father had given him
and to show his father companionship and gratitude. I am sure
that his father didn't put things this way to him, but he may have
expressed concern (widely shared) that James would miss out on
the logical opportunity to return to his familiar house and then to
his familiar job, which was being held open for when he was
sufficiently recovered to try.

James's ideas of returning to politics and website development,
particularly of his own website, with me, were simply beyond most
of his family and therapists' knowledge or understanding. There
was probably also a fear that I might dump James (which James
often expressed to me) leaving him spinning in space interstate
without a job in Queensland and no prospect of any other. James's
family, like most families, was so naïve to brain injury and its awful
outcomes that the need to engage a lawyer did not occur to them.
My occupation gave me a relatively sophisticated position, from
which it seemed to me that they were inclined to gratefully accept
any superficial advice that came their way from hospital staff.

I felt as if I might be imposing my will on James, so, although it
hurt me a lot – because it was like losing him all over again - I
decided that the right thing for me to do was to defer to James's
desire to go along with the general advice and also to please his
father as he saw it, rather than my wish to take him back to
Victoria with me. We would return to a part-time and virtual
relationship. I felt lonely in advance and rather angry with James
because his decision was not loyal to me and seemed infantile. But
I felt that because it was disadvantageous and insulting to me that it
must be the correct one – which shows my own warped
perspective on life.

Fortunately this warped perspective of mine only lasted a couple of days, during which I continued to research the pros and cons of remaining in Brisbane versus going to Victoria, on James's behalf.

Then again ...

Then, suddenly, it became absolutely clear that Victoria was a better place to go, and that Commonwealth Rehab Centres really were better than what was on offer through Queensland Workcover. I finally managed to convince James of this.

Dr R., the hospital registrar, also came to our rescue, after talking to Dr G., the consultant. "James," he said, "Sometimes you can retrieve something good out of a very bad situation. Going to Victoria probably will offer you more options. You can always come back if it doesn't work out. But consider how much you care about Sheila and want to live with her. Perhaps you should take this opportunity, even if it seems risky on the surface and think of it as a cloud with a silver lining."

That convinced James, thank heavens, although, in the early months of his coming to Frankston, he often got on the phone to his sister and two friends in Brisbane and also to Jill in Victoria, at all hours, to complain of my unexpectedly vicious nature. On my part, on a couple of occasions I went and slept in the spare bed and once I told him that he could go back to Brisbane as soon as his rehab had finished.

But nothing was straightforward. How someone who has no familiarity with hospitals or neurology would negotiate in this kind of situation mystifies me. We could not just depart for an informal trial in Victoria. We had to find a program there that would accept James and which James would find acceptable. And we had to find it within the five working days we had been allocated by Workcover to spend in Victoria before James was due back as an outpatient in Brisbane Mater Private. All this was tightly negotiated with Workcover Queensland.

Packing and leaving on a short trip to Melbourne – Mid July 2010

Organising to go interstate, even just for a week, caused clashes because it demanded prioritizing and time management skills from James that his very faulty memory and still fluctuating state of consciousness/attention span made unreliable, to say the least. That meant that I had to give him directions on matters which he believed he was still in total control of.

Although he could see that his situation had changed in many ways, his ability to analyse how and why suffered from problems of concentration and generally impaired ability to reflect and contextualise. He might be able to say that his thinking was affected by a head injury, but he could not really identify signs of this as they happened. His brain still told him, logically, that when things went wrong where he used to get them right, it must be someone else's fault, or something wrong with any 'system' involved. For instance, James demanded that I come in to Mater Private for an urgent counseling session about my lack of confidence in his ability to pack things in a timely fashion. We sat in a small room with the social worker there and I remember feeling a lot of resentment at James's accusations. That was even though I knew, intellectually, that, from his point of view, my behaviour was unreasonably bossy. Despite my knowledge that his brain injury was accidental and not personally aimed at causing me grief, I found myself resenting it as though it were a malevolent force inhabiting James and causing him to be unreasonable. I felt victimized by his injury and James felt victimized by me.

If I had not had a sophisticated understanding of brain injury, it would have been normal for me to have taken this personally, to have assumed that James was doing it on purpose. That could have led to my becoming frightened and despairing. Despite frequent disagreements, James was himself surprisingly forgiving during this period when what I thought of as his 'cultural software' was telling him that I was unreasonable. His 'cultural software' had become unreliable due to the damage to his senses and their processing of information coming in from his environment. I got the impression that the hardware – the 'intelligent ape' that we are all born with (as I think of it) – was still able to judge the situation and therefore to trust me. I slowly developed the impression that

James had an awareness of his neurological deficits and was adapting to them in a practical manner, but his more immediate and accessible personality had yet to catch up.

(Of course, I use the term 'ape' here in an evolutionary context, thinking of humans as a species of ape. I call our inner self the 'intelligent ape' because it seems to me to be a very old part of ourselves, built to adapt to all kinds of social and physical environments, but stable itself over time. I like to think that it goes back to a time before we were 'human'. The 'intelligent ape' idea is my name for something like a neurological self, that is, the 'me' that is in control of muscles, emotions and instinct. My idea is that it is the stuff we are born with, which develops as we mature, and which then interacts with what we acquire culturally, or 'learn' about our environment. The personality is composed - according to my model - of the hard-wired neurological self and the cultural self that grows up and interfaces with its social and physical environment.)

We were almost late for the flight to Melbourne, Victoria. James was still packing his bag when his father arrived to take us to the airport. I think James blamed me for this, accusing me of having told his father to come unreasonably early, whereas I told him he had started to pack unreasonably late. I had, in fact, packed most of his bag, but he wanted to grab items to take with him at the last moment and lost time trying to locate them. I had actually anticipated this so there was just enough time to get to the airport despite James.

The Airport and flying

Being at Brisbane Airport with James again was poignant. He had seen me off on flights to Melbourne so many times that we had a routine. The routine always involved going to the bookstore, where we would graze the titles and James would usually buy me a book and a copy of the *Financial Review* for the plane flight. We would start to read the *Fin Review* whilst drinking coffee in an airport café. Often we would finish our coffee whilst we stood in the boarding queue. James would always tell me to wait at the very end of the line and let all the other passengers go, and he would stand in line with me and put his arms around me and kiss me just before I went through the airline boarding doorway. When we travelled together we would both buy books and newspapers.

James was a somewhat profligate book buyer. If he liked a book he would buy copies for his friends or people he wanted it to influence. Despite its cost of nearly $50.00, he bought several copies of the second edition of *The Final Energy Crisis* (Pluto Press, UK,) a collection of articles by scientists, half of which I wrote and all of which I edited, giving it as presents to various members of his family, most of whom would have found its subject difficult and alien. Then again, two members of his family send him bibles and other religious tomes every birthday and Christmas. It seems to be a Sinnamon trait to present relatives with things they 'ought' to read.

Since his accident, however, James had been eerily unmoved by books. Whilst in the Royal Brisbane hospital he had come down to the bookstore to buy some Linux magazines, which are glossy productions about open-source politics, software engineering, and new software compatible with the Linux open-source operating system, but basically, he seemed to have lost his interest in actual books. He even seemed to disapprove of them when he thought about them. When I had brought him home from the hospital he had looked around his library and exclaimed in a strangely disgusted tone, "All these books. What a waste of money."

It was sad to look at the bright red bookshelves he had put up on his office wall just before the accident to organize his favorite recent acquisitions.

I naturally hoped to hear his interest in books revive, because that was such a strong part of his personality. I don't know what I would have done if it had never come back. (I read somewhere that the hardest time with a person who has a brain injury is not while they are struggling to improve, it is when they stop improving. Then you have to come to terms with what is left rather than with what you hoped for. As we passed the two year and four month mark, I thought I could see what this meant.)

The airport was an obvious time to retest James's new attitude to books. If I had not pointed out the bookstores he would not have noticed them, but I thought I detected slightly less antipathy. I dragged him in and tried to interest him in a variety of titles. He seemed to be making an effort to indulge me and to be trying to understand what it was I saw in these books. On this occasion I bought a book about Body Language because James's ability to read people was much diminished by the accident at that stage. I bought another one, ostensibly for me, about Mind Maps. Mind maps are very useful ways of organizing thoughts and plans visually.

We sat down and had a coffee and I reminded him of how he used to like books. While we were drinking he asked me if he looked okay and to check that he wasn't drawing attention to himself. I said he did look okay. By that I meant he looked okay for someone who has just had a massive head injury. In fact he looked surprisingly good, considering, but somewhat blank of expression. But then, I was comparing him to his appearance in hospital when he was still semi-paralysed. He was walking and talking now. If he was called upon to make a simple decision, however, you could tell there was something wrong. Most of the time I could foresee situations and act on his behalf, but not always. For instance, when he had to go through the metal detector separate from his luggage and by himself, he had difficulty following instructions. He set off alarms and emerged beltless with his pants sliding down around his hips in a panic about where his baggage had got to. It was hard for me to help there because of the security aspects. I didn't feel I could call out, "Excuse me, he's not a terrorist. He has a brain injury, that's why he's acting funny." I was also wary of the airline twigging that something was odd about him and not letting him on the plane.[30]

We sat together on the plane and James actually read about 25 pages of the book on body language.[31] He seemed so interested that I assumed the book must be really well written and began to wish he would fall asleep so I could read it. He told me later that he felt so nervous on the plane that he could think of nothing else to do except try furiously to focus on the text in front of him. Trying to pass the time and remember where he was and what he was supposed to do and to respond 'normally' to interaction with the airline stewards tested him to his limits. The trip seemed endless to him and he was physically very uncomfortable as a large tall man squeezed into a narrow seat. At the time, despite my many years of experience in psychiatry, I had no idea how utterly nerve-wracking this trip was for him. I was comparatively relaxed because I had managed to organize him and herd him into the plane and knew he could not go very far within its confines. I was happy to lose myself in my own book. His blandness of expression fooled me and I was very pleased to see him reading.

Although he handled the plane trip impeccably to all appearances, his performance became more erratic during the interminable second part of the journey. Getting from Red Hill in Queensland to Brisbane Airport and then from Brisbane Airport to Melbourne Airport is comparatively straight forward. The trip to the Brisbane airport by car is half an hour. The plane trip is only two and a half hours. Getting from Melbourne Airport to Frankston is a lot harder and takes much longer. If all goes well, the fastest way is to get a Skybus from the airport to Southern Cross station in the city. From there you can get a train direct to Frankston, although you may have to change at the station after Southern Cross to get it. When you get to Frankston you have to get a taxi or a bus up the hill to my house. Buses for our route from Frankston Station don't run after about 6p.m. There are usually queues for the taxis. You can also catch a special minibus from the Airport to Frankston, but that can take a long time and the schedule meant that this service was not available when James and I needed to get to Frankston on this occasion. As I said, if all goes well, you can travel in two shots to Frankston, then catch a taxi home. That is what we aimed to do.

Already on the Skybus bus from the airport James was showing signs of fatigue and insecurity. My natural way of passing time during arduous journeys is to grit my teeth and read books continually. This had once been James's method too. Now he kept interrupting me to ask questions I felt were unnecessary, such as, where we were exactly, how long the trip would take, was I angry with him, and did I realize that he was feeling lonely? I was also very tired and gave him the minimum of reassurance I could get away with before plunging back into my book.

We had some kind of argument on the platform whilst waiting for the Frankston train. The details escape me, but James stopped speaking to me. Once on the train we arranged our suitcases around us and prepared to ride it out. At Mordialloc the train stopped and urgent voices over loud speakers, speaking very rapidly, informed all passengers that they would have to complete their journeys on buses organized by the transport system.

It was dark and the weather was seriously inclement, although I cannot remember whether hot or cold or damp or dry. We had to cross uneven terrain rapidly, carrying our numerous bags, along with hundreds of other passengers. There was, however, the added difficulty of James's fatigue, emotional state, tendency to lose track of baggage, his relatively clumsy gait and poor physical strength and, worst of all, his difficulty in hearing and understanding instructions. For all I knew, he was so cross with me, he just might decide to make his way back to Brisbane.

It was hard to find the buses and I was afraid of losing track of James. We followed a crowd of struggling passengers up, down and around a station, then down a street. James had started speaking to me again in order to complain bitterly about the train service and I was arguing with him that there must be a reasonable explanation and telling him to hurry up. He was accusing me of making excuses for Melbourne transport, saying he would never take public transport again and how much he hated buses. James wore an indignant and ill-used expression as he lectured me on my poor attitude and illogical defense of the railways. I denied the accusations. Our dialogue degenerated into hisses.

We managed to get on the same bus and it took ages to get to Frankston as the bus followed a tortuous route from train station to train station around various associated road works and rail upgrades. James and I were again not speaking, which was a relief.

I cannot remember if we took a bus or if we got a taxi home. It was cold and the house was an incredible mess. Nubi the dog needed my attention and so did my parents. I felt disloyal to James. Our bed was not made. I had to get some dinner for us and there was little food. There was stuff all over the sitting room in my studio area where we ate. I felt that I had brought James into total chaos from a situation in Brisbane that was comparatively orderly and supportive. I felt ashamed and very upset. I told James this and apologized for bringing him into such an uncomfortable and inhospitable situation. I was surprised and somewhat restored when he put his arms around me and told me it didn't matter at all.

Choosing an Out Patient Rehab Hospital in Victoria

We had compiled a small list of rehab units in Victoria, which we intended to contact and hopefully to visit in the next week and a bit.

One of these was the Alfred Neurological Unit. Somehow I had managed to get hold of a Nurse in charge of this Unit and have a good chat while James was at the Mater Private. The nurse had been extremely helpful and had also talked with James, asking him questions about his career expectations and past experience. James felt uplifted by this chat. We therefore kept the Alfred in mind when we went down to Victoria.

Once there, however, we discovered that the Alfred Neurological Unit was almost impossible for patients with private sources of income to get into. James was in this mostly fortunate category because his injury came under Queensland Workcover.

Another specialist unit in a large hospital, The Epworth neurological unit, was impossible to get an appointment with in the time we had.

VicRehab, which seemed to be the only specialist neurological rehabilitation hospital in Melbourne turned out to be on the other side of the city and required catching a bus from my house or walking from my house for 20 minutes to Frankston station, then catching a train for four stations, then catching a bus which took an hour and 20 minutes to get to VicRehab. (I had a car but James would not be able to drive it, so public transport was a major consideration.)

We decided to apply to St John of God instead, which, it turned out, was only two short bus-rides away, or a walk and a bus-ride. James thought they sounded nice there, rather like the Mater Private. They didn't specialize in neurological rehabilitation, but they were a rehabilitation hospital that had a visiting neuropsychologist.

By great luck we were able to get an appointment at St John of God within a couple of days. On the appointed day, just as we were about to leave, the phone rang and a voice said, "I am just ringing to let you know that Dr-- will not be able to see you at 10 o'clock. She will be able to see you at midday, however." I said, thank you, hung up and told James that our appointment, for which we were just running out the door, had been put off until midday.

That gave us some time to go out and do some shopping before the appointment. When we returned home, the phone was ringing. It was St John of God asking where on earth we were. "But... but, you said the appointment had been moved to midday," I stammered.

"That wasn't us," said the woman on the phone. There would be no appointments available for weeks now.

"But we have to return to Brisbane in a few days!" I protested, feeling as though I was in a nightmare. James was slowly emotionally inflating at my side. "What?" he said. "What??" "Who rang you to tell you about a change of time?"

"I don't know," I said, tersely. Then I remembered that I had rung VicRehab the week before and had spoken to someone about an appointment, which I had subsequently not confirmed, due to our focus on St John of God. Perhaps it was VicRehab that had rung up to move the appointment. I looked through my diary and found a record of the conversation. I rang them.

"I don't want to go there!" cried James. "It's too far away."

"We have no choice," I said, through gritted teeth. I was still reeling at the bizarre coincidence where VicRehab had rung without identifying themselves, talking about a change of appointment time, just at the moment James and I had been about to leave for St John of God. Whose fault was it? According to James, it was all my fault. I should have asked who was ringing and which hospital and what doctor. Perhaps so, but should they not have identified themselves? Not just say, "Dr X cannot see you at 10 on Friday, but she can see you at midday." Did the brain-injured and their associates not have enough confusion in their daily lives without professional organizations adding to it?

The consequences, however, were much bigger than my mistake. Now James had no choice but to try to get into a hospital where he would have to commute about five hours a day in multiple forms of public transport, which he hated. Friday was the last day of the last week we had to find a rehab unit in Victoria.

I would never hear the end of it and I might never forgive myself.

I looked up VicRehab and began to realize, however, that it really did specialize in neuro-rehab, unlike St John of God. It began to seem like a good option for me, but the commuting would be a big problem for someone with the fatigue and disorganization of brain injury. I did not care to consider that it might be insurmountable.

When we finally got there for our midday appointment on that Friday a few days later, James was cheered to find out that the doctor he would see was Dr Kennedy. It seemed auspicious, given his love of one President Kennedy. He said the same to Dr Kennedy, who was quite impressed, since, having seen James's MRI report, she had expected something more obviously like a shambling wreck to wheel itself or lurch on crutches through the door.

The interview went very well. Dr Kennedy was right at home with brain injury and, to our great joy, informed us that James's insurers would be expected to pay for a taxi-ride to and from Frankston for James. This would amount to something like $60 or $80 each way. No more buses and trains. I was elated. James was a tad more circumspect; I wasn't going to be so easily pardoned for my crimes.

Getting lost again

I think one learns a lot about oneself through close association with a person recovering from brain injury. You are frequently called to justify your behaviour and it is not always easy to do so logically.

On the way back from VicRehab I was feeling cocky. Sitting on the train, I asked James if he wanted to get out at an earlier station and see if we could link up with another train line instead of a bus. I added that I was not sure that this was possible, but that, if he liked, we could just try and see. In my mind this would be a bit of an adventure. I do not know what James had understood of what I had proposed but he certainly didn't get the full meaning. Five minutes out of the train he was asking where the station we were looking for was. I said that I didn't know. I had some vague idea that it was over in a particular direction and that if we kept walking we would find it, but I admitted that I wasn't sure at all.

James was aghast. How could I set off without knowing *exactly* where I was going? How dare I get him into this?

In fact we became quite thoroughly lost. We walked for ages without seeing a service station where we might access a street-map. Although I knew that we would eventually recognize some landmark, James had no faith at all. Was I mad? Who did I think I was to drag him out here. He insisted I tell him where we were, but I didn't know. How far was it to where we were supposed to be going? *Exactly* how far?

I reiterated that I had explained to him already that it would be a bit of an adventure. James insisted that I had explained nothing of the sort. "Tell me the exact words! Go on, tell me exactly what you say you said to me. Of course you can't!" he accused.

We walked on and on along featureless pavements alongside busy roads in endless suburban streets that looked similar. When I stopped and asked the way I found the context within which people gave directions impossible to grasp. Melbourne suburbs are huge. It was very hot and we were very tired. James was exhausted and beside himself. I was a monstrously unreasonable and irresponsible person. He walked on in front of me to show his disapproval; indignation fueling his normally slowed pace. Taxis frequently drove by but every one was full. Eventually we found a station, but it was not on the line we were seeking. We staggered on to another station, a kilometer or two further on, which linked up with the one we needed. James barely spoke to me on the way home. That night he carried on for hours about apparently unrelated issues, upbraiding me for tiny things as if they were monstrous wrongs that I was attempting to conceal. He was utterly unable to forgive me my despicable playfulness with space to the detriment of his feelings. It was as though he was convinced that I had got us lost on purpose and that I must admit my 'fault' and atone for it in some way, which meant listening to him lecture me far into the night. I fled to the spare bedroom, where he followed me. Then he went and rang his sister Judith again. She said — again — that I was entitled to make mistakes.

James disagreed of course. The subject remained alive for a couple of days and nearly two years later, he still remembered. Another case of how strong emotion and being the focus of events, binds memories, even in adverse conscious-states.

Brief return to Brisbane then back to Victoria for the long haul

VicRehab could not take James on immediately. It took about three weeks before an agreement about treatment was finalized between Queensland Workcover and VicRehab, then another week or so before a rehab schedule was worked out.

During this time we returned to Brisbane where James had two more weeks at the Mater Private as an Out Patient. Over this time time we also packed up James's house, found someone to look after it, and made arrangements for him to come and live with me in Victoria for at least six months.

Packing to move

Our friends Rob and Julia said they would like to house-sit for James. This was a profound relief because James hated the responsibility of rental agreements and the house had too many things wrong with it to be a viable formal rental property. James also disapproved of renting, of exploitative landlordism. Since he currently still had 70% of his wages as income under Queensland Workcover and I wasn't charging rent, he could afford to avoid renting the house out and let friends take care of it.

We had about three weeks to pack and move everything that James felt he needed. Road freight was the only option we could afford. Brisbane is about 1,682 km from Melbourne by road.

The main things to move were computers, books, clothing and food.

James had three main computers he had been using: Tibrogargan,[32] Kazak,[33] and Grammos,[34] (the apple computer). Tibrogargan had a Linux operating system and was James's personal computer on which he did most of his work, stored his email and which harboured our home version of candobetter.net (then, candobetter.org, to be precise). Grammos was a fall-back computer for internet browsing and quick access to functions that were not automatically available on Linux. Kazak had once been a shared project between himself and W., a depressive electrical engineer. James had been building other computers, of which two or three were stacked around the dining room among boxes of computer parts. It was not obvious which worked and which were to be mined for parts. James could not remember. We both remembered that at least one was to be a test bed for upgrading *Candobetter*, whilst others had been intended for reselling. Our friend Ilan made a living from buying parts and assembling and programming computers and James had been thinking of doing the same thing.

As well as the computers currently in daily use, he wanted to take all the ones in the dining room and the boxes of parts to Melbourne, plus a lot of his books. I was cheered that he now valued books again, but I was pessimistic about his expectations to use all these extra computers and parts in order to return to his computer-building project as well as to his test-bed programming project. It all seemed so fraught now that James was not even able to find his clean socks or reliably dress himself.

With some horror, I anticipated sharing my house with about eight computers in various stages of being built, unfurled about the house, festooned with all sorts of discarded unlike items, such as clothing and books and newspapers, under layers of dust and unwashed coffee cups and snack-plates. Even though I was pleased to see James looking to books again, I also dreaded his filling my house up with mountains of them and I did not even want to think of the chaos if he began acquiring daily newspapers, with the intention of archiving cuttings. I was reassured to a large extent when I realized that most of the thousands of books in the house belonged to his father. Hearteningly, James was quite easily able to identify the books most 'currently' in use (as in before the accident). So was I, since we tended to read the same thing, with the exception of military histories.

Why did someone so preoccupied with virtual systems want books in 2010? Using the free open source Linux operating system instead of Microsoft kept James removed from the trends that big commercial systems floated and popularized with the majority of other computer users. James remained a hard copy person. In many senses he was a paradoxical nerd, used to writing code directly rather than through pretty graphical front-ends; reading in print, rather than on-line. James frequently surfed in text mode and so often missed my electronic illustrations for *Candobetter* unless I brought them to his attention. The illustrations were a source of pride for me because they were easy for me to do, but effective, and most sites did not have original illustrations.

It was not until about two years later that James, reminding me of an episode in the UK comedy series about male-female relations – *Coupling* -told me that another problem that had concerned him at the time was what to do with a cherished old collection of sexy magazines. He was afraid that they would come to light unpredictably and cause embarrassment. In the end, he tells me, he threw them out, with much regret. And I had attributed all of that angst to whether or not he should conserve his collections of publications like *Workers Liberty*. Such are the unexpected trials and tribulations of loss of independence in brain injury.

In the end I cajoled James into working out which computers and computer parts might still be viable and then he agreed to throw the rest out. He brought Kazak with him after offering it to W. and receiving a massive depressive unsolicited earful from W. about how James must give up any idea of further computer work now that he was brain injured.

It was too much for James to hear. James kept politely trying to get W. to give him the benefit of the doubt, twisting the phone cord in his hands as he listened to W. berate him with 'reality'. He came off the phone pale, disgusted and downcast. "How could he be so horrible to me? How could he be so unreasonable? I never want to talk to him again. What a madman!" In a sense W.'s prognosis for James had a reasonable basis, but his aggressive certainty did not. James's contacting of W. also shows how James had retained his characteristic sense of honesty. W. would never have asked about Kazak, but James went to the trouble of giving him first claim on a computer that he valued and which they had built together.

It was sad to see all that equipment out on the front porch, ready for the rubbish removal after nearly two weeks sharing time between rehab and sorting through the immediate treasures of his recent past life. I could see that throwing out some of his computers was for James a major sacrifice of hard-won assets that included work and time invested, as well as a concession to his loss of control over his life.

He could have stored some under the house, with his father's furniture and hardware collection (where he and his father had been building a downstairs apartment) but even he seemed to realize that if or when he returned long-term to Queensland, he would already be engaged in other projects and would not have the time to go over aging computer parts.

How did he make these decisions with so little memory and working consciousness? He seemed to operate with some awareness of his current deficits, but did he have any capacity to estimate the likely delay before he would be recovered enough to return to such projects? Did he also envision that he might never return to them, as W. had foretold?

At conscious level of interaction and decision-making this was not apparent, but my feeling is that, at some level, James was able to consider all these things, even then. There was and there is some overall global consciousness in James that still has access to a kind of brain index. It ticks away in the background. It does not communicate immediately with his consciousness and it is not accessible for problems like writing computer programs. It is not reliable in the short term, but it looks after him in the long term. In my paradigm of the wise old hardwired ape, perhaps it is organized along generic principles and the sorting is slow because it is hard for the hardwired ape to read the cultural software that had grown up with James but which James can no longer access reliably. The old ape sorts through this painstakingly, as though interpreting a foreign language.

We were in Queensland for two weeks and, as I mentioned in an earlier section, James spent much of those two weeks trying to clean and tidy the house ready for his house-sitters, even though the house was cleaner and tidier than it had ever been and the house-sitters said it was fine. It was a lot for him to get his head around. Consequently I laboured alone among the cobwebs and decades of dust in the front room, sorting the junk and filling new boxes with the things that James would take to Melbourne. Two things he would not be taking were an ancient mummified rat, as flat as a bookmark, and a collection of hand-made native spears from PNG. I carefully stored the latter but a tenant later found them and used some for curtain rods.

Two days before we left for Victoria, James insisted on retrieving his bicycle from the police station. This gave me an 'uh oh' feeling. The bicycle was, for his father and me, a hated reminder of what had happened to James's mind and body. If James recovered enough to balance on a bicycle, his risk of added brain injury in a simple uncoordinated fall would be high. We needed to get the bicycle, however, because we had been advised to photograph it and the helmet for forensic evidence.

On a very hot day in an outer suburb of Brisbane, we made two trips in James's ancient Mazda to a freight depot. We had one wooden pallet's worth. We were instructed to bind all James's possessions to the pallet with something like an acre of cling-plastic wrapping. This took us about an hour and a half, working together in the hot sun outside the roofed depot where the trucks were lined up. Kind drivers came over to offer us end-rolls of plastic and advise us on technique and quantity. I was surprised at how hard James worked over this short period and at how well he did, although it was I who made all the executive decisions and drove the car, of course. Naturally we had several small arguments about wrapping styles, but nothing major.

What would we do with the car? We had been going to drive it back to C. Street and park it, with instructions to Rob and Julia and James's father to turn the engine over every so often, so that we could use the car when we came back to Brisbane from time to time. Parting with that car was a final concrete loss for James; it was saying goodbye to every part of his independent former self. At the last moment I remembered how when I was a child my mother had transported her tiny Renault in the moving van along with the household contents and a piano from Melbourne to Sydney. I asked the clerk in charge of the freight depot the cost of freighting the car – a tiny little old Mazda. We were charged $75 to freight James's car. It felt like a reprieve, as though we had saved something of the 'old James' for future use. We were very cheered up by our decision.

As a driver saw us plastic-wrapping all the other items on a pallet for transport he came over and said that he could not understand why we didn't just put them all in the car with the bicycle on top on the roof rack. He had a point, but we had done things the wrong way round, so the computers and the bicycle went on a pallet in one truck and the car went on another.

PART TWO: OUT PATIENT IN VICTORIA

Orientating James to Victoria

A concern of those reviewing James's rehab options had been his lack of familiarity with Victoria. But his visits to me there had given him a strong acquaintance with Frankston and it turned out that he remembered it quite well.

Two years later, in August 2012, whilst we were visiting Brisbane, I noticed that his ability to find his way around Brisbane and Queensland without a map remained superior to his ability to navigate in Victoria, Melbourne and even Frankston. James lived in Brisbane for six years prior to his brain injury, but only visited Frankston in Melbourne for a couple of weeks at a time. His retained familiarity with Brisbane tells us that even though James's spatial orientation and his ability to lay down spatial memory were generally considered well-preserved, they were nothing like what they had been prior to the accident. Fortunately, however, he remains a very competent map reader.

Spatial orientation and organization around the house

The major task for settling into Frankston was orientating James to where things would go in his study and other space I shared with him in the house, which was divided into two and inhabited by my elderly parents at one end. Having my parents there was very helpful while James needed someone present all the time during the first six months or so after he left the hospital. He would become extremely anxious – even bursting into tears - if he came home from rehab in a taxi and found that there was no-one home.

I found that the best way to help James to memorise where things were supposed to be was to draw a diagram. My half of the house that I shared with my parents and which James now shared with me consisted of two bedrooms and a large work or 'studio' area on two levels, divided up by the levels and part walls, into three spaces. There was what we called 'the inner sanctum' where people could sit or watch television and which could be closed off with a room divider I had designed to keep the heat in during winter. It also contained a modular lounge that did as a spare bed.

On the same level as the inner sanctum and adjacent to the kitchen, but separated from it by double doors that you could bolt, was the dining room area, which contained a table, frequently half-hidden by papers and books. Below the dining room was an L-shaped area with picture windows looking down the main hill in Frankston to Port Phillip Bay. There were two main desk-surfaces here, plus some smaller ones, which I juggled depending on what felt comfortable and what tasks I was engaged in. There was a large easel and a painting and materials storage area flanked by two big metal-shelved units with a wooden platform over the top, forming a small mezzanine, on which I stored paintings and other large objects. James's also had the second bedroom as his study, since I had the larger part of the studio.

This diagram method worked for the large items of furniture and James remembered where he was supposed to put his socks and other clothes, although he didn't often get round to doing so, due to his poor organisation.

Dealing with misunderstandings due to impaired auditory processing

We had big misunderstandings resulting in blues every day or two. James's slow auditory processing meant that, particularly in the first year, he missed at least a third of what I said, but had no awareness of this. For this reason he was constantly hearing me say the opposite of what he expected and found me a mass of contradictions and contrariness. Not only did I not seem the angelic person who had rescued him from the white noise of the hospital; I seemed to him inhuman.

It was only with V., the Social Worker at VicRehab that we worked out what was happening. Most of our fights were due to James not hearing everything I said and my assuming that he had. It took us a long time to understand this and to keep it in mind. We also got helpful advice from his psychologist, Michael W. Our friend Jill was also very good at listening to us and interpreting.

James's auditory processing improved little, but he has learned to make allowances for it. Sadly, he misses out on a lot, particularly in social conversations, which he used to enjoy so very much.

How exactly did we deal with this?

When a simple conversation suddenly flared into a seemingly intractable fight, we followed this advice:

Face each other at a comfortable distance. Ask the other person what they think they heard. Make simple statements. Make 'I' statements. 'I' statements are ones where you say, "I think, I felt, I thought" to avoid saying "You said, you meant, you did," because, in a situation where sound-bites were actually dropped by the receiver, such accusations were likely to be wrong and to inflame the situation. Concentrate on how you felt, so that both people realize that feelings follow from perception and that perception may be wrong. Thus, "I feel confused because I thought we had agreed years ago on this political position," instead of, "What do you mean, you don't remember Bakunin? You gave me his biography!" A successful 'I statement' might lead to a response like, "I'm sorry. Could you say what you meant again slowly?"

Those were the kinds of disagreements we tended to have, which led V. to say, "Why can't you have normal arguments, such as who is doing the dishes?"

An example of a 'normal disagreement' might be: "I felt hurt because it sounded to me as if you thought I meant to leave the dishes for you, but I just needed reminding," instead of, "You're telling me I'm a horrible person because you've done all the dishes this week."

Another way of we got through these misunderstandings was to email each other about our perception, or to write in longhand. I still have a few exercise books worth of letters to me about misunderstandings from James, many of them concluding tragically that he was no longer welcome and wondering how he could possibly transport all his goods back to Brisbane. I would initially read these with my teeth gritted then reply grudgingly. When I realized, however, that James's strangely distorted accounts were simply records of what he had actually heard, I became much more optimistic about explaining myself.

Initially James's reactions to such misunderstandings, in addition to hurt feelings, could be catastrophic and childlike, but fortunately I could see that as a consequence of his relatively recent injuries and knew it would improve – although how much was uncertain. If I had not had an ability to judge these matters, his temporary deep interpersonal naivety could have caused an unnecessary permanent separation, with the added risk of enracinating an otherwise transitory behaviour. However, for all my knowledge, if we had not learned why we were having these constant ridiculous and wearisome arguments, we might have had to have gone our separate ways. Auditory and other perceptual processing problems are fundamental to many brain injuries and understanding them is basic to survival for the injured person and those close to them.

Knowing why we argued was liberating.

James became well aware of the need for me to make phone-calls for him to banks or to deal with automated replies or complex face-to-face contractual interactions, like mobile phone contracts, mechanical repairs, or legal work. We tended to put the matter simply to people as that James had difficulty hearing – which was true.

In the end, nothing has been more important than knowing and acknowledging that we have feelings. Somehow this humanizes us to the person we are arguing with. It is also, if one is angry, quite hard to identify one's feelings. Once identified and 'owned' however, the anger usually disappears and one regains a measure of self-control.

Perceptions affected by brain injury and psychological trauma

While still in Queensland, one night we visited a Tibetan restaurant we had enjoyed before the injury. James had been keen to go out but, once there, suddenly felt very tired. We had to move to another table because he felt threatened by the people on an adjacent table, probably because he misinterpreted what they were saying. Half way through his dinner he began to cry and we left early. For about a year after his injury he would still feel threatened by people walking behind him talking loudly, especially if they were males. Later he would just find it unpleasantly loud, especially if he was tired.

In Frankston we went walking in the parklands of Upper Sweetwater Creek with Nubi, the black staffi. By the side of the road, on the edge of the park, was an avenue of old pine trees, the pathway deep with pine-needles, pleasant to walk upon. I started down the path with Nubi, there, expecting James to follow. Some way down the track, when I looked around, James was standing with his back to me in the dark shadows at the beginning of the trail, in some kind of trouble. I discovered he was frightened by the evening shadows cast by the trees and that everything had an eerie and depressing aura to his eyes. He was experiencing the dusk as almost supernaturally threatening. On the basis of this way of seeing things at dusk, the doctor at VicRehab sent James to have his eyesight tested. Comprehensive testing revealed no particular pathology and we came to the conclusion that this was a post-traumatic stress reaction where James saw death and threats everywhere. He was like a person who had been under fire in a war and who takes a while to get out of battle-ready mode. He also needed to replace the glasses he had put away irretrievably in the sharps disposal container beside his bed in Brisbane Hospital. He had been using a vintage spare pair.

How does this gel with the impression we have of reduced concentration, attention and memory? Obviously the brain continues, damaged or not, to form perceptions. Near death experiences presumably engage some very deep processes in the mind.

James was aware of his stress. He spoke about the way he felt he had been close to death. For instance he described several times how he was preoccupied by the idea that, when he died, there might be minutes before his brain became entirely de-oxygenated, when he might have nothing else but to review all the things he had stuffed up in his life. He said this with such conviction and anxiety that I could only conclude that he had actually gone through some experience like this whilst stranded unconscious on that fateful by-way after the collision and before the ambulance came or perhaps also, thinking he was dying, again and again, in the hospital, from the time he was brought in and perhaps over those long days struggling in restraints as diffuse axonal damage set in and stealthily, unpredictably, took away memories and skills. Is that how it feels when parts of your brain die, bit by bit?[35]

Although he felt he had no memory of his time in hospital, except for waiting to see me, he did have a memory of fears that occurred then. He just didn't contextualise them at the time as happening in a hospital. Although his life was in danger and he was being treated, he did not understand where he was or the hospital system or the period of time this took, however he does have a memory of almost dying which seems very strong, and to resemble a kind of hell.

James's attitude to Nubi the dog became evident at this time. He saw Nubi as different from any other dog he had met and faintly inimical. As I mentioned earlier in the book, James was particularly bothered by the noises Nubi made when he ate and drank. I wondered if this was in part because he remembered the noises of his own stressed breathing. He had also found disgusting and upsetting a man in the next bed at the RBWH who breathed through a respirator, whereas most people would simply feel sorry for a person in that situation. Perhaps what James experienced now as disgust was what most of us would call fear.

Emotions and perceptions get rewired in brain injury.

Inability to cope with on-screen violence – even in cartoons

During James's traumatised period he could not bear to watch most films. He seemed unable to draw a line between simulated violence and suffering and real suffering. Around August 2010, shortly after his inpatient stay, he chose in a video store a war history drama about the starvation of Leningrad. Just as I was getting into it, James had to leave the room because the suffering was too close. I stopped the film because he also needed my company to feel safe.

Being an artist, I love three D digital graphics, and I was attracted to the colour-scheme and animation of Kung Fu Panda. Perhaps three months later, I got the DVD and suggested we watch it together. James was reluctant but gave the film a try. If you have seen this film, it begins in a picturesque Chinese village inhabited by diverse talking animals who wear clothes. The hero is a fat panda, the son of a skinny little stork who owns a noodle-soup business. Part of the story is the mystery of how a panda could become the son of a stork, but the plot mostly revolves around the elevation of the panda to high mystical status, before he finds the answer to the puzzle of his origins. The film is very funny. The animal characters are cute, clever and amusing. James, however, did not see them that way at all. He took one look at the Chinese period setting and said, "How those people must have suffered."

"But they are cartoon animals, James."

"Nonetheless, they still make me think of a terrible period in history."

After a few minutes, due to his ideas of the politics of the time in China (a mythical time in the film) he could not continue to watch it. It did not matter that the characters were electronic creations and the setting stylised and the end guaranteed to be happy, James was overwhelmed by memories of things he had read about the suffering of village-dwellers like these under Chinese dynastic rulers.

It seems to me that the attitude we develop to literature, plays and films, as abstractions, as imaginary stories, is a learned feature of our culture. Generally we protect children from violent imagery because we do not want to frighten them. As we grow up, we learn to distance ourselves from stories we see on the screen. Brain injury can strip this, along with other learned attitudes, either temporarily or permanently.

The first film James watched all the way through was *Twelve Good Men*, a 1936 film about a jury reaching their decision about the guilt of a man accused of murder. His father had taken him to the film many years previously. It was a film that makes valid and important points which have since been made many times, with more sophistication. How did James choose this film? I don't know. He later did not remember us watching it at his suggestion.

Another film we watched together at his suggestion was *Casualties of War* (1989) with actors Michael J. Fox and Sean Penn. This was the dramatisation of the kidnapping and abuse of a Vietnamese woman for use as a sex-slave by a small group of US soldiers. One soldier stood up against his peers.

Justice, courage and morality were probably values adopted as choices when James was in his teens. Maybe these and similar films he chose to review early in his recovery were part of retracing cultural developmental steps in his past.

James's catastrophic interpretation of films and his persistent fears at twilight were diagnosed as a brain injury associated depression which carried a high risk of suicide in a person of his premorbid abilities. For this reason he was referred urgently to a psychologist. The psychologists at VicRehab were booked out, so we had to find an outside one. James began seeing Dr Michael W., who gave both of us valuable support.

Networking

Traumatic Brain Injury Survivors Network Development Project

My major way of coping with any stress is to master it by reading and learning as much as I can about it, then using this information in any way I can. I searched the internet for brain injury e-lists in the hope of joining discussions and asking questions about James and my situation, however most of these e-lists were inactive when I looked. Eventually I came across the Traumatic Brain Injury Survivors Network, which was ideal because of its huge size and the fact that it took carers and brain-injured people. The huge size was important because it meant there was a big range of different kinds of injury. It also meant that, despite the general disability, there were enough people joining in for a high standard in aggregate both in quantity and quality of information in all kinds of media. Craig Sicilia, a brain injured person, started the whole thing off in conjunction with Brain Injury [digital] Radio, where he would interview 'survivors'. James did not identify his brain injury as a major problem at this early stage and so was not inclined to use the site. I did persuade him to respond to a call for Brain Injury Radio participants once, and he produced the letter below in November 2010. I've left in the typos to give an idea of his writing skill at the time.

James's Letter to the Brain Injury radio program

Dear Friend,

I am e-mailing to to express my interest in appearing on your show as a guest on 22 November.

In many ways, I am not a typical brain-injured person. I am luckier than most if, in some ways, unluckier than most. Whilst I suffer a number of the symptoms of a brain-injured person, I don't have a good many which feature in literature about the brain-injured.

I was made comatose by my brain injury in Brisbane, Australia when a car collided with me on 18 May on my pushbike on my way to work and I suffered post-traumatic amnesia (PTA) after I recovered from the coma. I have no recollection of the accident itself, or, indeed, any of my early period of treatment and little of most of my subsequent treatment during my total of 70 days as an inpatient.

I do have some recollection although of some of the events not long before the accident. For example, I was running late for work on that day, because my pushbike had a puncture. I remember walking my bicycle home where I repaired the puncture. I rung up work to tell them I would be late and set off again a second time. Some time after that a car colllided with me on what would normally be an unusually safe route to work for cyclists.

Although I am not an expert in the field of neuropsychology, I believe my recovery although not complete, could not have been anywhere near as good as it was without the care and attention for my time as an inpatient in a neurosurgery ward of a close friend, Sheila, in particular, and other good friends and family.

At present my therapists tell me I have good intellectual and language capacity. My visual and spatial processing is good.

However, my memory, comprehension of what is told to me, my ability to learn, my ability to adapt and my brain processing speed is slow. I also suffer from fatigue and can't work physically or intellectually nearly as long as I used to be able to.

Sheila got leave from her work as a psychiatric nurse a soon as she learned of my accident and came to my bedside. Her care for me involved vigorous mental stimulation in which I engaged her and others in intense intellectual discussions about politics and history, almost past their points of endurance. I particularly spoke of the late (and truly wonderful) President Kennedy of which I had read an excellent book (http://search.barnesandnoble.com/JFK-and-the-Unspeakable/James-W-Douglass/e/9781570757556) just prior to my accident. During my time as an inpatient I also read a number of times, at my father's suggestion, Kurt Vonnegut Jr's "Slaughterhouse 5", a book I had read many years before. (Apparently, I even engaged in a number of conversations with Billy Pilgrim, the main character from that book (and movie) at my bedside, who was entirely a figment of my imagination.)

At the moment, I have no recollection of any of this, but others have recounted these events to me.

Sheila also wrote for me and printed and bound in a black plastic folder "James Sinnamon's Orientation Booklet for his hospital stay". This book told me of my accident, why I was in hospital and how to cope with problems and how to do things I had forgotten how to, such as to make phone calls, particularly with the hospital phone system.

I am told I read this over and over again. Again I have no recollection of this. I got Day Leave from hospital, something that the brain injured there had never enjoyed before and was able to spend much of my time on release in close intimate friendship with my carer as well as in being engaged in further intellectual discussions, bush walks and other acivities.

Sheila also helped me to technically administer the web-site I had initiated about four years prior to my accident, to edit the material on it and to co-ordinate with numerous other contributors to that site. Of this, I have almost no recollection during my time as an outpatient. In case, it may be of interest, the web-site is http://candobetter.org

At the same time as I was being so well cared for by my friends and family, I am told of another brain injured patient who mostly sat in his bed, sad and alone. Evidently he had no close friends or family to care for him.

If my life had turned out differently, that sad person could so easily have been me. Without the care and stimulation I got from my freinds, I don't think my recovery would have been anywhere near as good. I fear for how that other brain injured patient is getting on in life now.

This has made me realise just how people without close or intimate friends are deprived and not just when they need help recovering from brain damage. That so many seem to be deprived of good friends is a sad indictment of what sort of society Australia has become in recent decades.

Whilst it should not be any government's role to interfere in peoples' lives and to impose unwanted friendships, a decent caring government should proactively do what it can to ensure that the need for close and intimate friendships is met as widely as possible.

In spite of the good care I received from family and friends, they could not prevent much of my early stay in hospital from being a terrifying and traumatic ordeal for me. I can only guess at how much worse it would have been for others who did not have the care from close friends that I did.

One bad memory was of sharing a ward with 5 other patients. Some spoke loudly, shouted and wailed all night, often making rest impossible for me. Often I simply got out of bed and left and went to to another room (perhaps a meal room, social room, study or library) where I could get at least get peace and quiet and read. I badly missed the lack of close friends and familiy, outside of visiting hours, which did not begin until 10:00AM.

Every morning, I woke, sometimes even earlier than 6:00AM and would wait until 10:00AM when Sheila was able to begin her day's visit. The wait was quite an ordeal for me.

If you think any of this, or other details I have not discussed here, may be of interest to your listeners, please get in touch.

Thank You
James Sinnamon, Australia.

Why TBI really is epidemic

The TBI Network promoted the concept of TBI as a 'hidden epidemic'. At first this seemed to me the kind of cliché that any special-interest group is likely to market itself with. Then I began to realise that, just in relation to automobile trips, TBI incidents must be growing almost exponentially day by day as a reflection of the density and complexity of traffic, the frequency and distance of trips and, above all, the speed of travel. If you pause to consider it, walking is a relatively risk free mode of transport. The horse is somewhat riskier and the rise of horse-riding must have increased the statistical rate of TBI, but the automobile must be the greatest contributor in history – apart from war. Hand-guns in countries where they are frequently carried, and various big hard heavy machines would surely also be a major factor.

And then there is war. The impact of explosions, single and repeated, as well as the effects of poisonous chemicals and heavy metals in war must be a major contributor to war injuries, many of which may well remain undiagnosed, mistaken simply for 'states of mind' or 'compensation neurosis'.

Nubi – How James's injury affected a dog

This is a big chapter, mostly about a dog. Why would I spend so long telling you about a dog, when this book is supposed to be about how a brain injury affected a man? Well you will learn a bit about the place of dogs in our lives and how a distant car accident a head injury affected one dog in particular.[36] Through that dog's story, you will get an indirect view of how James's injury affected my life at the time as well, and the lives of the other people who shared our house – my parents – about whom I don't write much in this book either.

If he derived timely support from Dr Michael W., James's accidental perceptual rewiring and the massive reorganisation of my life after his accident had tragic results for Nubi, my black Staffordshire dog. Nubi had somehow lost his owners and then learned to fend for himself in a forest, before being captured by the RSPCA. At the lost dogs home they taught him to walk and obey simple orders and one day in 2009, when he was two and a half years old, I came in with Miss Bianca Dog. Miss Dog, aged 17, was the surviving sibling of Nero Dog. They were a pair of brother and sister Jack Russell [accidentally] crossed with Staffordshire – an unexpectedly delightful combination. In 2008 Nero, aged 16, suffering renal failure in addition to traumatic arthritis from old road trauma was euthanized.

Miss Dog and the rest of us were bereft by Nero's death. Knowing her life to be nearing its end, I decided, in the hope of cheering her up, and buffering my own grief, to get another dog.

When Miss Dog and Nubi were introduced at the lost dogs' home, they both behaved politely towards each other. Nubi showed more ability to interact with me than did other dogs at the shelter.

There were many signs nonetheless that Nubi had a traumatic past. Whilst I was paying the RSPCA the money to purchase and register him, he chewed through his new leash and attempted to abscond. The shelter then sold me one of the very stout chains that the staff there used instead of leashes, and he went home on a chain-leash. I learned that he would also never enter an unfamiliar doorway. It was as though he feared what he might encounter inside. Initially he had to be carried across the threshold of my home in Frankston, like a bashful and reluctant bride.

When he saw our big back yard with its undulating lawn, trees, shed, hedges, bushes, orchard, flower beds and ponds, Nubi walked about with his jaw hanging. It was a sunny day and I sat on a stone flower-bed wall, watching him explore. After a little while, he came back to where I was sitting, looked up at me and laid his head and his upper body on my lap and licked my hand. It was clear to me that he was expressing gratitude and relief at having a home.

On the first night home, Nubi's breathing became laboured. As a Staffordshire he had a gaping mouth and a tendency to drool. The next day he seemed better. I took him for a walk and was surprised at how well-trained he was, at how little he seemed to pull. On our way back from the park, about 0.5 km from the house, Nubi lay down on someone's front lawn and refused to budge. After about 5 minutes he got up and slowly proceeded up the hill, only to recline in a gutter. This time he lay there for about ten minutes. He seemed apologetic. Could he be ill?

The next day I took him to the RSPCA shelter vet and discovered that he had kennel cough, which seems to be a catch-all diagnosis for contagious upper respiratory infections. (I had to carry him over the threshold at the vet's as well. He was a very large dog and made me drag him through the doorway by holding him under the arms and heaving.) I discovered at the consultation that he had recently been castrated at that very surgery and really felt for him.

The kennel cough took two treatment episodes. The second time I took him to our usual vet, whom I had confidence in. Slowly Nubi regained his health. As he regained his health he became a lot bouncier, began to pull on the leash and to jump on poor Miss Dog. I learned that Nubi was a dog of fairly unsubtle humour, with a good streak of violence.

When I left for work Nubi would demonstrate 'separation anxiety' and try to join me, throwing himself against the gate. So I would shut the front door behind me to prevent this. Often he would still manage to push past me, and lie down on the ground so that I would have to pick him up and drag him bodily back inside the house. If I did not put him inside the house, he would try to push through the front gate or jump the fence.

We realised that he was jumping the fence quite often, after a while, since he turned up at three veterinary surgeries where well-meaning members of the public would take him to have his microchip read.

He would ultimately find his way home if left to his own devices, but his independent habits were unusual among Frankston dogs. When he decided to visit someone's home and dogs and then took a swim in their pond, they tended to think he was lost. We would get phone calls – "Do you know where your dog is?" I would think that I knew but was proven wrong on several occasions.

A friend told me that the front fence of an acquaintance with two long-legged Staffordshire dogs like Nubi had finished up looking like Stalag 17. Until I had spent a day adding a couple of feet of wire to our front fence, I had taken this to be an exaggeration.

When I began to add to the fence, Nubi sat watching my activity sociably. After a while, though, he worked out what I was doing. He got my attention, looked at me, and whined at the injustice. He then climbed on a huge pile of firewood under the carport. Nubi was obviously sizing up the leap required from the wood pile to clear the neighbour's side fence and make a get-away via their yard. I got up there with him and felt that he would never make it.

I apologised to Nubi, explaining that the laws of the land were not my doing. Then I apologised because he could not understand English. Nubi was turning out to be a problem simply because he was an independent soul and I felt sorry about having to trap him in a garden, big as it was. It was obvious to me that Nubi had learned to live as a real dog, in the wild. He had probably sired puppies too. I remembered the freedom that dogs had had when I was a child. They took themselves for walks down the street and had their own society, although they knew where their homes were. I didn't hear any stories about dogs attacking children, but I knew to behave with restraint towards animals and I think most children did then. The animals were not confined and dominated by humans. They had lives of their own, worked out their own hierarchies and problems, had some opportunity to mature and develop confidence.

After Nubi worked out that he could not jump from the woodpile and escape via the neighbour's yard, he started throwing himself bodily against the front fence. It seemed he intended to break it. He failed to do so and was finally distracted with dinner. One day, however, I crossed the street to go to a garage-sale and heard a lot of noise behind me. Turning, I watched as Nubi scaled the fence like a cat and heaved himself over the two feet of stout wire netting at the top, joining me in the middle of the road, expecting to attend the garage sale.

To stop Nubi climbing the fence like a cat, or battering it down with his body, I added a kind of inside wire-fence buffer.

If nothing else, it would give burglars pause for thought.

Although Nubi had lived in the wild, his behaviour made it clear that he missed a human family that he had grown up with. He was fascinated by teenage boys, especially those wearing baseball caps on backwards. When we went for walks he would stop and watch boys walking or cycling as they approached from a distance, his whole body riveted in anticipation, until the boys got right up close, and his posture would slump as he realised they were not his lost human family.

If I could have returned him to that family, I would have. He obviously pined for them so much.

After I acquired Nubi and before James's accident, I made a couple of trips to Brisbane to see James, leaving Nubi in the care of my parents, both of them 84 at that time. My parents shared the other side of our big house, which my mother and I had bought together. It was because we wholly owned the house that I could afford to work half-time and write, politic and paint the rest of the time.

When I came back, Nubi was really pleased to see me. In the sitting room we would recline on a modular lounge together as I watched TV and Nubi would roll all over me in an affectionately dominant way. He was reasonably gentle with me, but not so gentle with Bianca, of whom he was jealous. The house became divided between the two dogs, with Bianca staying in the half where my parents were, and Nubi hanging out with me. Often I would attempt to redress this situation, but Nubi was sneakily vengeful and would lay in wait for Miss Dog.

Nubi began by loving other dogs. At first he was gentle with them, but he grew very domineering over time.

Then James was knocked off his bicycle by a car and I disappeared more and more, for longer and longer, to Brisbane.

I left my 85 year old parents to run the household and keep Miss Dog and Nubi in line. Well, Miss Dog was the most perfect of animals, gentle and humorous, so it was really Nubi who had to be kept in line. Miss Dog was, however, becoming so elderly that she spent most of her time in a chair and only went for walks down to the end of the block. These walks gradually got shorter and shorter, until she was content to just go outside the front gate and sniff for messages. Miss Dog had in her day been a magnificent athlete with a cheerful and confident demeanour, who would drown out any excited human discussion with a volley of participatory barks. The loss of her brother, who had been her constant companion, took the wind out of her sails. You could see that she wanted to be friends with Nubi but did not know what to make of him.

More and more, neither did we. Still he was a loveable animal with many amusing quirks, despite his violent streak.

Miss Dog was euthanized, also for renal failure, in 2010, whilst I was in Brisbane with James after his accident I will never forget her and regretted not being at her side.

My father kept taking Nubi for walks, but Nubi, without daily input from me, became insecure and unruly. As I mentioned earlier in this book, one day when my father kindly took him to one of the local parks, Nubi knocked my father over, grabbed my father's baseball hat, and the leash, and disappeared over the horizon. He reappeared briefly to intimidate another dog and its humans, before taking off again. No blood was shed, but the people with the dog he had intimidated took a complaint to the council.

My father walked back up the hill to our house, where he found Nubi waiting outside the gate.

It was at that point that my father told me in Brisbane that he was having problems handling Nubi and that I probably should come home sooner rather than later.

When I next came home I took Nubi out for a big long walk in a bush area with marshes and streams − a great place for a dog. Nubi, who loved chasing sticks, began barking at me to pick up every stick we passed. He actually leapt up at me and menaced me briefly with his teeth snapping and cold eyes. My response brought him to his senses and he behaved himself after that. However I had seen a glimpse of something out of control. I hoped that the problem would not grow.

Fairly soon after that I was able to bring James back to Frankston. This was my chance to rebond with Nubi and settle him down.

Mysteriously, rubber and plastic dog toys began to appear in the front garden. There was a squeaky hamburger, a green squeaky dinosaur, a squeaky plastic bone, a knobbly ball with a rattle. Every other day there was a new toy. Where were they coming from? I suspected that Nubi had found another way out of the garden, but, since he was coming back and seemed to be discrete, I chose not to think too hard about this and sort of hoped someone was throwing them across the fence to him or that he was picking them up when we went for walks and I simply hadn't noticed. I harboured the cowardly hope that Nubi was working out his own problems, that he had some traffic sense, and that no news was good news.

One evening James returned from a walk down to the beach (ten minutes away). He had taken Nubi with him and Nubi had somehow got away from James and had crossed and recrossed the Nepean Highway. James said, "It wasn't possible for him not to have been hit by a car." Nubi showed no sign of injury, so we both assumed that miraculously he had escaped harm.

The next morning, however, Nubi showed signs of severe pain. He felt hot and his abdomen appeared tender. It was a Sunday. Because of the temperature and because he appeared seriously ill, I took him to an emergency vet that charged per hour at about the same rate as a taxi. Nubi was there for 8 hours and ran up $1000. Right at the end of his stay there, which he spent mostly under observation, an incoming vet at change of shift noticed that Nubi was experiencing pain when his neck was moved. We finally worked out that he had some kind of neck or back injury. I got him out of that flag-fall operation with some painkillers to give him and the next day I took him to our ordinary vet, who I discovered had been open on the Sunday too. There was some question that there might be an injury inside Nubi's mouth and it was decided that he would have to be put under a general anaesthetic for a proper examination, due to the enormous pain any movement caused him.

We brought him back the next day and checked him in for his G.A. and examination. Several hours later we came to get him. He was sitting in a large cage on the floor in the recovery area and he didn't realise I had entered the room. He looked utterly desolate, like a man who had finished up back in jail, after a short time on parole.

It was a pleasure indeed to go up to him and take him out of his cage. Despite his pain and dopiness from the anaesthetic, a look of amazed joy spread across his face. Jan F. our vet showed me an X-ray and said that Nubi had a couple of spiny outgrowths on parts of his vertebral column, perhaps due to trauma in the past. "You're going to have an expensive dog, there," he said, expressing sympathy. "It could have happened when he was throwing himself against the fence or just playing. These dogs take risks all the time, unfortunately. Or, when he was young, he could have been involved in professional dog-fights, with a lot of impact. You just don't know."

The dog-fighting scenario had occurred to me as well. Nubi seemed pretty hair-trigger at times with other dogs. Then again, not being with him enough to settle him and make him more secure due to attending James in Brisbane had not helped.

The vet said that, really, our best option was to make Nubi take it easy and to hope that the whole thing would settle down.

With the pain-killers, Nubi got cocky and pretty soon he was in agony again.

I had previous experience of a dog with a bad back. After Nero (Bianca's brother) had been hit and seriously injured by a car, he had developed severe back problems. These had been very effectively treated by a dog-chiropractor, a vet called Dr Alex Hauler at the Canine Sports Clinic in Dandenong. We found out about that clinic from Peter C., a friend who was also a vet. Peter, on hearing about Nero's problems, related a couple of amazing stories about dogs whose hind legs became paralysed, suddenly regaining the ability to walk. The Canine Sports Clinic was a really professional operation that catered mainly to gigantic greyhounds. I gathered that seeing a couple of Jack Russels crossed with Staffy was something of a professional favour to our vet friend.

As we sat in the waiting room on that visit years ago we were in awe of the parade of huge, muscular racing animals that preceded our consultation. Nero and Bianca were also very stimulated by the experience.

Once in the consulting room, they became more leery though. Dr Hauler asked me to hold Nero's front legs and head. His assistant took Nero's shoulders, and Dr Hauler took hold of Nero's back legs. The look on Nero's face was profound. Suddenly Dr Hauler rotated Nero's hips, like a door knob on its spindle, while Nero's front body was kept immobile. There was a blink of pure shock on Nero's face, almost instantly replaced by relief. Then we did Bianca, who had mild loss of strength in her back legs.

These treatments were very effective and kept both our dogs in excellent condition for months, after which time we would take them back to the Canine Sports Clinic, which acquired an electronic player-piano composed of dogs singing individual notes. This was big hit with our dogs, who were real singers – especially Nero.

On our last visit, Dr Hauler seemed a little irritable and preoccupied. He was at pains to recommend a gel one could obtain from chemists, that was helpful for dogs in pain. He urged me to be careful not to allow Nero and Bianca to overdo things. "They are not young dogs, but they don't have a concept of stopping. You have to rein them in. Bianca's back legs will deteriorate if you keep playing high impact games with her, notably ball chasing." Bianca was entirely addicted to ball-games. This was bad news.

In the year before she was euthanized, Bianca lost her passion for balls and finally seemed to prefer not to be reminded of them. Nearly completely blind, she had also lost her sense of smell. All her smart black spots now faded to pure white, but still the loveliest dog, she greeted me with almost human sounds of enthusiasm when I came to her, trusting me to take her out to the gate for a sniff and a short walk.

Anyway, the next time we went to the Canine Sports Clinic, the name had been taken down and the front door was locked. It was as if we had only imagined the greyhounds and the singing dog piano. We discovered that dear Dr Hauler had died of a brain tumour. I now realised that he knew his time was near when he took such pains to give me the name of a particular pain-killing jell and told me so sternly to make sure my dogs didn't overdo it. He didn't care to tell me that he would probably not be there for much longer.

Now I thought of taking Nubi to a chiropractor, but Dr Hauler was dead. Another chiropractor, who did mostly humans but dogs on the side, had closed down his practice. Finally I tracked a vet chiropractor who did home visits. Tall, blond and rangy, she looked like the kind of woman who might be cast as a country vet in an Australian soap. She was mainly a horse vet and chiropractor but kept her hand in with dogs.

We put Nubi in my parent's lounge-room at the other end of the house for the vet consultation. Assisting were James and our self-appointed gardener and amateur dog-whisperer, Leanne, who has an obsessive compulsive disorder and uses gardening for therapy. Nubi looked extremely morose and held himself stiffly, moving as little as possible. The vet got down on her knees, nearly at his nose-level, and had a feel of his neck, reassuring him verbally as she did so. She thought he seemed pretty bad and didn't feel very confident, but she did do some kind of manipulation. To everyone's surprise, especially Nubi's, it worked. He was suddenly animated, wagging his tail and licking everyone. The vet said she would come back in two weeks for a repeat treatment.

Nubi maintained his improvement for a few days until we went down to Sweetwater Creek, ten minutes walk below the house, and he exuberantly raced down a ravine, swam across the creek and up the other side, then swam back across the creek and returned to me at a gallop up the steep bank. On the way back home he began to stiffen up.

By the time of the vet's next visit he was so bad that she declined to touch him. She then recommended another chiropractor in Geelong (70 or so km away). We made a pilgrimage there but this chiropractor was also a kind of mystical healer and we didn't see any improvement for Nubi.

Ultimately I questioned whether Nubi had received a blow when he had crossed the highway. I decided there was undoubtedly a big muscular component, whatever the cause. I also surmised possible overuse, subsequent to some of Nubi's marathon stick-chasing expeditions. I had recognised this phenomenon with Miss Dog in her last years when, after a very long walk, this usually indefatigable little dog had a bout of paralysis lasting days, where she simply could not walk. It resolved very slowly over weeks. It seemed to me that two months was a not unreasonable convalescence for such conditions. Getting Nubi to take it easy, as long as he was in pain, was not very difficult. He was inactive and his appetite decreased. He lost weight and muscle mass, but gradually he began to look happier.

In Frankston you can take dogs down to the beach after 7p.m.. Around Christmas 2010 or early 2011 of the Australian summer, James and I took Nubi down to the beach, for some hydrotherapy. We mixed in the shallows with other dogs and people. A couple of people with a small Staffy came up to us and said, to their dog, "Is it? Is that Bob? Yes, I believe it is." The Staffy approached Nubi in a friendly way, but Nubi, however, looked guilty and tried to dissociate himself from the small Staffy and its humans.

"It *is* him," said the small Staffy owners to each other. "Did you know that your dog visits a red cattle dog in Thames Street," they asked James and me. "We know him," said the man, "because my wife's sister lives there with her husband, children and their dog, Max. They call your dog, "Bob." He jumps the fence and comes to play with their dog."

"So that's where the toys are all coming from!" I said.

"Yes. Max has loads of toys. He doesn't mind. But we haven't seen this one for a while. We wondered if he had moved."

"No, he hasn't moved, but he has been very, very ill. He is only just beginning to recover."

"So you're Bob!" we said, to Nubi, who had a very minimal expression on his face, as if he thought any answer would get him into trouble. He clearly understood the situation and would have preferred to have kept his arrangement private. For a little while we would call out to him, "Bob!" and shake our heads, then pat him. He really looked as if he had a hard time knowing what expression to wear, so we stopped the teasing.

The red cattle dog, Max, lived in the house diagonally behind us. An elderly Labrador, Zac, with a single monotonous bark, lived in the house on the left of ours. Nubi was welcome at Zac's place and could easily jump the fence there, but their back yard was secure, we had thought. Zac's owner admitted, however, that he had seen Nubi jumping their back fence, presumably to visit the dog in the next garden, Max. Zac's owner was also a bit of a dog-libertarian, so hadn't said anything.

I had figured that Nubi had worked things out. But now he was ill and depressed. I decided to walk him round to Max's place and see if I could cheer him up. James and I slowly walked together round the corner to the house in question. There was a slim young red cattle dog in the front garden, which was unfenced. A young girl came out and, when I explained why we were there, she urged Max to greet his old friend, which he did and Nubi responded pleasantly. On another occasion I met the female owner of the house, who told us that Max and Nubi (who she knew as 'Bob' of course) used to play tag for hours and run around inside the house, jumping on beds and tearing around generally. Max's owner found this behaviour amusing, fortunately.

On yet another occasion when we took Nubi round there, Max suddenly attacked Nubi and Nubi was unable to defend himself and appeared shaken.

Nubi had not regained weight but I only became aware of how diminished he was the day when my mother came in with another black Staffordshire. I was busy trying to write and found the interruption irritating.

"Here's Nubi," said my mother.

"I'm busy," said I.

"Well, I just brought him in because I thought you might be worried."

"Why would I be worried?"

"Well, the young girl who owns Max brought Nubi back here. She said they are going out and she didn't want him to be unaccounted for, so she brought him round and handed him over to me in the garden."

"What? But Nubi didn't go out. He's too ill."

Since Nubi was lying on the couch by my side, this black dog my mother had brought in could not be Nubi. Nubi was also puzzled at what was happening and got up weakly. The other black dog seemed younger. He had a white blaze on his chest, where Nubi was jet black. The young dog was beginning to look a little doubtful as if he wondered what he had got himself into.

"Mum, how could you mistake this dog for Nubi? Look at them together."

"Yes, well, now I look, I can see that Nubi is thinner and more miserable. Goodness, I wonder what the neighbours thought they were doing. I suppose they really thought that this dog was Nubi."

"So, who on earth is he?" I wondered. "I'll take him back to them and tell them they have made a mistake." With a martyred sigh at the interruption, and fearing that I might have to go to some lengths to find out who this dog was, I put the visitor on one of Nubi's leashes and led him out the door. Nubi followed us to the door slowly with an expression of mild curiosity animating his generally hopeless demeanour.

No-one was home at Max's house. I thought I would have to take the new dog home and look after him overnight then make inquiries or, unfortunately, ring the dog-pound.

"Maybe if I walk around the neighbourhood with you, you will take me to your home," I suggested to the dog, who looked back at me pleasantly, although I thought there was just a hint of disquiet there.

I began to walk slowly round the block. On our left a street led down to the creek. It seemed to me that the dog looked in that direction. I let the leash slacken and the dog turned down that road, leading me. About five doors down he turned into a house where a couple of men were doing renovations to a patio. They didn't notice me until I said, "Excuse me, is this your dog?" Before I had finished the question, the older man dropped his tools and said, "Boss! Where have you been?"

"He's been at my house. Actually, he visited a dog called Max and Max's owners thought that he was my dog, Nubi, and brought him back to my house, and my mother mistook him for Nubi and took him in..."

Boss's owner understood immediately. "That dog will be the death of me," he said. "He goes visiting everywhere."

"It's a good cover, being a black Staffordshire," I observed. "They can cover for each other."

The man called his wife out to meet me and a dark haired woman came out in an apron and extended her hand. Boss was wagging his tail and disappeared inside the front door.

Soon after the visit from Boss, Nubi seemed to turn the corner. He began piling on muscle and doubled in weight, hungry as a horse. His jet black coat gleamed again and he looked ready to jump fences, so my father and James made every effort to take him out when I was at work.

One day my father said that Nubi had had an altercation with a white dog and its owner. Dad couldn't describe the white dog, but no harm had been done.

"Nubi's getting more and more unruly," my father said, a few weeks later. "It's getting so I can't control him. I really don't think I had better take him out again."

"Okay," I said, "I'll make sure that James does when I can't.

But the very next day my father relented and took Nubi down to Sweetwater Creek. I was working at the hospital, when I received a phone call from James to say that my father and Nubi had been reported to the council because Nubi had attacked a woman's dog and bitten the woman on the thigh.

This was very worrying news. Nubi was on a path that could easily end on death row. We were at our wits' end. It was obvious that I had failed Nubi in a number of ways. James's attitude where he sometimes locked Nubi out of his room or made him stop licking himself seemed an added stressor for Nubi. Nubi had originally been brought back to a home where he was in competition with an elderly favourite dog. Now that Miss Dog was gone, however, he probably felt he was in competition with James, because James was so unaccountably critical of him. James simply could not stand the noise of Nubi drinking from his bowl or licking himself. I had never paid attention to any of these noises, since they were natural and inevitable. I might as well object to the sounds of birds or the wind. For James, however, it was really serious. There was some kind of 'synthesia' going where stimulation of one sensory pathway led to a reaction in another sensory or cognitive pathway. Somehow the sound of Nubi dealing with his own saliva and tongue uncontrollably evoked disturbing images and thoughts for James with a painful auditory response. The repetitive motion involved in licking also bothered him if he could see it. There was not much I could do about this except try to help both of them adjust. I wondered about getting professional dog-training help. James had already consulted his psychologist about this sensory problem. In fact he brought the topic up much earlier at VicRehab with his speech therapist, relationship therapist, and neuropsychologist. They had taught him distraction techniques and to explore underlying feelings, but they acknowledged that he had real neurologically based difficulty with some sounds. For all I knew, the disagreeable sensations that James was avoiding could even have been harbingers of fits, called 'auras', which he avoided by avoiding the stimulus. Four years and eight months later, James remains sensitized to licking sounds, although less so.

Just as these complications were piling up, my mother met an old man down the street who said he was going back to Italy to die. He was leading a very cute little Pomeranian dog at the time and when my mother admired the Pomeranian, he said that the dog would have to go to the pound if he could find no-one to take it. My mother told him to bring the dog to the house in a couple of days, on Monday, and that we would take it in.

I was aghast. On Monday no dog or man showed up, so I relaxed.

On Tuesday the doorbell rang and there was the man and the Pomeranian. The Pomeranian's name was Lucky. Lucky looked like a dog who had no choice but to put up with the old man's vagaries and look out for himself as much as he could. We learned that the old man's wife had died a year or so ago and that he felt he was getting in the way of his son and daughter in law, who had a new baby.

So we took Lucky in. The old man said good-bye and we closed the door. Unsurprisingly, Lucky seemed diminished by this act. At the first opportunity, within minutes of the man's departure, he slipped past someone who was going out the door, and took off too fast to catch. We intercepted him as he was returning to his old master's home and just before the old man arrived back himself. Standing in the street, the old man wished Lucky a 'good life' and bid him adieu. Poor Lucky was beside himself.

As kindly as possible we made Lucky aware of his lack of choice. Lucky, after initially jumping on the much larger Nubi and simulating sex in a very domineering way, took up residence on our bed. From this superior altitude he posed a constant positional challenge to Nubi, who was too large a dog to feel comfortable sharing a bed with two humans. In fact he was a creature of very simple habits. He slept on the floor or the couch and really disliked bedcovers.

After a while it became obvious to me that Lucky stayed on our bed more than was normal. I had to consider the idea that he was actually afraid of Nubi.

Then we got a phone call from the council about the incident where Nubi had bitten the woman and attacked her white dog, which was described as a Samoyed.

The writing was on the wall. Nubi was 'this far' off being declared a 'dangerous dog' forever to be confined to the backyard and to wear a special collar, and never to be allowed off-leash. Some hope.

For a large fee I got in a man from Bark-Busters. The results were initially equivocal. Nubi found the Bark Buster's man quite outrageous. You could tell from Nubi's behaviour. He kow-towed but there was a look of disgust and rebellion in his eye. It was hard for me to keep up the tough discipline. It was exhausting. Nubi was a lot smarter than me.

James would walk Lucky and Nubi to work with me and come with them to meet me most nights. One day we walked to the hospital together through the park on the way. I let Nubi and Lucky off the leash because the way was clear and there were no dogs in view. On the way back James did the same. Lucky and Nubi raced round a corner and out onto a back street. There must thousands of dogs in the area, but they encountered the same woman with her white Samoyed. Urged on by Lucky, Nubi sank his teeth into the Samoyed's neck. James tried to pull him off but Nubi's jaw had locked. Eventually James prised his jaw apart.

I went to see the woman a few days later. She was very understanding but she said she had had a dangerous dog herself which had had to be put down. I could not help wondering what it was about her dog that had made it a repeated target of attacks. In fact this last incident was one of three involving Nubi with the same dog. I actually recognised the Samoyed's owner from the local gym where I had, by curious coincidence, heard her describing how Nubi had attacked her dog and bitten her in Sweetwater Creek. I had pricked up my ears because I had recognised the description of my father. The woman also had said, in the gym, that her dog had been attacked by other dogs on several occasions, and seemed afraid of dogs. "I just can't explain it," she said.

I did not tell the woman I had overheard this conversation. She told me how her dog had suffered from being bitten, how she had suffered, and how she was afraid that Nubi might turn on a child one day. I had to admit this was not beyond the bounds of possibility.

I knew I was beaten. I couldn't send Nubi to the country to a farm because he would chase sheep and chickens. He just could not be trusted. In addition to the lack of good support from me while I was in Brisbane, something else had happened to push him over the edge and we would never know what. James's hostility to Nubi washing himself and drinking was another source of tension for the dog. James was not nasty to him, but would stop him from licking himself. Nubi found this hard to take. Lucky's presence might be an exacerbating factor as well. But before James or Lucky, Nubi had been left in the care of my very elderly parents while I had been absent, and before that he had been a wild dog for a while. Before that, perhaps he had been a fighting dog. I wondered if he had even been forced to fight a Samoyed. I would never know.

That afternoon, the Council Animal Officer rang and she told me that both James and my father would be charged and required to appear before a magistrate for failure to control a dangerous animal. My father was 86 at this time and James was just out of hospital with severe brain injury. The costs could run into tens of thousands of dollars.

I was in a very bad mood that night. I felt guilty about Nubi and cornered. Because I had not raised Nubi from a puppy I did not have the same deep love I had felt for Nero and Bianca. I think that if they had been implicated in dangerous dog activities I might have moved to another town and hidden their crimes. Even if I could have done that for Nubi, I did not have enough love to carry us through. I was caught up in responsibility for an entire household; my parents, James, Nubi and now Lucky. The strain was becoming unbearable.

The next morning I made a phone call to the vet, and another to the Council Animal Officer.

"If I have the dog euthanized, would that change anything?" I asked.

"Perhaps I could try to persuade the Samoyed's owner not to press charges. She is really most concerned about the possibility that your dog could attack a child."

"I have made an appointment for the vet to come round this afternoon and euthanize Nubi and last night I gave the Samoyed's owner money for her vet and doctor's bills."

"Good on you," said the animal officer. "What time are you expecting the vet?"

"The vet is coming here at 2p.m.," I said.

"Which vet?" asked the Animal officer. I told her and I knew she would check to see that I had gone through with my expressed intention.

There were a couple of hours before the vet was due. I gave Nubi a nice snack of chicken, put his harness on, and walked him down to Sweetwater Creek. It was a week-day and we saw no-one. We walked along the familiar dirt paths, and along the boardwalk amongst the paperbark trees. The canopy from big eucalypts on the ravine edge rose up above the Melaleuca, giving a jungle effect. I let Nubi off his leash so that he could jump in his favourite tea-tree leaf tannin-dark pool that formed in the overflow from the creek after rain. He swam and rolled luxuriously. He always loved the water and except in deepest winter, took every opportunity to dunk himself in any water-holding hollow, but the tannin pool was his favorite. Now every time I walk past it, I think of him. Nubi lacked the cultural software to have an inkling of what was in store for him. Every day was a new day and the tannin pool was a sumptuous swimming environment and a delicious drink as well.

The vet rang to say that she would prefer to come an hour earlier. That was fine by me; I was not keen to play the innocent to Nubi for any longer than I had to. The vet and her assistant were both women. They told me how they would administer a sedative and then the 'green dream' which is a lethal dose of barbiturate.

"We should warn you that, sometimes, the animal will appear to die, but suddenly there will be some activity. You may think he is coming back to life, but he isn't."

Nubi was in a good mood, very relaxed. He lay on the two-seater couch with his head in my lap, attended by two admiring female visitors, as far as he was concerned. They patted him and paid him compliments and the vet put a needle in the pad of fat at the back of his neck. Nubi registered no discomfort at all. He did not even seem aware of the needle. Soon he was even more relaxed, utterly confident. The vet's assistant shaved his foreleg. The vet inserted the needle and slowly, firmly pushed the green barbiturate into Nubi's blood-stream. My huge tears splashed on his healthy coat. Nubi's eyes were dreamy. I could feel his heartbeat. Suddenly his eyes widened and he shot to his feet. Then he fell down and was dead.

James was in the garden, digging a grave. I had not invited him to the murder.

He came in after the vet and her assistant had left. I asked him to carry Nubi to his grave in the left hand corner of the back garden, near where he used to visit Zac's garden. (Zac had since died of old age.)

James put Nubi in the hole and I left him to fill it.

Sometime later I was loading photos from James's electronic camera onto my computer and I found perhaps fifty photos of Nubi's body in his grave before James had covered him with earth. James later said that it was a way of dealing with his grief.

For some weeks, James resented Lucky, who he saw as profiting from Nubi's death. And I resented James, for his irrational attitude to Nubi's mouth sounds, which I felt had been a factor that could have pushed Nubi over the edge. I also resented my mother for bringing Lucky home. And I resented myself for resenting everyone.

I did not miss Nubi immediately, but perhaps one year later, I found myself walking Lucky in the old haunts where I had taken Nero and Bianca, and later, Nubi. I found that, not only did I clearly visualise Nero and Bianca, but I also remembered Nubi. Oddly I did not feel that he would have resented me for what I had done. Looking at photographs and movies of Nubi, in our many bushland and seaside walks, I realised that the two years or so that he spent with me, apart from his terrible illness, had contained much happiness for Nubi and for me.

VicRehab

I began this book by keeping a kind of diary of events and situations and priorities but during the most intensive period of James' out-patient rehabilitation, at VicRehab, I was too busy to make many useful notes. And, although I participated at home, and shared some sessions with him at VicRehab, James attended mostly by himself, and has very little memory of this time.

VicRehab was a large former government brain injury hospital, known as TAC (for Traffic Accident Commission) Rehabilitation Centre which had been privatized in 1997 and divided into inpatient rehabilitation and outpatient rehabilitation. It seemed vast because it was all built on one level. I recognized the architectural type from my early nursing experience in large state institutions. In the inpatient rehabilitation building, which had been modernized, a lot of people were still in early recovery and spent substantial time in bed, tended by nurses and doctors. All the walkways in this bed-patient area were glassed in.

Connecting the two sections was a long covered walkway with lawn on either side, forming a kind of linked courtyard.

The outpatient centre was very solidly built with concrete walls and floors and ill-fitting doors on very solid toilets, long concrete walkways, suitable terrain for wheelchairs. This part had been refitted with a large polished wood-floored gym and various rooms for therapists, on a smaller scale, simply furnished. The woodworking and occcupational therapy rooms, which included a model kitchen, were also in the old part. There was a cafeteria franchise that sold rather expensive cakes, sandwiches and coffees but was a good place to sit and wait out of the weather. There were external tables and chairs on a patio created by the wide verandah on the courtyard side of the outpatient building. Wheelchair-bound patients often hung out there on sunny days.

At the front was a circular drive for taxis to pull up and where therapists could park their cars and enter via the reception area, bounded by a grassy lawn with a couple of well-established weeping ornamental trees. At the back of the hospital was a large parking lot with small recently planted trees and little shade.

The Therapists

The main people he would see there were a speech therapist, D., a social worker, V., a neuropsychologist, S., the occupational therapist, C., a woodworking therapist and physical rehabilitation therapists at two different gyms on the premises. Once a month he saw Dr K., who was a rehabilitation specialist.

C., the occupational therapist, a young quietly presented woman, who later walked the Kokoda trail during her vacation, managed to establish an immediate rapport with James. He came home after his first meeting with her and mentioned how she had made him feel good about himself.

The other therapists never made quite the same personal impression on James, although he spent more time with them and they had a huge impact on his recovery. There was a policy of moving occupational therapists around, apparently, and about half way through his time at VicRehab C was replaced by another O.T. with whom he did not feel nearly as comfortable.

A therapeutic session coordinating system at the newly privatized VicRehab was an interesting form of economic rationalisation of the traditional hospital format of outpatient rehabilitation where patients came in on regular days to a regular staff who had regular time tables. After each therapeutic session the next session would be scheduled by the therapist by writing it on a form each patient carried with them and deposited with coordinators located in the reception area, at the end of the day. It seemed a rather hazardous system to use with brain-injured patients to me. The coordinators tried to eliminate double or impossible bookings, but therapists and patients still sometimes failed to coincide because either could get mixed up about the times. My impression was that the associated stress was a constant item on the agenda at staff meetings. Every service at VicRehab was individually costed. So, if James came in for one hour in the morning but had two hours free until an afternoon session, he could not just go and use the gym, because he was only entitled to so many hours gym a week. James found this inefficient from his point of view because it meant he had nothing to do except buy expensive coffees or read if his sessions were not back to back. Being brain-injured he was easily at a loss. There was apparently a sort of sitting room for outpatients but he was never sure of the rules. After a while he learned his way to some shops nearby and would walk there and do a bit of shopping.

Morning scrambles

VicRehab's unofficial code of dress for patients was tracksuit gear because of the gym sessions and the need for simplicity of dress in a diversely handicapped clientele.

My memory of James during this time was the morning scramble before he left about three mornings a week in a taxi wearing tracksuit gear. There was continuous psychological stress for him because, as far as he was concerned, he really should just be getting on with the programming tasks he had set himself for *Candobetter* before the accident.

For a person with a new brain injury, getting out of bed, getting breakfast and getting dressed and out of the house is a major undertaking. Once he was sure that he had a session on, then James would fly about the house looking for his tracksuit pants, running shoes and suitable tops. He then had to locate his backpack, wallet, glasses, combs, pens and papers. As he found one item, he would pick it up and carry it with him as he searched for the remaining items. Quite often he would put the first item down somewhere. Everyone is capable of doing this but for a brain-injured person the amount of unconscious picking up and putting down can be huge. Although they may have the best intentions, they leave a trail like a toddler. I used to do my best to round these things up ready for him before we went to bed, but it was as if they took on a life of their own and hid themselves in unusual places all over the house after we went to sleep. Everything electronic and half our clothing seemed to be black or dark coloured and tended to blend into the many shadows in our house. I took on the role of a personal butler over this period, which was a little out of character, since I am not known for my tidiness or sartorial preparedness.

I was always curious to know what had transpired in James's sessions, but James generally had very little recollection of this because most of it went over his head. He was more inclined to focus on the behaviour of the taxi driver who had driven him home. I would hear whether the taxi-driver and James had anything in common and how much the fare was. James was sometimes asked to sign chits that had not been filled out and he was concerned about whether he should stand up to the taxi-drivers or not.

Because I only worked one week in two, I was able to attend the Social Worker's relationship sessions every second week with James, and occasionally all his other sessions – speech therapy, neuropsyche, occupational therapy and woodwork. Although there was quite a lot of quarrelling at home due to misunderstandings, attending VicRehab had a normalizing effect. The organization presented a controlled and simplified environment. I found the theory and practice of the therapy interesting rather than mystifying because I understood the basics and I was familiar with hospital environments.

Physical Rehab

The staff in the rehabilitation gym were predictably fit looking and mostly fairly young. Set apart by their uniforms and their domain from the psychological staff, they were also taller on average, more physically agile and often seemed rather stern and stressed.

There were two main gyms with several huge spaces each in the rehabilitation centre. One was for neurological physiotherapy and the other was for more general muscle building, as far as I could understand. Patients started off with very obvious disabilities, both physical and mental but I saw big improvements over time. They were of both sexes and all ages – people who had suffered strokes or traumatic accidents of the most catastrophic kind. These were the people who, in Brisbane Hospital's neurosurgical ward, I had thought of as the world's quintessentially unlucky people, lost to themselves and society. At VicRehab, however, for those who could participate in rehabilitation, the improvements were obvious over time. At first the patients doggedly pursued their exercise regime, which was often terribly simple and tiny, focusing perhaps only on one part of their body and obviously occupying all their attention. Towards the end of James's stay I heard a pair of James's rehab contemporaries joking together about their starting points in physical therapy and where they were now, demonstrating insight into their improvements and remaining disabilities. Their faces were expressive and they were optimistic and alert to each other and their surroundings. When I had first noticed them at the gym they had looked robotic, unreactive, with lost faces.

Although the hospital gym was orders of magnitude ahead of commercial gyms, I was surprised at the relative lack of supervision of patients there after they seemed to have learned their exercise machine circuits. James injured himself because of this and it was only because I happened to be present that the injury was finally picked up.

James had been following an exercise course that involved sit-ups using a piece of equipment with padded parts under his legs. Because of his peripheral neuritis (of unknown cause and present before the accident, as previously mentioned) he was unable to feel much sensation in his legs. In this case it seems that he misinterpreted discomfort on the reverse of his calves as ordinary muscle fatigue which he thought must be normal in these exercises. In fact he had worn his skin away there, exposing the muscle layer. This came to my attention as I sat beside the machine. Suddenly he exclaimed, "It's no use! It's just too hard. I can't stand the pain." I asked him where the pain was and looked at his calves and realized what had happened. When we told the gym supervisor, she upbraided James and said that he should have stopped using the equipment when he had experienced pain. As usual, the knowledge all professionals theoretically had of how brain injured people could misinterpret sensation simply failed to be applied in practice despite their knowing that James had altered sensation in that area anyhow.

In defense of the gym staff, I concede that it is very hard to remember, imagine, anticipate and look out for everything that can go wrong with a brain injured person. Without the specialized gyms and their staff at VicRehab, James's recovery would have been much more limited. There probably should be more awareness of altered sensation and staff should probably examine patients head to foot after every circuit.

Since his accident James has had two serious infections in his foot due to simple blisters which he did not feel until the infection became generalized. If I had examined his feet every day this would not have happened.

The fact that James had neurological damage to his whole body, causing weakness in the muscles, meant that he was especially vulnerable to muscle injury. When his allocated time in the VicRehab gym ran out, he was referred to a commercial gym, although we were warned by one of the rehab workers that none of the commercial gyms take much care of disabled clients. It doesn't fit their image. James came away with something later diagnosed as severe bursitis, which became a chronic handicap that prevented him from fully raising his left arm. This was probably caused by doing bench presses, even though I tried to look out for him. But what did I know about bench presses? The one-size-fits all slogan-style advice and instructions at this gym were totally inadequate. The supervision was nominal. The gym didn't take care of anyone. Non-brain injured people also come out of such gyms with serious injuries. I look back with horror at the way I pushed weights using my knees when I joined as James's required carer. Today I have damaged knees. The hard sell of body building and toning relies on generating a kind of religious enthusiasm long enough to get a contract signed. During that period, before people drop out, they subject their bodies in an act of faith to harsh stresses in arbitrary routines with unfamiliar equipment. Were it not for James's injury, I would never have had contact with such places, let alone have signed up!

If you can build a box, you can build anything

The Woodworking Teacher was a carpenter who had taught secondary school before moving into the rehab area. He would say that, basically, if you can learn to build a box, you can build anything. Tall, solid, middle-aged and balding, his territory was a couple of rooms furnished with carpentry benches and tools in one and a circular saw and shelves full of timber in the other. With their saw-dusty odours and direct access to the outside, they made a nice change from the clinical environment,. Near the front doorway there was a display board with photographs of patients' completed projects – mostly items of furniture. The teacher guided two to three patients at a time in woodworking projects, with the assistance of a rostered nurse. I spent a few sessions with James there and helped him select materials for his projects at hardware stores. There was a young boy of about 15 who also attended the woodworking sessions when James was there. The boy wore a helmet, presumably because of the danger of fits. He had one mildly atrophied arm almost entirely immobilized by neurological trauma, which he used as a stabilising support for his projects while he manipulated, hammered or planed them. He chatted a lot with the teacher, at a level much younger than his years. They mostly talked about his visits to his family and his ideas about impressing his girlfriend. He had been there a while and had completed several projects, some of which were displayed in the collection of photographs. I felt very glad that James, although his balance, strength and thinking were affected, had no obvious stigmata of brain injury, mental or physical. It would have been so much harder if he had. I imagined the distress of the young boy's family as they began to realize the limitations of his recovery. It was a reminder that, the younger you are, the less learning you have to fall back on.

James was good at and enjoyed the woodworking. With help, he made a wardrobe and a cupboard, to my specifications. I was the one who quickly thought up new projects for the house and wrote down the dimensions. James was always too preoccupied with coming to grips with his computer problems to remember to prepare projects for woodwork without reminding, although, in principle, he was keen to be prepared. He built a small yellow cupboard which we still use in the kitchen and a very large wardrobe which we still use in the bedroom, and which I eventually painted.

Structured supportive environment

VicRehab was, for the most part, mildly stimulating and generally reassuring. While I was there James continued to improve quite rapidly, so the whole experience was relatively optimistic. I would drive myself there and James and I often went for lunch at a local shopping centre and shopping in the Vietnamese shops in Springvale road on the way home. Melbourne is a huge, sprawling city and VicRehab was a long car drive from Frankston. We would chat and I would test James to see if he was learning the way home. James always carried a street directory to VicRehab and studied it both in the taxis and when I drove him. He was trying to learn his way around Melbourne. It was taking him a long time. He would familiarize himself with one route but when our habits changed later, the memory of that route would fade. He could follow a map well, but he could not easily remember its contents or orientate himself without the map. James's spontaneous practice of keeping a street directory with him was probably a really good instinctive kind of self-retraining for a skill impairment that showed up on neuropsychological tests as an inability to reproduce a recently copied pattern from memory. This impairment also showed up in computer work, as a reduced ability to retrace his steps to investigate a problem.

The importance of diaries to the brain-injured

From the literature I read voraciously and widely about brain injury rehabilitation and prognosis, I gleaned that the most important variable that an injured person might control was to compensate for damaged planning abilities by effective note-taking and diary use. I also learned that few brain-injured people develop really good skills in this.

The Speech therapist, Social Worker, Neuropsychologist and Occupational Therapist worked closely together to reinforce priorities in James's head and Choosing A Diary was an important subject of two or even three speech therapy sessions.

James and I had already allocated transparency folders and a number of A4 notebooks to things like "IT technology", "Test bed", "Vital documents" but we were trying now to standardize so that James would use one diary for appointments and another notebook for every kind of reflection or detailed records of work done. The diary would have fixed printed dates so that future appointments could be scheduled in it. The notebook would allow James to write in dates for each entry, forming an historical sequence. It was very difficult for James to see the merit in this note-book plan where everything went into one notebook under dates. His preference was still to open a new notebook for every new subject or project. He had the same approach to filing bills and work on loose paper. Despite his love of computers, he retained a love of paper and dreams of orderly ranks of filing cabinets. James had a curatorial mind.

I can remember that, some time before James's accident, our friend Jill and I amused ourselves on one occasion picturing each others' brains. Because Jill often seemed to forget things that were obvious to me, but could recover these memories under prodding from different angles, I compared her brain to two cylinders, one inside the other, each with similar sized holes in rows around them, which needed to be lined up in a random process to let light in for Jill's brain to be able to see the connections the holes made when they lined up. When this process was achieved you could see connections all over the place and the space inside the cylinders glowed.

James's brain, Jill and I agreed, could resemble a vast silo with a number of dwarfed high tech fittings and metal Escher staircases leading down into the gloom at the base of the silo. Shafts of light from windows high up made glowing dust-motes and highlighted thousands of toy soldiers arranged in historical battles, world maps on the walls and globes on wooden desks on a vast factory floor divided by metal filing cabinets into activities right down to a microscopic level. Nothing had ever been thrown out. When you looked up, you could see that the whole silo was crisscrossed with blue RJ45 cables[37] leading to batteries of computers and that there were vines coming in the windows. We might have added that there was a radio loudly tuned to public news radio station.

I cannot remember what my brain was supposed to look like.

As far as James was concerned, nothing much had changed in his view of the world except that he was now often tired, could not seem to achieve much lately, and had an awareness of once easily accessible remembered material having disappeared from his mind. He still tried to keep up with political events worldwide, through a variety of independent internet sites and current events and news radio. But it was as though a lot of the RJ45s had become disconnected, truncated and were no longer receiving or only transmitting intermittently. As for information storage, an unknown amount of the hard disks on the batteries of computers in his brain had developed randomly corrupt files and were working very slowly and unreliably. The clock and timing functions, which had once been ultra-reliable, now required external regulating.

There were signs that he was aware of this. He went out and bought several electric wall-clocks and installed them at strategic points around the house, acquired filing cabinets, and stocked up on notebooks and diaries, pens, pencils, pencil sharpeners and, of course, computers. His tendency to buy in bulk drew mildly alarmed reactions from me, since I was always trying to increase space. The corridors of our home were narrow and clogged up easily.

Almost all my records were now electronic and I really didn't need to conserve ancient papers I had placed years ago in the metal filing cabinets in the back shed, which possums had mostly shredded anyway. With deep satisfaction, I heaved the filing cabinets on the nature strip for a hard rubbish collection. I was piqued when, oblivious to my activities, James independently trawled the same hard rubbish collection and returned with two more filing cabinets and put them in exactly the same place. He subsequently forgot all about them.

There was so much choice on the internet and in actual stores of different kinds of diaries that, after the VicRehab team stimulated the idea, a fascinating search continued for weeks. James and I each finished up with multiple diaries. Weight and handiness was an important consideration. After he bought a couple of huge expensive desktop ones, we found a treasure trove of low-cost copies of smaller, lightweight ones in a variety of formats at a store owned by Cambodian importers who had a wonderful general store in Frankston. The upshot was that we finished up with several very similar looking diaries and notebooks each, with predictable confusion. The bulk of them were brown or black leather-look which, like dark clothes and black electronic equipment, faded into dark furniture and shadows. James pasted white labels with his name on a diary and sequence numbers on his notebooks. Eventually I bought a bright coloured opaque plastic removable cover for James to keep on whichever was his current notebook and a large glittering holographic square to identify his diary. These methods seemed to work fairly well, but multiple diary creep persisted along with diary misplacement.[38]

Of course, if James had been a Microsoft computer user, he might have simply adapted his life around a smart phone or ipad. Or he could have used Google to provide an online diary, as my father did. But we were both averse to being tracked on the internet or to paying high monthly costs for smart apps. So, for the time-being, James stuck to paper diaries.

Neuropsychology was the therapeutic lynchpin of neurorehabilitation. James's neuropsychologist, S. was small and slim. She laughed easily and was given to colourful patterned dresses and her office was decorated with interesting prints. Her therapy was based on objective testing of James's perception, function and intelligence.

She tested his verbal skills, knowledge of word meanings, verbal reasoning, ability to generate lists of words under pressure, visual and intellectual skills including problem solving and planning, attention and concentration, memory and learning. James's processing speed was found to be notably slow. His verbal memory and learning were average, not nearly as good as they had been, obviously. His visual reasoning was quite good as was his problem solving. Under pressure, however, those retained logical skills went out the window and he had become rather impulsive. Fatigue further affected his judgement.

We both now learned that disputes were now more likely to happen, especially while he was still recovering. We should say, stop, don't go there and suggest thinking about things in the morning. James was feeling insecure about who he was and out of control of his life. He tended to rely on me for everything he could no longer reliably do himself and needed to become more independent by taking more care with things he now found difficult.

Family meeting with staff

James's father, Ian, flew down to Victoria for a meeting with VicRehab. We had lunch together at a local shopping mall beforehand and were almost late. The meeting contained no surprises for me, but was reassuring and informative for Ian. James tended to try to argue that the deficits that his therapists described to us at the meeting were part of normal variations in normal people and could not necessarily be related to a brain injury. It was clear that his insight was partial and fluctuant. We heard later that the staff were amused when James, his father, and I, all pulled out writing pads and took notes during the whole family meeting. They joked about this being a family trait.

James and the 2008-2012 Queensland Elections

James ran for Lord Mayor of Brisbane in 2008 in order to highlight the lack of consultation with the people of Brisbane in privatizing public assets and in massive public-private infrastructure works to cope with government-engineered population growth. This was his first campaign, done in his scarce spare time, with no budget at all. It became possible because he had met a like-minded soul – me – and because we both knew how to use the electronic media and had our own alternative outlet. Coinciding with our debut as website publishers was the fact that Australia's grotesque defamation laws had been modified in 2006, thus making real debate and reporting possible for the first time in Australia.[39] We used *Candobetter* to promote James's efforts and policies and also to publicise the fact that State radio, tv and print media outlets refused to cover new candidates and kept coverage to a very narrow range of policies. We thus began a modest but public record of these soft Orwellian circumstances.

James decided to try to keep on contesting elections. I committed to supporting and encouraging his efforts.

James next ran for the March 21ˢᵗ 2009 Queensland State Election. He was running against Andrew Fraser, the State Treasurer, who had held the seat of Mt Coot-tha since 2004. James actually lived in this electorate. His objective was to force the press to ask the ALP Government to reveal any plans of privatizing assets, in view of widespread public opposition to ongoing privatization . This time we started to plan early in order to have a better chance of alerting the electorate to his candidature and his policies. Towards this, in 2009, he interviewed Andrew Fraser about his policies. I filmed the interview, but had to subtitle it because the treasurer sat (surely not by coincidence) under an extremely noisy industrial size fan. The content of this interview was very revealing but the noisy circumstances and the angle of photography don't do James justice. The interview is in three parts. The first part is here:

https://www.youtube.com/watch?v=EaMYQRGM5Yk.

James was also recorded speaking at the electoral launch of another Mt Coot-tha independent candidate, Dave Zwolenski, an aspiring actor, who was actually featuring in his own comedy about the difficulties of independent candidates.   James's talent for off-the-cuff speaking and his pleasant charm are much more apparent in this record, transcribed in the footnote and available on film here - https://www.youtube.com/watch?v=yC4qGs0epUg - than in the interview with Andrew Fraser. Zwolenski's publicity ploy was to run to the Town Hall, stripping naked in stages. His policy was not to tell a lie. The only political point of this was that the Queensland parliament had recently made it legal for politicians to lie. James gives a good speech off-the-cuff at Zwolenski's launch, not having been told that Zwolenski was making a comedy, in which James would eventually only appear briefly with his shirt untucked and saying something erudite about Abraham Lincoln and lies. The film of James's speech at Zwolenski's launch was made available to us after our request. James innocently tried to get Zwolenski to enlarge his policies to include opposing privatization. If he had been able to do this he might have withdrawn himself and put his efforts into supporting Zwolenski's campaign, since it was apparently better auspiced than his own. He was unable to interest Zwolenski in this for reasons which became obvious after the television program.

Zwolenski scored 202 votes and James scored 163 votes in the 2009 elections, something that surprised neither. Incidentally, I personally reckon that if James had run to the town hall naked, he would easily have got more votes than Zwolenski, but that's just my opinion.

With the full cooperation of the commercial press and the ABC, no other candidate raised the issue of privatization effectively in this Federal election, even though the public were very aware of it. Exactly as James predicted in his speech at Dave Zwolenski's launch, only after the 2009 election, did Queensland Premier Bligh inform the Queensland public that she had 'discovered' a budget deficit that required her to flog off $19 billion of assets.

This excerpt contains James's predictions about the Bligh Government's secret plans for privatisation in his speech at Dave Zwolenski's launch:

> *"It's obvious that we are not a democracy in the sense of government of the people by the people for the people. The reason for that is if you just think of how many decisions are made by politicians that are opposed by the majority and ask how that is democracy. The obvious example is privatization. They have sold, since the Beattie Government came to power – SGIO - that was the State Government Insurance office, against an explicit promise by Beattie not to fully privatize it. Then, later on, they privatized TAB, and they privatized Dalrymple Bay coal-loader. Probably a lot of other things that they privatized even I haven't heard about because another comes up every couple of days, like the Gladstone power station flogged off in 1994. They flogged off a couple of the airports last year. And the list goes on. In spite of the fact that every public opinion poll shows that the majority of public opinion is emphatically opposed. So there's something wrong. And one thing I've tried to do is to force the politicians that probably have secret plans to go and flog off Queensland rail and our water grid and probably our electricity generators, in the coming years. I've forced them to actually defend their election policies during an election campaign. I've actually challenged Laurence Springborg and Anna Bligh and the treasurer, whom we are both standing against, to publicly defend – either commit*

themselves to not privatizing or publicly defend their arguments for privatization. And I have got no response. They will not debate the issue. They just want to wait until they're safely in power and go ahead and do it, regardless of what we think. I'm also deeply worried about the insane policy of digging up as much coal as we can in Queensland. We dig up a record amount of coal. We're contributing far more than our fair share to the coming global ecological holocaust, and yet our Premier Anna Bligh wants to dig up, to triple our coal exports, by 2030, which is just totally utterly mad. We have to at least stabilize and aim to reduce it, not increase it, whilst we are watching the polar ice-caps melt."

The rest of the speech is contained in the endnote.[40]

Of course the irony persisted, in 2012, when the issue of privatization would, as James predicted, bring down the Queensland Labor Government (Premier Bligh). Ironic too that Zwolenski had made such a joke of independent candidates and had not publicized in his film anything serious about James's platform on privatization. Most ironic that James was subsequently put out of the election race and active politics forever by a bicycle encounter with two boys in a four wheel drive in May 2010. We are glad, however, that Zwolenski gave James the opportunity to speak and filmed his speech.

These two films we published of James speaking were intended to be the first of many, and would of course be supplemented with ongoing articles on *candobetter* and on other sites and correspondence with politicians, other political actors (like the unions and grass-roots community groups) and the mass media – notably the ABC. We intended to consolidate and slowly increase our activity and impact leading up to the 2010 Federal election.

So, our next goal was the August 21st 2010 Federal election in Queensland, in the seat of Brisbane, with privatization the main issue again. James began a campaign to get the unions to flag it, but they only kidnapped it. That is, they appeared to take it up, and then took it nowhere, leaving people stranded. We wrote some strong articles about this, attracting flack from the people and organisations we criticised. The Greens were no better. It was obvious that Queenslanders (as Australians in all states) were getting more and more upset about the issue, but they had nowhere to go. I personally believe that James did have a chance of being elected in the 2010 election due to the freakish situation of a press that refused to discuss something that nearly every Queenslander really wanted to discuss. The long-time member for the Queensland State seat of Brisbane, Arch Beavis, who had held it from 1990, lost his seat to Liberal Teresa Gambaro in 2010.

If the Greens, various 'Socialists', or even the ailing Democrats had run against privatization, they would have been in with a good chance because this was not a Liberal seat. But no-one ran on the issue! The minor parties mostly ran on the highly divisive issue of asylum seekers, which was a gift from the mainstream press to the two major political parties who used it as a classic wedge issue. No-one running even suggested that we should stop supporting the wars that created the refugees and asylum seekers.

James would have run, with my full support and a small but consolidated campaign, but he was, of course, in no condition to do so. If he had won or got a very high vote in contesting Arch Beavis's seat this would have made him well-known in Queensland.

I consider the interruption to his career a tragedy for Queensland and for Australian politics as well as a personal tragedy. The issue of privatization in all parts of Australia underpins a continuing distancing of citizens from control over the financing of and the distribution and price of water, power, housing, finances, ports, scientific research, education, roads, public transport and other major public assets, which the associated public institutions provided. Australia does not have a constitution or other code that guarantees any particular civil rights at law. Our public institutions were really the only structures in Australia through which citizens could assert their rights to certain basic necessities and standards.[41]

James still says he would not have had a chance of winning, but he does not remember our plans and how well they were working. We were building up a large reserve of original articles on privatization, media influence, and systemic political party corruption. We were acquiring many new readers. See http://candobetter.net/StopQueenslandFireSale and http://candobetter.net/privatisation and related tags on *Candobetter*. There was no other alternative press that permitted interaction and uncensored debate on a wide and coherent platform. And we were contributing to other online forums and relevant community activities. Furthermore, since this time there has been an increasing trend to vote for independents in Australia due to peoples' growing understanding that the traditional complex of the Liberal and Labor parties, the Greens, the 'Socialists' and the Democrats is captive to the corporate sector and its corporate press, notably the Murdoch Press and the Fairfax Press.

James was an easy and fluent speaker on political matters, with a wide knowledge of Australian and international history, politics and geography. He was a friendly person who enjoyed talking to all kinds of people. He could debate. He was kind and attractive, although it is true that his shirts tended to come untucked and he did not have many 'important' connections, but he had set himself up to become known and to acquire supporters. I believe that he did, indeed, have the makings of a successful but sincere politician.

Since James had been disabled by the near fatal traffic accident in May 2010, he was not able to stand as a candidate in the State elections in 2012, nor was he able to on 31 January 2015, when the issue of privatization was finally a policy issue for the Labor Party.

Leading up to the 2015 election, called early by Liberal-National Premier Campbell Newman who had himself become very unpopular because of selling off assets, the Liberal National Party was saying that Queensland must sell public assets to pay off debt. The Labor Party was however finally opposing further asset sales. Along with this position they also proposed to allocate two thirds of all Government Owned Corporation (GOC) revenue to paying down the State debt. The Liberal National Party said this would reduce capacity for service delivery as any dollar that is put to debt reduction cannot be spent on service delivery. The Labor Party's plan still left one third of GOC revenue for service delivery. How much of such revenue could still be allocated if the GOC's were sold, as the LNP proposed to do? Zero? If the debt were reduced, GOC revenue share allocated to service delivery could always be increased. How much could such allocation ever be increased if the GOC's were sold? Again a big fat zero. It takes a duplicitous media to avoid lampooning the party pushing such a stupid argument.[42]

So the line was that, if we didn't sell Queensland's remaining assets (to private corporate interests in which our commercial press corporations or their corporate friends probably have financial interests) we would not be able to pay for the infrastructure demand generated by population growth. As usual, no corporate or government voice acknowledged that the population growth had been engineered through direct state sponsorship and encouragement of interstate and overseas immigration and is unwanted by the public. As already mentioned, it was this state-engineered population growth that generated all the excuses for forcing unpleasant changes on the electorate. This issue was taken up finally by William Bourke's Sustainable Population Party and found some expression among other independents, however it needed much wider exposure and remains one of *Candobetter*'s chief areas of political consciousness-raising.

A purpose of this book is to record one man's sincere political efforts and to encourage others of like mind to do likewise. We encourage readers to get to know our website and themselves to write, forward and comment on articles there. One should not be too hard on Dave Zwolenski. He was a professional actor striving to raise his professional profile just as James was trying to raise his political profile, both of them in difficult environments. I wonder, indeed, if Dave were to do another program on the ironic events that followed his and James's political careers, whether he might attract a measure of new interest.

Computing and Brain Injury: Catastrophe: Candobetter.org finally crashes

For some time there had been an irritating notice telling us that it was not possible to change the 'header' of the *Candobetter* site. James had no idea of what this referred to.

When *candobetter.org* (as it was called then) crashed three times in one day, somewhere round the middle of November 2010, James took it off-line (i.e. unpublished it) with a notice to readers that it would be off-line for a few hours pending transfer to a new server.

It remained off-line for over a month because we could not get it back up.

James was still attending VicRehab as an out-patient about three times a week. Almost all the rest of his time and energy were taken up by Rehab homework, but he was tormented by the need to get *Candobetter* up again. So was I. As I have said elsewhere, *Candobetter* was our mutual project; it was a large part of our relationship. I thought that we could adapt to a situation where there was no longer any *Candobetter*, but it would not be easy. The idea of just abandoning the site without trying to rescue it was out of the question. Because the site had not provided an income to James prior to the accident, it was assigned almost no importance by those who were in charge of funding James's rehabilitation. They saw their role as promoting James's return to a role as a cleaner. They had no contractual obligation and to them it therefore made no sense to try to reanimate James's information technology skills. As the object of that contract, James was expected to relinquish any expectations that could not be met within it.

The fact that I had put years of my life into building up the site was not even taken into consideration in James's rehabilitation or by the compensation system, at common or statutory law, even though for me it obviously involved many full working weeks which might otherwise have been salaried.

James's last information technology employment contract (2004), which had been a software engineering research position at the Australian National University, had been lost to another candidate. At VicRehab, some possibly even thought that information technology skills were a figment of his imagination. In the implicit assumption that the employment market makes correct determinations, others may have thought that because he had had to seek employment in another field, he must have been incompetent in the first. In fact, the research and develop.m.ent field was a musical chairs one, where the risk of spending time without a chair was high, especially for older candidates. It was extraordinary bad luck that James was severely injured just at a time when he was independently building on his IT knowledge and experience in internet publishing. It is an indictment of the real inefficiency of our society that our legal system only recognises direct financial losses and reduces citizens roles to their immediate monetary value. It was even worse luck that the Queensland compensation system would only fund occupational rehabilitation in Queensland and then generally only with the previous employer. Although James by now wanted to live permanently with me in Victoria, by default, he could only be treated as if he were going back to work in Queensland. But, although the system could compel him to return to a particular job, it could not compel him to live in a particular state.

One point of view at the Rehab centre was that, since he was under Workcover Queensland for an accident on the way to a job as a Patient Support Officer (which in his case was a cleaning position) he had no business devoting energy to computer work. Another Rehab centre point of view was that injuries had to be treated 'holistically' and so, if James cared so much about his internet site, some time should be set aside for him to deal with it. A further point of view was that, if James was really worried about the site and he was prevented from attending to it, the stress would be counterproductive. In the against side, *Candobetter* was not a money-earning venture. The fact that it might have been or might have had this potential was too remote a contingency. The fact that James and I had sacrificed earnings and time to build it did not count.

Polite but tense discussions therefore ensued about the ethics of James spending time trying to rehabilitate *Candobetter*, while he was attending VicRehab. He was attending Speech Therapy sessions which re-educated him to basic grammar, word-finding, and logic. (As he progressed in these I asked if he could be given incrementally complex exercises in mathematics and science, but the program simply did not run to these.) As well as the speech therapy, he also had sessions in facial expression recognition, occupational therapy, did woodwork, and attended a neurologically oriented gym program and an ordinary gym program. On alternate weeks he also attended by himself and with me a social worker to assist his handling of our relationship.

(In July 2012 James had no recollection of these rehabilitation programs and people. This lack of memory formation months after he was discharged as out of Post Traumatic Amnesia calls into question the notion of consciousness itself, if measured in levels of real attention and awareness on a critical level.)

The rehab program did not stop with attendance at these sessions. James had to practice his balance lessons, do exercises, do homework for his speech therapist and put her advice and the occupational therapist's advice into practice, such as tidying his room, doing a share of the housework, organizing a diary, doing a daily and weekly plan. The objective was to help him to organize himself now that his memory was unreliable and that ability to prioritise was affected by brain injury. Because of the effects of his injury, especially the fatigue, these tasks took almost every bit of his energy and time. Working on the computer meant time not spent working on his formal rehab program.

James and I asked the speech therapist if, instead of following her generic program, she could reorientate the program to try to help James to organize himself enough to try to fix *Candobetter*.

D., the speech therapist, who was also a children's book author, was one of the ones who thought that treatment should be holistic and she could see that *Candobetter's* crash was eating away at James. With her help, and with James's cooperation, we worked out a way to try to help James organize what he had to do, subject to a time-limit of two to three weeks. This would also help her and me to see what was involved as well, and to monitor whether James was actually making headway or simply going round in circles. The speech therapist was skilled and tried out several approaches with James's cooperation. With James, I designed forms with the names, IP addresses, passwords and purposes for each computer, plus visual cues for each computer. We also made spaces for the Australian virtual host site – OCG.net[43] and for the proposed new mirror site at LVPS Hosting. With James's help, we tried to identify goals and aims for each computer and what had been done, was in process of being done, what was intended and what was just James surmising.

James was expected to plan how to achieve his aims in steps then let us know his progress, step by step, and to record this on the forms. These records were necessary because of the risk that James would forget a step or redo a step if interrupted as well as because of technical problems. The exercise would also be useful to assess whether he was able to lay down complex memories and whether enough skill had survived for him to be able to achieve these technical objectives.

Pinning James down to following some plan we could assess required a lot of energy from all concerned, in coaxing, cajoling, and threats to stop supporting his efforts. For James much effort was required to resist impulse and cooperate with us. Although he initially showed insight into the idea that his memory and fatigue problems could affect his ability to perform the tasks he had assigned himself, once he got involved in a task, he would forget the context of brain injury and resent attempts to monitor what he was doing. It helped that his speech therapist cooperated by setting tasks which I would remind him of and about which she would require him to account at his next session.

Whilst sometimes resistive and frustrated by D. and me, James was conspicuously careful and diplomatic in his dealings with the technical support staff at OCG, the group that provided space to house the *Candobetter* site. He was notably professional in the manner in which he framed technical questions so as not to be misunderstood or to waste the other person's time. It was easier to make himself understood when talking to IT people than with me and D.. It also had to be very good for his brain to think the problems through and express them, then to try out solutions. Even though his memory would usually erase what he had done, the surviving machinery was being oiled.

James's commercial contract with OCG had not included technical assistance, but because his site was so large and had a lot of traffic, they gave him a bit more attention than they otherwise might have. A technician there was aware of James's head injury, although he did not know the extent of it. It was this kind technician who, whilst James was in hospital, repaired the site after a hacking attempt.

James would ask for this technician via the OCG help-line. Unfortunately any technician might answer the request for help and customer reliance on particular individuals seemed to be discouraged. Nonetheless, the kind technician managed to intervene several times in a useful way. He even spent some of his time-off exchanging emails with James.

Transfer attempts

Initial attempts James made to transfer *Candobetter* from Rejkavik to another OCG-based virtual webserver called Basil failed. (For readers curious about names, Basil had actually been meant to be Basel, like the Swiss town, in James's series of Scandinavian countries and snowy places, but was misspelled by a technician who took it for a man's name.)

The problems started with James initially not being able to get anything onto Basil. Diagnosis of the problem relied on occasional help from the OCG technician. Assessment of the situation was further complicated because James had difficulty working out whether the problems lay with him or with the virtual server. After a few exhausting and anxious days when he tried to understand what was going on, he realized that the new virtual server was not empty as promised. There was already material on it from who knows where. This had to be cleared. After it was supposed to have been cleared, there were still problems where it seemed that the space simply wasn't big enough. While James was wrestling with this problem in his brain-injured state, the OCG.com hosting business was taken over by a new and larger organization and James could no longer get in touch with the person who had initially been helpful.

Migrating and Upgrading Candobetter to the Canadian Server

Before the Australian version of *Candobetter* crashed, we had been aware of its instability. We attributed this partly to the growing rate of traffic. We had long planned to create a mirror site to take some of the pressure off the Australian site. A month before *Candobetter* crashed James had managed to copy all the files on our site over to my LVPS hosting site. This was not enough, however, to make the new site functional. You could read the articles, but you could not use any of the links, make comments, or contribute new articles.

Candobetter, as a content-managed site, was normally very interactive, permitting authors to publish, edit and update their work without consulting James or me for routine matters. If publication had relied on us, we would have been completely overwhelmed, since there were over 2000 articles on the site, all of them interactive.

Now that our Australian site was non-functional, the problem was not just how to update the second site from the old site (using free open source software called 'rsync'),[44] but how to move *Candobetter* in its entirety to the Canadian site and then get it to refunction as a fully interactive site, in a process known as 'migration' of a site.

Why was this so difficult? Trying to find out how to migrate a site took James all his time and energy. Conceptualising the problems and then explaining them to me, in addition, was beyond him. I was too busy with writing, editing and nursing to take the time to understand the field and thereby be able to meet him half way. Nor was there anyone at his rehab or among our contacts who could explain any more to me than James could. It turned out that none of the other news-website owners we knew had any actual hands-on experience. They knew less than I did.

Complexity: The virtual machinery involved was inherently very complex and therefore the task of coordinating it and moving it required sustained and strong attention, concentration and stamina – all of which were in short supply for James.

Servers are computers with hardware and software architecture adapted to help them to host and publish internet sites.[45] Servers can be divided up to serve several clients. A server computer can run one or more 'virtual servers', that is, complex digital 'machines' (i.e. 'programs') acting as if they were themselves separate computers. They are meant to run completely separate from each other so that they can be used by different clients without those clients having any access to another client's virtual server. The 'format' is the environment that supports the virtual machines, which include within them programs and applications that allow code to be written and published as web pages or digital movies, social networking sites or in other internet forms that the public can perceive as information and interact with.

Web-pages, especially interactive ones, are produced by a large variety of programs, in different code-languages, of which some examples are PHP, Python, Java, Javascript and C or C++. Some languages are commercial, but many are open source[46] and freeware.[47] Most are constantly evolving independently. Website developers, like James, adapt and change the software programs and applications they start with so that their pages become further individuated.

The hardware environment and the software format may be sourced from many different designers and manufacturers. The software in particular, which includes things like 'drivers' that interface directly with the hardware, comes in several versions with multiple updates, patches and other variations. These variations, whilst useful in their own right, complicate the website migration process because the format (or operating environment) of the new computer server needs to be as close to identical as possible to that of the old server.[48] Although there are many server providers to choose from, providing a variety of operating environments at a variety of prices, you may not be able to get what you want.

With *Candobetter* the correct operating environment would have been 'Debian Linux', that is a Debian distribution of Linux operating system, using Drupal Content Management Version 4.2 and similarly dated versions of Mysql and PHP. PHP is a programming language and Drupal is an application written in PHP.

Up-to-dateness of versions: Before the accident James had been preparing to update the versions of the Debian Linux, Drupal, Mysql and PHP which provided the architecture for *Candobetter*. Working almost full-time at the hospital had made it hard for him to find the time to do this as it was inherently quite a difficult and time consuming job. Learning and using Drupal is usually described in the trade as 'a steep learning curve', but rewarding and versatile. Upgrading and experimenting with new software was part of the reason that James had purchased more than one computer, because he needed a 'test-bed' to trial his various upgrades before republishing *Candobetter* in upgraded architecture on the internet. He intended to upgrade the software on one computer then copy it to the other, which would view it as a website. In that way he would see what worked and what did not. It was not as simple as downloading a new version of Microsoft – although even there you still get incompatibilities with hardware and software – hence Microsoft's certification protocols. Working with open source software allows web-designers and software engineers to try new things out and contribute to learning within their community. The longer you leave software without updating it, the more problems you will have if you try to go from an early version to a much later version. For this reason you may need to upgrade gently, from an old version to the next version and all the way up through several versions. Each new version of the software components will present new problems or challenges interacting with other software, and so on.[49]

Candobetter was also a particularly difficult site to 'migrate' because of its size and complexity. I learned that it is very difficult to transfer a large and complex site. Transferring a single page is much easier, since if one software application in the original page does not work in a new environment, the task of adapting the content to a substitute is relatively simple. In the end, you can always copy, cut and paste the content of a website (i.e. the words and pictures) into a new format.

D. and I tried to help James organize into 'chunks' his attempts to migrate the site and to upgrade so that he would not forget what he was doing before he had a chance to complete it and so that we could work out what he was actually accomplishing, if anything.

He had a habit of working on several sites - the local one on Tibrogargan, the virtual server in a Melbourne university that housed *Candobetter*, and the one I rented in Canada, where we were trying to migrate *Candobetter* - doing comparisons and trying to transfer from one to the other. He could easily forget which site he was working on at any time and would become amazed and then worried that his computer was being hacked when he accidentally tried to use one password for another. At some levels sites he was working on would throw him off every five minutes for reasons of security and so mix-ups with passwords made this even worse. James's professional concern for security meant that he frequently changed passwords. To his credit, he was quite good at retaining his new passwords for a day or two, but after an interruption of a week or more, while he was occupied doing something else, or switched computers or programs, he would forget what they were and would ask me. I would write them down as fast as I heard them and into a little book with a highly identifiable cover by touch and sight, so that we both could find it, because he was constantly losing the exercise books and little bits of paper that he wrote the passwords down in cryptically. We would also argue about password systems because occasionally I would latch on to an idea – such as a zoological theme which I found easy to remember - and he would defend another which he found easy to remember – typically a geographical theme. (As noted, we had websites named Basil for Basel, Rejkjavik. Tibrogargan was the odd one out, not being snowy. It is the name of one of the glasshouse mountains in South East Queensland, which looks like a huge granite gorilla head. The name is Aboriginal and has nothing to do with gorillas.)

James admitted to his speech therapist that he did not seem to be making progress. D. had been making inquiries and said that she had heard that migrating sites was a job that was usually done by a team, over several days. She was right.

Candobetter is saved 21-12-2010

It had been three weeks and we were getting nowhere. Like programmed lab-rats we would repeatedly go to *Candobetter* where there was only a static page with a message about the site being under repair. For a while a lot of other people continued to visit the site as well. Gradually visits dropped down to zero. James had lost his opportunity to seek election to parliament in the 2010 Federal elections and he would probably also not be able contest the 2012 State elections.

Without *Candobetter*, we had lost our roles as creative activists with a potentially major media outlet for publishing a particular brand of alternative points of view for as wide a number of people who wanted to contribute. Without *Candobetter* our social network with others who shared our political activism would gradually melt away. We would become relatively isolated and powerless, like most people. Six years of work and dedication down the drain. Because of a motor vehicle accident. *Candobetter* was never far from our minds, like a desperately ill relative on whom we were resisting pulling the plug.

James's old server people were not able to help him, but the new LVPS hosting people, who were far less expensive, went out of their way to be of service. Noticing that our site remained in distress for several days in one of our later bids to move the site from its original v-server to theirs, they repeatedly asked, "Can we help?" James was skeptical that they could, but in the end they did.

At first James resisted their help. I was only paying about $19.00 a month for the site and that did not include management, but it did include something called technical support. James was very worried about me asking for any help because of the potential costs, which we thought might even run into thousands, since hourly pay for someone to do the work that James was trying to do was quite high. These reservations will give you an idea of the minimal budget to which we were accustomed. They also reflected how the accident had removed James's memory of more recent decreases in pricing for these services. Server providers from developed and developing countries now competed for the same customers. This situation was often detrimental to Australian information technology employees, but it worked in our favour in this case.[50]

One day we got another email from LVPS hosting helpdesk asking could they PLEASE help us. Could we please tell them what problems we were having? They must have been mystified. James did let them know that he had been ill and could not work the long hard hours he had been able to previously, but he did not reveal that he had a newly acquired, serious brain injury.

I had to wait until James was completely stymied before he would give them the information to enter the site. The minute he gave me the go-ahead, I told them, "Yes, please help us!"

For reasons still unknown to us – but they seemed like exquisite kindness to me – the helpdesk people at LVPS Hosting formed some kind of team with different people on different shifts and worked day and night for several days to 'migrate' our site from OCG.com to the Canadian site. The whole process took about two weeks.

At the end of their shifts we would receive emailed messages in somewhat fragmented English asking for us to try it out and tell us if there were any problems. We found lots of problems which we told them about. They would say that they were not experts but then would promise to try to fix the problems and they did.

We were saved and *Candobetter* was saved – now as candobetter.net – by some angels we only ever knew as 'help desk' from LVPS Hosting in Canada on 21-12-2010.

Out-Patient Rehab draws to a close

Psychologist – Dr Michael W.

As I mentioned earlier in the book in relation to his catastrophic interpretation of films, notably Kung Fu Panda, James showed signs of post traumatic depression and was referred to a psychologist. Vic Rehab's psychologist was so inundated with other patients that James had to find a psychologist outside the organisation. He found one in Frankston who proved to be supportive, interested and enthusiastic.

The Frankston psychologist, Michael W., due to similar family dysfunctionalities, had also once belonged to the same 'socialist' political group that had taken over James's life from his teens to his early twenties. Michael W. could therefore speak the same language and understand James's profound regret for having given so much to a movement he later realised was led by people who used it for corrupt financial takeover opportunities. After leaving a traumatic family home at eighteen years of age, a poorly socialised James (who was nevertheless superbly self-educated in political history and theory) had joined this group in the early 1980s because it appeared to offer friendship, meaning, direction, engagement and stimulating literature. As well as taking a portion of James's income, the political group had directed him to change jobs and states on several occasions and to work with others to disrupt workplaces. According to James, this was purportedly done to educate workers to organise collectively in their common interests against employers who were supposedly all ruthlessly and unconscionably exploiting their workers, paying them less than the true value of their labour. Theoretically this was to lead to greater union and class struggles on a broader scale. James particularly regrets these disruptions because it was a time when working conditions were quite good in Australia and he wonders if the group did not actually assist the subsequent deterioration in Australian working conditions in general. The frequent changing of workplaces, industries and states effectively ruined excellent career and training prospects in several occupations. James eventually left the group because he felt that the people dominating it were deriving income corruptly by using it to take over other voluntary organisations for their funding sources, and dissipating the energy of the young people who joined them in meaningless activities which actually detracted from the purported democratic political objectives of the organisation.

Between that time and his traumatic brain injury, James continued to engage in facilitating political engagement where he saw important causes and this was one of *Candobetter's* main roles.

James never lost that commitment to political engagement. As described at the beginning of this book, he was preoccupied with systems and the suffering of others while he was still bed-ridden with Post Traumatic Amnesia in the neurosurgical ward in Brisbane Hospital. It was therefore helpful for him to have a psychologist with an active past and continuing interest in this area.

Apart from this, Michael W.'s expertise in relationship counselling helped prevent either of us sinking into despair or our relationship from running onto the rocks. We continued to see him until three and a half years after James's accident, when we had to stop, due to costs. Nonetheless, going together to a familiar counsellor remains a very important resort in times of crisis.

Tertiary Rehab and work prospects

In James's last week at VicRehab, somewhere around March 2011, I bought a woodworking table and a horse with clamps for sawing, in the hope that both of us would find the time and energy to make use of what he had learned from woodwork classes at VicRehab. Not until July 2012 did James make use of it, with a jig-saw, to carve a piece of wood to fit a hole a branch had clawed into an external wall of the house. He had forgotten that I had bought the woodworking table for him, and had since assumed that it belonged to my father, who used it to cut firewood.

I wrote on 23 March 2011 that the problem for the future now was how and when James would get occupational rehabilitation; if and when he would get a lump sum in compensation under statutory law from Queensland Workcover, and if and when he would be compensated under civil law by the insurers of the boy's car which had collided with his bicycle. Amounts involved would depend on the apportioning of fault under Queensland law. The civil case would not be resolved until about three years and eight months after the initial injury.

James's long period of wearing mainly tracksuits, which had begun around July 2010 with Mater Private, ended in late February 2011, as VicRehab declared that they had done as much for him as they were able. Normally he would have graduated to some vocational therapy, but as long as he remained in Victoria, that was not on the cards because of Workcover Queensland rules. Dr K referred to this lack of vocational therapy as a 'dreadful, shocking circumstance' and wished us all the best.

Commonwealth Employment Service, interview with Meredith

Apart from the better opportunities for rehabilitation in Victoria, part of our decision to bring James to Victoria was based on the idea that Commonwealth Rehabilitation Services (CRS) would be available at some time to help James with vocational rehabilitation in Victoria. It proved to be impossible to access these services while he was still receiving Queensland Workcover payments. CRS expected payment by James's insurer for any work they did, but Queensland Workcover did not have vocational rehabilitation in states outside Queensland as part of its brief. If

James tried to use the Commonwealth Rehabilitation service, he would legally signal to them that he was looking for work (whether or not he found work) and they would stop paying him his salary for Brisbane Hospital. Even undergoing testing to evaluate his potential for specific kinds of employment, if performed by CRS, threatened his means of support while he was ill. Much later Queensland Workcover would fund six hours with a CRS Vocational Psychologist, but only as part of a neurological assessment of function prescribed by his psychiatrist. We were really glad that they managed to do this for James.

Private Gym

Vic Rehab referred James to a local gym, which was situated right next door to an All You Can Eat, in what looked like a clever business arrangement.. Initially VicRehab wished to free me from taking on more chauffeur duties for James by specifying the need for a professional carer to accompany him. Queensland Workcover would only acknowledge one carer for each of their liabilities, however, even though I was unpaid, so I also had to drive him to his gym work outs.

I thought that by standing there monitoring his performance and safety, which was my unpaid role, I was attracting attention to him. What he was doing also looked kind of fun, so I joined the gym for a year, at the cost of about $800.00, which included a training evaluation and health check every three months. At the first health check they decided I needed to lose 15kg. 5 or even 10 kg I might have thought, but 15 kg struck me as excessive. Training by itself would not guarantee weight/fat-loss. I would have to eat less. I actually got fatter whilst attending the gym over several months.

Initially charming, which was how they hypnotised me into joining, the gym staff lost their polish shortly after I signed up and their attitude from then on seemed consistently peevish. I got the idea that working there was rather grim, involving a rapid through-put of underpaid trainees from local college courses. There seemed to be a star system among the regular staff, with their ranking closely aligned with bodily perfection and avoidance of eye-contact with clients. The imperfect and the trainees worked most of the weekends and manned the gym after hours. All the staff had been trained to use upbeat language and used the word 'awesome' whenever they could.

James and I were given a printed A4 sheet of paper with diagrams of our exercise routine next to charts to record our progress. James mislaid his and was treated like a bad child, only grudgingly being given a second chart. The gym began renovations, but failed to keep its clientele informed. We would come in and half the equipment would have disappeared, with the exercise circuits incomplete and no staff present to orientate us. The last two times we were unable to complete our courses. James developed a shoulder injury and we drifted away gratefully, glad of an excuse.

PART THREE: RETURNING TO THE HUMAN RACE

Rapid improvement, insight and phantom memories – March 2011

Early in March 2011 James began to develop some real insight and I got a glimpse of his real feelings about his changed self. For instance, he told me and his psychologist that, although it seemed silly to him, he grieved for his lost memories, which he still felt, although he had lost them, like a phantom foot or something.

One evening, after spending the whole day trying to establish a programming test bed on the computer he called Tibrogagan (having given up doing so on the Swedish B3 machine[51] he had bought a couple of months prior and on his new laptop), he came wearily into the bedroom and sat on the bed. Looking straight ahead and away from me, he said, "Sheila, do you think I'm alright?" I asked him what he meant. He said that he meant it in two ways, but principally, that he was worried that he wasn't up to creating a test bed. "I don't have enough stamina. My concentration isn't up to it. It may be too hard."

I suggested to him that he needed to remember what his VicRehab therapists had told him, about breaking tasks up into small pieces, and that he needed to have definite steps to fulfil, one after the other. I offered to look at his plans the next morning.

This approach cheered him up and when I next referred to his difficulties, he denied having any.

He also said to me one morning that, although he had not died on the 18th of May 2010, he felt as if he had. "It is as if the person I was before the 18th of May did die in the accident, and then another person, very similar, was born on that day." I asked him what the difference was. "The new James is a less sharp version of the old, without the same concentration."

It seemed to me at the time that this was what I felt myself, although I also hoped that things would continue to improve.

On 25th of March 2011 I noted that the period between ten and ten and a half months after the accident was perhaps the month of the greatest improvement to date. James's injury was unlikely to reach any assessable final state before 18 months have gone by, so the acceleration in improvement increased the parameters of my expectations. On the one hand I began to imagine that, despite the obvious severity of his injuries, James might somehow improve enough to almost have his old life back – certainly to have his computer skills back. On the other hand I feared that he might only become well enough to work as a drudge in some insecure occupation. My preference was that the seriousness of his undoubted injury would be compensated enough for him to be able to do with his time what he wanted – to run *Candobetter* and pursue whatever political and intellectual fruits might still come of what he had laid down before the injury.

Better to be a good political citizen and involved in one's community than to spend about two days a week working in an industry of no particular appeal or real value in a job that others might actually want. Because James's energy levels were about 20% what they once were, work for work's sake would leave him nothing left to devote to our mutual political, environmental and publishing engagement on the internet.

How our political and personal association began

James and I met at the Woodford Festival in Queensland in 2004. In my life outside nursing, as an environmental sociologist, I was part of a 'peak oil' panel which included Prof Ian Lowe, Don Henry of the Australian Conservation Foundation, and another NGO representative. That evening I gave another talk about energy resources in the 'Greenhouse', a big tent at the festival.

Audience members took sides on the subject and some people congregated outside the tent to ask me questions as I exited. The planes of their sweating faces caught the lantern light. One of those flickering faces belonged to James, who towered 33cm above me, his face a mystery in the distorting shadows. After some questions about oil-depletion and population numbers, he blurted out how, suddenly the day before, to his utter dismay, he had learned that his two year contract at the Australian National University was not going to be renewed, thus ruining his doctoral ambitions. I sympathised and told him of my own sadness at how my MA by research thesis on different land-use planning systems and their effect on population growth had been buried, despite – or because of - its continuing predictive capacity.

I didn't remain at the festival for long after my talk, but became fascinated by the nearby Glasshouse Mountains. I spent a very hot afternoon walking around Mt Tibrogargan, fascinated by its hard reolite features and the bright green scrub surrounding it. I also came across a truly enormous orb spider hanging from a tree. It was far bigger than any spider of any kind I had ever seen. I tried to film it but could not hold steady due to viciously attacking mosquitoes. I did salvage a series of stills from the film, but disappointingly none provided any effective context to judge the size of this monster spider.

Because of my talk and our short encounter afterwards, James joined an NGO that tried to educate people about overpopulation and the environment. He would often ring me in Victoria from Queensland and chat. I really had not got a good look at him at the festival and when next I met him in Canberra at an NGO planning meeting, I was quite impressed. He had a charming smile, golden skin, and intense Prussian blue eyes. Losing his desk job had slimmed him down and his shorts revealed superbly muscled legs. I know people often exaggerate when they describe people as looking younger than their age, but James really looked ten years younger than his 45 years. Even his real age still meant he was seven years younger than me. I felt that the age difference removed me from serious consideration and thus gave me licence to lightly pay him an open compliment.

I actually said, "Wow! I had no idea you were so handsome!" But James did not see any problems with the age-difference and he took this as encouragement.

A few months later I was planning a holiday in north Queensland, filming big fig, cypress and kauri trees in the Atherton tablelands. James invited me to spend a few days in Brisbane on my way back.

He said he felt, however that he needed to warn me about some aspects of his living conditions. He explained that his house was rather full of his father's possessions, but that he did not feel he could move or throw them out because he was renting the house from his father. I seemed always to be doing something when James rang me and found it difficult to concentrate on the house problem he was trying to convey in this case. One thing that did get my attention though was when James asked me if I liked possums. When I said that I did, he asked me how I felt about possums in the house. When I asked him what he meant, he told me there was one in the kitchen as he spoke. He described its progress round the kitchen, heading for the toaster, which it checked for toast. It wasn't a pet possum, however, it lived outside, in a bit of jungle at the back of the house. When I said that it sounded like fun having a possum in the kitchen, James said that was what he thought, but he was just checking because he knew some people would disapprove.

He met me at the airport with a brilliant smile and was wearing shorts again. He insisted that I drive his car so that I would be familiar with Queensland. He organised two film nights for me to show environment and population films. The first one went unbelievably wrong due to malfunctioning equipment in a library where there was no assistance available. The weather was atrocious, with crushing humid heat despite continuous violent downpour. I responded to these conditions with a rare and disabling migraine, making an irreversible fool of myself. James showed nothing but kindness and helpfulness. I thought the next film-show, in the town of Maleny, went quite well. (Years later, as I have learned from experience, I realise that the films were too long, too slow, too many and too arcane.)

Back in Melbourne, I would often receive phone calls from James about his Drupal content-managed website, *Candobetter*, asking my opinion on content and style, which was hard for me to give because I was still always concentrating on something else. He then asked me if he could post my thesis about the 'growth lobby' and different land-use planning systems on it. He wanted it to become an interactive site where people would engage politically with problems of the day. He asked for my help in providing other material for it, but I had other focuses back in Melbourne, including covering political events and doing my own research. I had started a book about the effect of different political systems on population numbers, which would eventually become a four volume work. After he had published my thesis on the website, he would keep me briefed on how many downloads it had. Gradually I realised that lots of people were reading my work. James had read it as well. I started to pay attention and made a real effort to repay James's recognition and publicity for my work. I wrote an article and James asked me if I could illustrate it. I remember that he said, "Just a few lines to evoke a subject is enough." I did an electronic cartoon of two Mexicans sitting outside an adobe hut in the Australian desert for an article entitled, "Working man's vegetable plot under attack again," about a property develop.m.ent growth lobby apologist's recommendation that Australians alter their concept of beauty from a verdant back yard to a desert rockery, so as to enable a population twice Australia's current size. Suddenly I was hooked: not only could I write effectively and quickly, but I enjoyed the freedom of fast electronic drawing.

The website developed its direction largely from my thesis, as a website for reform in democracy, environment, population, land-use planning and energy policy. I learned a lot from James's knowledge of political and military history and benefited from his knowledge of scientific data and mathematics and his huge and reliable memory for facts, measures and names.

From that time we both became committed to *Candobetter* and gradually to each other. We learned about each other's strengths in knowledge and perspective on politics. Instead of a few days, I spent two months with James in Brisbane. Although it was five minutes drive and 20 minutes walk to the heart of Brisbane CBD, the house was shielded by a field of bamboo that came up to the windows and provided shelter for various birds and other animals, including Asian house geckos, whose slow evening insect-hunts on the window-panes fascinated me as we ate dinner at a dining room table surrounded by piles of books. Brush tail possums did a circuit between the Black Bean tree and the roof and the Morton Bay fig, where the ring-tails lived. The night was alive with pattering feet in and on the vaulted galvanised iron roof as fruit bats shrieked and whistled in a fiesta of continuous glorious summer nights. I realise that these conditions would irritate many people but I liked the nomadic menagerie and the way it animated the house and the night. Buried in a field of feral bamboo (planted by one of a series of earlier tenants), at the top of a remnant rainforest, was like being on a secret island off the grid, although well-supplied with computers. As well as James's own book collection, which contained interesting new subjects, James's father, Ian, had archived numerous art and architecture books there. I enjoyed his musing about art and architecture through tea and dark fruitcake when he visited.

I returned to my job in Victoria with great reluctance. After that James and I commuted, with me doing most of the travelling because my part-time job permitted it, and we corresponded by email several times daily.

We thus came together because of our mutual concern about overpopulation, oil depletion and the overdependence of modern society on fossil fuels. For a decade or more I had led a kind of double life: working five days a fortnight as a psychiatric nurse and much of the rest of the time writing a thesis in environmental sociology and then writing two editions of a book about fossil fuel depletion, which was published by Pluto Press in the UK. My thesis was about how different land-use planning systems in Australia and France led to different outcomes in population growth and economic systems. The book was *The Final Energy Crisis*, Pluto Press, 2005 and it was co-edited by economist Andrew McKillop and I. Although this was a period where a number of books were coming out on 'peak oil' and 'oil depletion', many were one-man tracts of a fairly predictable nature. *The Final Energy Crisis* was a collection of articles by scientists about different aspects of the problem. I was the sole editor of a second edition which came out in 2008[52] with 22 chapters of which only four remained from the first edition. The second edition was divided into sections: Part I, 'Measuring our predicament' reviewed the theory of finite fossil fuel resources; Part II was 'Geodestinies', the politics of oil production and distribution; Part III was about the global picture – false solutions, hopes and fears, including conventional nuclear power and fast breeder reactors and the European fusion project, which a particle physicist analysed for the book. Part IV looked at how different countries might cope without large quantities of fossil fuel, contrasting their historical pre-industrial histories or using examples of countries, like North Korea and Cuba, which had both lost access to cheap oil when the Soviet Union broke apart. The main countries studied by different authors were Australia, France, Korea and Japan – in contrast to the majority of books on the subject which focus on the United States. I wrote all the chapter introductions, transitions, and about six of the articles.

James was a huge help in proof-reading for technical and scientific data because he had a very reliable knowledge of measurements and scientific definitions.[53] His excellent grip on maths and his superb knowledge of geography and history made him alert to miscalculations, misspellings and mistakes of place, person and events. Although he complained of being a slow reader, he was actually a very close and careful reader, so this took time. Now that he was injured he really did read slowly and could not read much at a time, so I could not rely on him as a proof-reader anymore, to my dismay. Even if he retained the ability to pick up mistakes, he did not have the energy to read the volume of information necessary.

Planning for the future

The prospect of making money out of our work on the internet was something we had kept in mind before the accident. Among other things our rapidly climbing numbers of readers and authors before the site had crashed had placed us in a good position to start up an internet-based book publishing company, using modern cloud technology and local just-in-time printing, with global outlets. Although we didn't want to be a for-profit organisation, I had firmly intended for James to be compensated as the creator, maintainer, and owner of his successful internet site with some kind of salary.

The issue of money-making became all the more crucial when James was injured.

Now I devised a fall-back plan. If we both got too snowed-under by circumstances to continue doing what we set out to do, then perhaps we could simply concentrate on renting his house and living in mine and tending my vegetable garden and orchard. Unfortunately, because I had spent so much time travelling back and forth to Brisbane, the trees had been neglected during a vicious drought and the vegetable plot was a joke. James was very resistant to renting his house, because he hated landlordism, but he was also worried that someone might break into it. VicRehab leaned on him sensibly to maximise his financial options and convinced him to find someone to pay him rent for the house on the basis that if he found himself without an income, social security would make him sell the house if he could not show that he derived an income from it. In such a case he might as well live from rent than lose his house and draw unemployment benefits.

Driving lessons, driving tests

Towards the end of his stay at VicRehab, James raised the question of returning to driving. His neuropsychologist felt that his reactions were so slow that this would probably not be possible, but his rehabilitation doctor referred him on to the in-house occupational therapist who specialized in driving competency. I think her name was Susan. She presented as sterner than the other occupational therapists, perhaps because of the need to impress the seriousness of driving upon her clients and their relatives. I did not envy her the responsibility she had for judging whether seriously brain-injured people were safe to drive.

James had to sit a box-ticking test where he was presented with a number of drawings with captions to gauge his comprehension of the law and traffic regulations in different situations. He answered them quickly and efficiently and got all but one very difficult one right. I was looking over his shoulder and I did not get as good a score as he. Susan said that his performance was good and that most people failed the difficult question. She told him that she would now schedule an actual driving test. She seemed to indicate that, based on his written performance, he probably would not have too much difficulty in passing it, but reassured him that, if he didn't, he could do another test. She also added that sometimes people who could not pass manual driving tests were able to pass automatic driving tests. James's car was a manual.

I was elated at the idea that James might soon be able to chauffeur himself because I was becoming worn out with driving and unable to attend properly to my writing or much else. It felt as if I was on a treadmill. Nonetheless, like the VicRehab neuropsychologist, I really wondered if James had sufficient coordination, attentiveness, and rapidity of response to be safe.

James was himself mainly haunted by the idea that he might reveal his ineptness at parallel parking – something he had preferred to avoid before the accident. He really wanted to find somewhere to test himself before the official test.

In our busy schedule, I found time to drive him in his car to a friend's farm and we drove around her homestead on dirt tracks, avoiding a flock of hens, turning, backing etc. I sensed that this was not really a very good simulation of driving on-road and in traffic. I let James drive the car out of the farm gate to the other side of the empty unpaved country road and park it for me to take over and drive us home. James had difficulty performing this relatively simple task whilst steering, braking and changing gears. I was glad that the only car on the road was still quite a long way off.

Despite her experience, it seemed that Susan had not been warned of a likely disconnect between James's intellectual performance on the paper test and the reality of his slow cognition and impaired coordination. Driving is really quite demanding of higher brain and bodily coordination.

The driving test was conducted by Susan and Peter. Peter was a professional driving instructor who had a lot of experience with brain-injured clients. On the day of the driving test James dealt with his nervousness by greeting Susan and Peter somewhat overconfidently. He was joking urbanely as the three left me in the clinic waiting room and proceeded to the driving instructor's car. They were gone for about three quarters of an hour. Apparently the test drive had been quite frightening. Susan's reaction was to appear even sterner, rather than flurried, although it seemed to me that she looked pale. James was still talking with the driving instructor whilst Susan conveyed to me that James had driven far too fast and had coped very poorly with intersections or the freeway.

It seemed to me that she was probably suppressing the desire to ban James from driving into perpetuity, but she said that she would recommend six driving lessons and then another six. If that didn't work she would consider more driving lessons and then review the situation, but there was no guarantee that he would ultimately get his licence back. She would support James having lessons in a manual car but she rather thought that - if he ever got his licence back - he would probably have to settle for an automatic car-driving licence.

The driving lessons continued over several months.

Peter, the driving instructor who had conducted the first test, lived near us and became James's regular driving instructor. I admired his gentleness, politeness, patience, humour and frankness with James. He was a very gifted driving instructor. I also admired his courage, but he pointed out that he had his own set of brakes and could take over the wheel if necessary.

After the first six lessons James went for a second driving test. Although he was nervous about the outcome, he really didn't appreciate how poor his driving was and thus was surprised when he failed this test too.

James persistently approached intersections too fast and slowed and accelerated unpredictably, no matter how many times he was told about this. Steering and observing road conditions required more attention than he could safely allocate whilst also changing up and down gears with his hand and manipulating the clutch with his size twelve plus feet, which sometimes overlapped between accelerator and clutch. If he also needed to roll down the window or adjust a mirror, the car would drift to one side as he did so.

Although these things were obvious to us, James only noticed them at the time they were pointed out. He could not understand why he failed two driving tests. He would acknowledge certain problems that had arisen, but he would rationalize these, hinting that Susan and Peter had unrealistic standards.

He was coming up for a third driving test and Peter warned him – ever so kindly – that he might not pass this one either. Susan, who had the last word in these matters, said she would only allow him to continue to try and get his license back in an automatic car. This meant a significant and symbolic loss. James had kept that little manual car for years. It was a feature of his pre-accident life. We had brought it down from Queensland to Victoria.

James and I now agreed to swap cars. I would take on his little old manual Mazda and he would (if he ever got his licence) take over my big old Ford Falcon station wagon.

Still James could not take in how different his abilities were and he kept on making the same mistakes. I started to ride in the back of the driving instructor's car to help try to get James to develop some insight.

Between the two of us Peter and I started to make inroads. Although James's condition was improving anyhow, insight does not necessarily follow neurological improvement.

One day, however, James came to me, mortified, to say that he had just realized, "What a fool I am! I have been trying to show you all how I can drive, but I suddenly realized that you have all been telling me all the time that I have changed and I no longer have the same capacity. Oh why didn't I listen to you? I hope it isn't too late."

From that time on there was more hope and James did improve. Although it might seem cruel and confronting to the reader, Peter and I would talk to James about 'the old James' and 'the new James' with regard to his driving, and that seemed to get through to him.

James learned – very slowly – that he had to drive about 5 kilometer per hour (kph) slower than before the accident to reasonably guarantee safety.

He got his license on the second test in an automatic car on 8 July 2011 – about six months after his first test.

And after that, until 2013 and 2014, when we replaced them with newer second hand cars, he drove my Ford and I drove his Mazda.

His getting a license back made a huge difference to my life and his. Until he could drive I had to drive him almost everywhere. Although Workcover Queensland had provided taxis to Vic Rehab twice a week, any other appointments, shopping, recreation – all required me to chauffeur. My parents, who lived in the same house, also could no longer drive. When James was again able to drive, they were able to allocate their needs between the two of us.

James's driving instructor was a very patient man who seemed to have a very good understanding, developed from experience, of the capacity of people with brain injury. Without someone like that, perhaps James would not have got his licence back.

I wrote on 20 July 2012 that James was then driving very reliably. He tolerated well ongoing unsolicited observations from me about his driving when I was a passenger, but I noticed that my need to make these was diminishing. My mother, who is a very nervous passenger, thinks that James is a good driver.

Income: Guarding against sexually transmitted poverty

Sometimes professionals assumed that I was James's wife.

I realised quite early that if James should ever need to survive on unemployment benefits or some other form of social security, if he were identified as my partner, his payments might be reduced or he might not qualify at all and I would be expected to support him on my salary.

The pre-accident circumstances in which James and I had lived separately in separate states working separate jobs with separate bank accounts became very important when it came to assessing James's and my incomes for social security and income tax purposes.

Before the accident he supported himself in one Australian state and I worked half-time at the local hospital, wrote books, edited *Candobetter*, was an environmental activist, and shared a house with my aged parents and dogs in another. Why should his living with me make me liable to support him if he could no longer support himself, when both of us had paid taxes, superannuation and insurance to cover situations like the one we now found ourselves in?

Ultimately I was able to maintain my status as single because James and I would have continued to live in separate states, were it not for his injury. That is, the only reason he was living with me was that he needed care subsequent to a traumatic brain injury.

So many marriages/partnerships are said to 'break up' after one of the partners gets a brain injury. Brain injury is also associated with poverty. You can see why.

The statistics probably do not reflect the fact that many couples may make the decision to separate on paper so that the well one does not have to take total financial responsibility for the other one and become a virtual slave for the state.

Renting James's house while he lives in mine

James had been very fortunate to find two trustworthy friends to take care of his house while he was undergoing rehabilitation in Victoria. After six months, our friends, Rob and Julia indicated that other commitments meant they could not continue to house-sit. James was relieved to be able to organize to rent the house for $200 a week to another friend. The advantage for the friend was low rent in high-rent Brisbane and for James it was to have someone reliable to look after the house, with the understanding that we could come and stay there when we needed to.

We thought that, if James ever got a 'pay-out' by suing the insurers of the car involved in the collision with his bicycle, it would be good to use this to repair and renovate his house in Brisbane, which is over 100 years old and needed floors repaired, a balcony reconstructed and new plumbing, plus a coat of paint inside and outside. Ideally it would also be restumped. The roof was in good order, although the gutters were full of leaves.

In fact such a pay-out would need to be astronomical. Full restoration of a house over one hundred years old can easily take $150,000 to $200,000. To get that money 'back' in rent would take ten or more years. During those years there would be no income from rent and income from capital would be diminished by the repair amount. Simple repairs, even without restoring the original style and timber, need to conform to ever-stricter codes. Insurance is very costly and finds many reasons in old houses not to reimburse the full loss of accidental damage. It is wise to remember that old adage (which I only recently learned) not to insure anything you can afford to lose.

At some stage in the future, we thought we might also go and live there and repair and renovate my house in Victoria, where we are living now. James would like to conserve the option of us one day living in Brisbane.

The house is James's only asset and we intend to try to protect it. Should we ever separate, then he will not be without his own place, nor me without mine.

That Australian social security has this habit of pressuring its recipients to capitalize on their remaining 'assets' if they live with a partner, makes life precarious. Marriages and carer associations are not life-time guaranteed and if you break up and have no home of your own, you are a hundred times more insecure than you were before – of little use to yourself and a slave to employment and landlords. If you have a brain injury, you are likely to wind up living in a state of neglect in a miserable share-house, boarding house, or special accommodation.

Indigenous cultures preserve land within families and clans. In continental European societies – where Roman law prevails - governments have an obligation to provide housing for their citizens. By that standard, Australia is not civilized. Cultures under Roman law – notably the Napoleonic system of Continental Europe - heavily tax any land and assets that go out of family hands. The law also guarantees equal inheritance rights to male and female children and mandates that children must receive their parents' assets. This means that men and women in countries with Roman law have a good chance of entering marriage with their own assets. Furthermore, if they divorce, those assets will not go to the new partner; they will still go to the children, or, if there are no children, to grandparents, siblings, cousins. For this reason serial marriage is not much of a problem in France. Affairs do not produce the same mayhem as they do in countries where spouses may demand compensation in the form of the family home and children of the first family may be left in poverty.

Thyroid glands and Vitamins

In early May 2011, James noticed that his performance at the local gym, where he had been referred by VicRehab, was deteriorating. I also began to feel that he was not as quick on the uptake, that his verbal comprehension had again slowed.

We visited a clinic where a GP had taken blood tests about two months prior. Between that time and this, James had seen another GP in the same clinic, on an occasion when the first GP was unavailable. He had gone to find out his blood test results, but the second GP was unable to locate any record of tests. This time, since James was feeling so weak, we went back to the same clinic and saw the first GP, who was able to put his hands on the blood test results immediately. The results disclosed low thyroxin, low vitamin B3, low vitamin D, low iron. This made me think that James may have been suffering from these problems for some time, and that they had recently worsened, causing him to feel noticeably weaker.

I was surprised about the Vitamin D because he had been taking supplements from the time he was prescribed this by the Mater Private Hospital. I had got him to keep them up because I had read that a large proportion of brain injured patients are low on vitamin D and that there is some reason to believe that Vitamin D supplements increase rate of recovery. Given my busyness and James's poor memory, it was possible, however, that James had not taken the supplements for a period prior to the blood test, either because he had run out or because he had fallen out of the routine.

The GP ordered repeats of the tests to see where James was at that time, prior to commencing any treatment.

I was not inclined to delay and took James to a pharmacy where we got the chemist to choose vitamin supplements in the areas James was low in. Without a prescription it was not possible to treat the thyroid condition, and anyway, it could be dangerous without knowing the most recent level of thyroxin in the blood.

James's thyroid results had normalised by the second test.

Another blood test was done in May 2012 and showed low thyroxin. It was suggested that James return three weeks from then for another test. That test was somehow overlooked, so James only went for a repeat test in late June. The results for this were lost, so he had to go for another.

I had an argument with the generally courteous GP at the clinic about my mother's thyroid testing while I was there with James because I was worried that something similar might happen to him. The doctor seemed to think that my attitude was hysterical, but I did not want to find myself the only responsible person in a mentally and physically deteriorating household. My father at the age of 88 was still very useful in taking care of my mother but it was beginning to wear him down.

My mother had suffered a slow mental and physical decline over a period of years, then suddenly, after a gastric infection and then a tooth infection with very high temperatures, she completely lost her memory, overnight in March 2012. Prior to this she had had a period of delirium. We do not know how long she had had infections but she was also discovered to have an underactive thyroid. Perhaps the thyroid was damaged by the high temperatures or perhaps a low thyroid had made her vulnerable to these events. The GP, due probably to the volume of patients he was required by the clinic to see, the very small time allocated per patient, and the consequent brief, infrequent and irregular contact with my mother, failed to appreciate the significance of these events occurring together. He also completely underestimated both the degree of my mother's mental deterioration and its suddenness because he did not ask her enough questions and relied on incomplete mini-mental status exams. He wanted to wait to test the thyroid again, but I convinced him to treat her immediately.

I said, "She's losing her hair, she has a heart problem, she has all the signs, why don't you start it now?"

"You're right of course. I will," he said.

My mother's response even to a low dose of thyroxin was overnight. She had been spending whole days in bed, but the morning after an evening dose of thyroxin, she rose at 8am and showered herself. She stopped being delusional. She became able to retain the name of the day and to do her 'serial 7s' (a concentration test). Her memory began to come back slowly although it is doubtful now whether she will make a full recovery.[54] The doctor slowly increased the thyroxin in order to spare her heart. At one stage, however, I accompanied my mother to see him (because she was not fit to go on her own) and he did not remember that he had already prescribed her thyroxin. Meanwhile she was still in a very poor mental state and required a lot of support from home, so we would have appreciated being able to have confidence in a more alert appraisal by the GP.

The Australian General Practitioner system was problematic for us. Rapid population growth – a situation cynically engineered by Australian state and Federal governments and the 'Growth Lobby' (consisting of property developers, financiers, building material suppliers and a host of hangers-on) – meant that services were overwhelmed. My mother and I had been accustomed to seeing a particular GP for years, but appointment waiting times blew out there to 4 weeks – useless for any acute care problem or basic need, such as a new prescription or an unfit for work certificate. Then that GP began to reduce her pensioner-patient load, which meant that my mother simply could not see her. My mother tried to find a caring female GP but found that her age and her vagueness meant that she had fallen so far down the pecking order that she saw a different doctor each time. So, no professional monitored her closely. After this, our whole family – including James – began to attend a GP Superclinic that seemed mostly to employ elderly experienced doctors. The problem was that these doctors saw so many patients so fast – presumably so that they or the Clinic could make a profit - that it was not possible for them to stay on top of complex individual patient histories.

James generally got good service from the specialists he went to see about his injury, but the waiting times were a couple of months. The specialists and Workcover Queensland expected a GP to write certificates and generally monitor James's health, but his brain-injured condition meant that it was not easy for him to give a useful report on his health and he was too easily fobbed off if I did not go with him.

I was working half time as a nurse and nearly full time writing and editing books and the website, and I found the need to monitor and accompany both James and my mother to the GP overwhelming at times. We should all have been able to rely on doctors with sufficient time to know a limited number of patients, but that seemed to have become impossible.

The production line model just does not work for medical practice.

On 23 July 2012 James got the results of the previous thyroid function test. These showed that his thyroid gland had become more underactive. His GP prescribed him thyroxin. Since thyroid underactivity can impact on hearing and verbal processing, I expected some noticeable improvement, but I did not want to leave the problem in the hands of a GP. Brain injury can make people more vulnerable and more prone to hormonal imbalances. The simple reason for this is that the hypothalamus and the pituitary gland (located in the brain) control most hormonal activity.[55] We asked to be referred to a specialist. Whilst waiting the two months or so before a specialist appointment was available, we returned to Brisbane for legal appointments. James decided to go and see his old doctor, Dr S. at Paddington Clinic. This turned out to be an excellent idea. Dr S. was one of those doctors who is able to remember specific patients. She not only produced all his old thyroid results, but she was able and willing to write letters testifying to his loss of function which we would use in seeking damages and later in seeking pensionable status.

Eventually we saw Dr S.M, an endocrinologist, about James's thyroid. He was a very careful listener. He realised the importance of assessing James for any damage through trauma to his pituitary and related hormonal system. He was also vigilant about average blood glucose levels and cholesterol – both of which are affected by thyroid disease. After he received James's results he diagnosed him as having Hasimoto's disease.

"Hashimoto thyroiditis is a condition that affects the function of the thyroid, which is a butterfly-shaped gland in the

lower neck. The thyroid makes hormones that help regulate a wide variety of critical body functions. For example, thyroid hormones influence growth and development, body temperature, heart rate, menstrual cycles, and weight. Hashimoto thyroiditis is a form of chronic inflammation that can damage the thyroid, reducing its ability to produce hormones.

One of the first signs of Hashimoto thyroiditis is an enlargement of the thyroid called a goiter. Depending on its size, the enlarged thyroid can cause the neck to look swollen and may interfere with breathing and swallowing. As damage to the thyroid continues, the gland can shrink over a period of years and the goiter may eventually disappear.

Other signs and symptoms resulting from an underactive thyroid can include excessive tiredness (fatigue), weight gain or difficulty losing weight, hair that is thin and dry, a slow heart rate, joint or muscle pain, and constipation. People with this condition may also have a pale, puffy face and feel cold even when others around them are warm. Affected women can have heavy or irregular menstrual periods and difficulty conceiving a child (impaired fertility). Difficulty concentrating and depression can also be signs of a shortage of thyroid hormones.

Hashimoto thyroiditis usually appears in mid-adulthood, although it can occur earlier or later in life. Its signs and symptoms tend to develop gradually over months or years.

Hashimoto thyroiditis is thought to result from a combination of genetic and environmental factors. Some of these factors have been identified, but many remain unknown. Hashimoto thyroiditis is classified as an autoimmune disorder, one of a large group of conditions that occur when the immune system attacks the body's own tissues and organs. In people with Hashimoto thyroiditis, white blood cells called lymphocytes accumulate abnormally in the thyroid, which can damage it. The lymphocytes make immune system proteins called antibodies that attack and destroy thyroid cells. When too many thyroid cells become damaged or die, the thyroid can no longer make enough hormones to regulate body functions. This shortage of thyroid hormones underlies the signs and symptoms of Hashimoto thyroiditis. However, some people with thyroid antibodies never develop hypothyroidism or experience any related signs or symptoms.

People with Hashimoto thyroiditis have an increased risk of developing other autoimmune disorders, including vitiligo, rheumatoid arthritis, Addison disease, type 1 diabetes, multiple sclerosis, and pernicious anemia.

Variations in several genes have been studied as possible risk factors for Hashimoto thyroiditis. Some of these genes are part of a family called the human leukocyte antigen (HLA) complex. The HLA complex helps the immune system distinguish the body's own proteins from proteins made by foreign invaders (such as viruses and bacteria). Other genes that have been associated with Hashimoto thyroiditis help regulate the immune system or are involved in normal thyroid function. Most of the genetic variations that have been discovered are thought to have a small impact on a person's overall risk of developing this condition.[56]

As the reference cited above intimates, the course of Hashimoto's disease varies in different people. Discussion with Dr S.M. identified that James's thyroid had probably had a period when it waxed and waned, reacting to its depletion by furiously manufacturing thyroxine until now, when it had gone into irreversible decline. The decline might be slow or fast. I wondered whether James's reports of gaining weight, working very hard and being very depressed in his last software engineering research position could have reflected one of those periods when his throxine production had waned. I also remembered one holiday when he had seemed to burst with confidence and muscle, and his hair had seemed thicker than usual. However, after that time he had begun to lose his hair and to look tired, pale and worried, although he continued to cycle to work and maintain muscle. He had also seemed to find it difficult to write with ease at this time.

James has responded well to gradually increasing doses of thyroid hormone and high doses of Vitamin D (deficiency of which plays a big role in autoimmune diseases, notably thyroid and glucose levels). Each time he has seen Dr S.M. the doctor has said, "If you are stable after this dose, I will refer you back to your GP."

But most times he has had to increase James's dose of thyroxine and Vitamin D, so James has so far avoided returning to the hit or miss treatment of a local GP, thank heavens.

Bizarrely, in early 2015 James has still not found a GP with whom he can reliably form a useful therapeutic relationship. We used his Brisbane doctor's letter to inform a pension application. James still tends to see her when he goes to Brisbane and the experience is always useful. James's move from Queensland to Victoria meant that he was treated under Queensland law whist in Victoria. Victorian GP's are used to relying on the Transport Accident Commission (TAC) provision of neurological and rehab specialists and reviewers, but these resources were not available to them with James. This meant that they found themselves on unfamiliar ground, legally and medically.

16 Months Post-Injury: First Assessment for Permanent and Stable Injury

In September 2011 Workcover Queensland paid for James and me to travel back to Brisbane for him to be assessed by a neuropsychologist, Dr Brona O.

This was an early assessment with the aim of predicting final recovery under a section of the Queensland Workcover Act which allows for early payouts to people who move permanently interstate. The outcome was that the neuropsychologist and the doctor involved estimated that James had six to twelve months more important recovery to do.

I scanned the report very quickly, afraid of what I might see. I took in that James was assessed as having an very high IQ pre-injury, and then I skimmed over his deficits:

'In the present assessment Mr Sinnamon's overall Full-Scale IQ summary score fell in the Above Average performance range (FIQ = 118, 114-122, 88th percentile), and was consistent with the reading-based estimate of his premorbid intellectual functional level however there was considerable variability amongst his index scores. For instance, while his Verbal Comprehension (VCI = 98th percentile)

and Perceptual Reasoning ability (PRI = 92nd percentile) were impressive, both his Working Memory (WMI = 63rd percentile) and visuomotor Processing Speed (PSI = 42nd percentile) were significantly and unusually weaker than these scores."

The report also described him as retaining:

"Perceptual Reasoning in the 95th percentile and Perceptual Analysis and Synthesis in the 91st percentile and Word Knowledge in the 99th percentile and Semantic Knowledge in the 99th percentile."

The 98th percentile is 135 and above, so the 99th percentile is above 135. The measurements don't get more specific.

What this seemed to tell me incontrovertibly was that James was no longer nearly as intelligent as he had been at the same time that I learned that his intelligence had been unusually high before the accident. I experienced this as a terrible blow. I remember that I went walking along the creek with the little Pomeranian dog, Lucky, digesting this information with difficulty. I had been so close to James for so long that it was as if it were my brain that was now officially mutilated. I had vivid mental pictures of an axe going through my skull as I walked along the track and I felt sick.

I did retain the fact that James was considered to have another six months to a year to improve and that he was diagnosed with a mild psychiatric condition which might respond to treatment. Since I didn't really think he did have any psychiatric condition, all that was left was the idea that there was still some time to boot up what was left. Under this stress I had the idea of music therapy, which I will go into further later.

About a week after I received the report, I showed it to our friend Jill, with James's permission, while we were waiting for dinner to cook. Maybe that's not the reader's idea of a pre-dinner activity, but Jill is a very clear thinker. Among several eclectic qualifications, she is also an audiologist, with a good understanding of the brain and auditory processing, and has known James nearly as long as I have. James decided to look at the report himself. We printed out copies for Jill and for James and James slowly read through. I kept waiting for him to scream and faint or react in some catastrophic way, all the time hoping that his reading skills and knowledge of neuropsychological terminology would prevent him taking in the full meaning of the report.

I was partly right. James was gratified and consoled by the information in writing that he had had such a huge IQ and that he still had high results in some areas. He simply skimmed over the 'deficits'. I took him back to them and he read the list. Then he said, "But I knew that anyhow."

Jill said to me later, "The deficits seem to be mostly in the details."

And I replied, "But his work requires attention to detail and memory."

3.4.2 Summary of Current Functioning in Specific Cognitive Domains

Further analysis of Mr James's neuropsychological profile (see Appendix I) revealed that while most of his cognitive skills were impressive and well preserved, there was evidence of likely residual relative declines (considered to be scores falling at the 50th percentile and below) and deficits (bold) in his:

1. Processing speed and efficiency in visual and auditory attention tasks

(especially in complex, novel and sustained attention tasks) e.g.

  o Grapho-motor speed = 37th percentile, *W
  o Single target scanning = 50th percentile, 0 omissions
  o Dual Symbol Search = 50th percentile, 1 error
  o Numerical sequence tracing = 37th percentile, 0 errors
  o Color naming = 50th percentile, 0 errors
  o Word reading = 37th percentile, 1 error
  o Copying complex visuospatial design = 30/36, 2nd to 5th percentile
  o Tower mean first move and time per move ratios = 50th percentile

2. Immediate auditory attention and working memory, e.g
  o CVLT List A = 6 words, 31st percentile; CVLT

List B = 5 words, 31st percentile

   o Auditory attention to numbers = 50th percentile (5 forwards and backwards)

   o Letter-Number reordering = 37th percentile

3. Immediate and delayed recall of auditory and visual information, e.g.

   o Immediate Story recall = 50th percentile (with marked confabulation)

   o Delayed free and cued retrieval of CVLT word list = 50th percentile

   o Delayed free recall of multiple visual designs = 50th percentile

4. Novel sequential problem solving, self monitoring, response regulation, and mental flexibility, e.g.

   o Tower building = 50th percentile, Move Accuracy ratio = 25th percentile

   o Word recall = 6 repeats, 31st percentile; 7 intrusions = 31st percentile

   o Verbal fluency = 3 repeats, 50th percentile

   o Category word switching = 50th percentile, 1 error

   o Number-letter switching = 63rd percentile, 2 errors"

The full report is in an appendix at the end of this book.

Phone consult with Workcover Neuropsychologist

At my request, Workcover Queensland paid for me to have 15 minutes on the phone with James's neuropsychologist in order to get a handle on the results of her day long testing of him and what it might mean for him.

Very cheeringly, at first, she said that she had found nothing specific in her tests that she thought would prevent James from going back to program design. When pressed, however, she admitted that she didn't know what program design actually entailed. James described it as administering different programs to make them work together. By this he seemed to mean coordinating and writing, tweaking, rewriting programs to run with each other. This entails quite a lot of planning, judgement, working memory and self-monitoring – all the things that James's test showed him to be deficient in. The neuropsychologist agreed that there might be problems in those areas and she reinforced what we already knew about compensating for these difficulties. James had learned about most of these strategies at VicRehab, but he did not remember much of his lessons or his teachers.

Knowing that James took written information in better, I tried to record and summarize what she had said to James in a short email, in terms I thought he would understand. I didn't expect him to take it all in after one read and I was not even sure that he would read it, but it was there for the record and for him to refer to whenever. This book also has that function. James's injury means that he simply does not have the time or energy to take in all the information that he needs.

I recorded that to improve planning of tasks with reduced RAM[57] (ability to hold info and patterns in his working memory) it was helpful to:
- simplify tasks as much as possible
- break tasks down into segments
- make plans with stages
- check and revise aims as one goes along, with each stage
- check details more carefully than one would once have done

I still hoped that James might be able to take some responsibility for the technical and management side of *Candobetter*. Even if his retained intellect made this theoretically possible, he would need to become aware of changes in his computing abilities and to learn to compensate for it to become practically possible. This seemed like the problem of driving more slowly and not trying to show that you already know how to drive. Since James's hard-won understanding about how he had to take a different approach to driving post-accident, I was trying to prompt the idea that computer coding would also require a more considered approach than had been necessary pre-accident, when a lot of what he did, involving looking and doing, had become automatic and very reliable.

It was also very important to allow more time for each task and be aware that continuous work would fatigue him and that his performance would drop with fatigue.

To combat the impact of auditory processing slowness, he should
- supplement all learning with writing or some visual aid
- try using music for mneumonic purposes.

The neuropsychologist agreed with my idea of trying music to regenerate neurones and connections and she liked my idea of getting James to follow orchestral scores. Singing could only help too.

On the bright side, neuropsychologists do not test reading and writing ability much in these tests and my feeling was that James learned much better from reading stuff, (and this was apparent very early in his rehab).

Weird car tire replacement experience

Was lack of attention to detail responsible for the very strange thing that James did on the 24th of October 2011? When he had only been about three months out of hospital he had changed a punctured tire on my car in the dark and rain. A few months later he managed the same thing again by the side of the road in the lengthening shadows of the evening. Then, on the 24th of October, a tire went flat with a screw in it and he put the spare tire on ... BACK TO FRONT and screwed the nuts on upside down! He asked my father to look at what he had done and my father picked the nuts but did not see – in the dark – that the tire itself was back to front with the inside facing out. They both got in the car and it made, to quote my father, "a god-awful shrieking sound that made you sure that the car was being wrecked." Nonetheless, James drove it from one side of the road to park it on the other. "Why didn't he stop?" my father exclaimed. "Surely anyone else would have stopped dead at that noise for fear of wrecking the car?"

James later said that he heard the noise but chose to assume that it was less awful than it sounded. I think his slowness of processing may also have influenced this decision in that he may not have been able to respond quickly enough, so he was more inclined to continue with what he had intended to do. Fortunately, even though the wheel entirely refused to turn, no damage was done to the rim or the car itself. The tire was ripped free of its valve.

When James met me at the gate when I returned home from work that night and told me what had happened, I was shocked. I was worried primarily that he might have hurt the car, since it was still sitting by the side of the road, pending a visit to the garage the next day. James thought that I was angry with him and that I condemned him for his mistake. He insisted that he hadn't meant any harm. This is a typical concern of James's – that I might think that he meant harm. That is almost never my concern with James. He is not a nasty person at all. No, I was just upset that my car might have been destroyed and upset also that James had had such a strange lapse. I was worried that it meant that there were certain unpredictable circumstances where he might come to harm – always a concern with brain injury.

When we took the tire down to the car mechanic the next day, the mechanic said that he had never, in forty odd years with cars, seen anyone put a tire on back to front.

Ongoing Therapy

Aspergers?

We began seeing Michael W., the psychologist at Mirrool Counselling in Frankston, early in 2011. Michael wondered if James pre-accident might have been a very high functioning Aspergers Syndrome, due to his remarkable gifts of memory and logic, his current poor capacity for empathy and his apparent love of little metal objects and machines. I had described James's reaching down on tottering feet to pick up a lone paperclip on the otherwise featureless hospital floor whilst still in the neurosurgical ward - and. In November 2011 Michael said to me in one of my personal sessions with him (which I had cut down to about once a month to stem costs) that he felt that James was now clearly showing the capacity to empathise.

"Or, if not exactly empathy, he does appear to think about how people feel and to make allowances for this, which amounts, in my mind, to empathy." Therefore he was not a person with Aspergers syndrome, after all. He had just looked that way due to the blunting of perception, interest, imagination and responsiveness caused by brain injury.

"Isn't he recovering marvellously?" I commented to Michael.

"He seems to be recovering exponentially!" Michael responded enthusiastically. "Every time I see him he has progressed in leaps and bounds!"

We had had a back to back session and James had just left to walk Lucky, the Pomeranian, although I had invited him to share my session. He wanted to get exercise, to give Lucky exercise, and to pass by the doctor's to get a medical certificate. This was an example of how he was able to organize priorities now.

He was less engaged in organising more complex priorities, however, such as those involved in upgrading the website management software. He was still very inclined to overestimate his energy and capacity to carry out several tasks, so he would begin quite complex projects – like photocopying multiple documents for collation and sending to multiple recipients - be interrupted, either by his own fatigue or some external commitment, and would never get back to them. One of his greatest failures was his refiling project begun around January 2012, where he attempted to substitute a more complex series of boxes and categories for the simple circular file with pre-labeled compartments that I had started for his documents. He had by that time forgotten the two second hand filing cabinets he had bought and placed in the garden shed for a major archival project January 2011. There were signs of these unfinished projects covering the dining-room table, his two desks, and littering his couch before floating down onto the carpet to accumulate in drifts like leaves from more lofty projects onto a forest floor. Organising complex priorities involved very high-level functioning because he had to take into account his 'deficits' and his 'potentials', which required more insight than he could maintain.

On the other hand he seemed to be able to complete tasks that could be achieved within a single day or in self-defining stages. Thus, perhaps as early as July 2010, he was able to remove the doors from the built-in wardrobe in his study and put up a curtain track and then hang curtains. (Removing the outwards-opening wardrobe doors made more room). A few weeks earlier, in a marathon effort lasting a couple of hours, he managed to assemble DIY bookshelves whilst following instructions of a complexity that defeated me, although I was assisting him. Such efforts were interspersed by weeks of lesser and diffuse activity. He was also able to string blue RJ45 Ethernet cables around the ceilings of the studio we also used as our offices. (Functionality trumped aesthetics.) He liked to get cables off the ground and did the same thing over time with all the phone lines and a number of electric cords and powerboards. When he felt he needed to replace our old Wireless Access Point (WAP) with a new router/WAP because he wanted to have the choice of using either Ethernet cables instead of wireless, although it took him several weeks, he was able to join internet lists and ask about device brands, then find a retail outlet, purchase a router, then bring it back and get it functioning... eventually.

James's attention and retained facility with organizing cables – in effect, his hard-wiring for hard-wiring - was one of the things that had caused his psychologist to think he might be a person with high functioning Aspergers syndrome. By the same token, it could be explained by his professional consciousness of safe cabling and the need to have a high-quality network. He had concerns about the security of wireless networks and he also preferred to minimize his exposure to different kinds of waves, including mobile phone waves, as a general precaution.

Around late February 2011 he was also able to apply himself to a fairly detailed logical and critical response to the insurer report for the driver and passenger of the car which had hit him in May 2010. It probably helped that he was responding to paragraphed statements in an orderly document and, of course, I went over it with him.

Do it yourself Music therapy

I had downloaded and printed out the orchestral score for parts of Beathoven's 7th symphony, which was one that I had been taught to follow in the school music class when I was about 10 years old. Whilst it is quite difficult to pick out the melody as it jumps among the numerous instruments of an orchestra, focusing on a score adds considerably to musical enjoyment. I was trying to stimulate James's brain to form new pathways by learning new complex music. Since brain injured people often don't learn to like new music, this was a double challenge. It also required coordination of eye and ear in real-time. Although James was still very scatty and tired, he gave this project a lot of attention most mornings trying to follow the score for longer as the music played. He seemed to have an immediate grasp for mathematic patterns in the system for symbolizing notes and timing that I lacked. He kept asking me technical details about musical notation which I couldn't answer. Our friend Jill could supply some answers which I could not. James, not a person to allow a situation to remain simple, also had a go at redesigning musical notation. There was a lot else that James and I needed to attend to at the same time and I was unable to continue to give him the assistance he needed to pursue this after about two weeks, due to my own time constraints. This was the case with a lot of interventions to help his recovery, however two weeks was enough to make a difference, and as it happened, we went on to singing, which still involved reading music.

Social singing

Jill and I both enjoy singing. Jill is a member of more than one choir, knows the words of many songs and performs in public quite often in her choirs. She is a really good amateur singer. I had a couple of singing teachers when I was quite young, including one who tried to teach me to sing opera at the conservatorium of music in Melbourne. I never took my singing seriously enough to discipline myself though. Although I knew I could sing somewhat freakishly loudly and hit some high notes, I experienced these talents as quite unreliable. I also rarely remembered the words to songs. Nevertheless I do enjoy singing and have little self-consciousness about bellowing out songs in the shower, in the car and wherever I experience the urge. For a long time I was the joyful owner of two very musical dogs, Nero and Bianca, who would sing at the drop of a hat with me and Jill, until they both died in their late teens. Jill and I often also sang after a meal together and now James showed an interest in being included in this activity. Keen to encourage him, we listened attentively as he tried singing a song.

Jill had never seen *Happy Feet*, the film about the penguin who had rhythm but could not sing, but I had and the noise that came out of James's mouth was very similar. James even looked like the penguin in *Happy Feet*, displaying a tendency to weakly flap his arms by his sides and stretch out his neck to hit notes. Hitting notes was too generous a term, however, since he would rarely actually hit one fair and square. Futhermore, wherever his voice landed, it did not stay for long, but wavered between several ill-defined points, more like a squawk than a note, but with less force. James sounded and looked as if he had a throat or a chest obstruction. It was therefore surprising when he came to a halt after a couple of verses (for which he got the words right) and asked us seriously what we thought. Obviously he thought he sounded quite good. Jill and I exchanged startled looks before composing our faces and telling him he wasn't entirely in tune. "Are you having trouble breathing? Does your throat feel as if it is closing?" I asked. "No," was the answer, surprised.

I wondered whether there might be some undiagnosed damage to James's airway structure in combination with neurological changes to his motor abilities to produce sound as well as the problems we knew he had receiving sound. Although it did seemed obvious to me that he did have problems coordinating sound, no structural problem of the voice-box, for instance, has been identified.

Jill had a lot of ideas about how to improve James's singing and so we all worked on our voices together. James made some progress, but I frankly doubted that he would ever be able to sing songs and be treated politely, let alone be taken seriously.

We wondered if James's singing voice, or lack of it, had any relationship with his auditory processing problems and his dislike of certain noises and the typical brain-injured preference for familiar music. I had the idea that perhaps we could use singing for a kind of biofeedback, stealthily surprising James's brain by exploiting an unexpected pathway.

Rather like children, Jill, James and I would spend quite a bit of our sessions arguing about which songs we would sing, since we all had distinct preferences. Jill and I both also like to sing in French, which was quite unfair on James, who did not know any French. Jill liked what I call 'swing' and which I loathe. James enjoyed songs where the singers screamed and shouted tonelessly (to my ears). Nonetheless, when we agreed on a piece, we enjoyed singing it.

Jill taught us how to 'sonic bathe'. This was a great success, with the new dog, Lucky, as well, who found being sung to hypnotic. Having once had two singing dogs, brother and sister, I kept hoping that Lucky would eventually sing, but he seemed not to be able to imagine such a thing.

To sonic bathe you choose one member of your group as the object of your sonic attentions. You then sing sounds towards them. The sounds are completely spontaneous, and seemed to come to each of us very easily. We made sounds like whales, birds, clicks, whistles, and percussive sounds as well as more conventional musical sounds. Each singer was affected by the others but their response was completely unscripted; it might be harmony or change of rhythm or a contrast. Dropping the voice to a whisper was another interesting variation, which could be sibilant and percussive. One tended to repeat sounds one could easily produce so false notes were not a problem. It was quite an instinctive experience and produced pleasant and interesting emotion.

Music and brain injury

My knowledge of music and brain injury was limited to the idea that music was very fundamental to the brain and another source of associations and ordering which might assist memory and learning from an unconventional direction. James's appearance of discomfort as he sang (which he was unaware of) also made me wonder whether, if his singing improved, his posture and expressiveness might, and who knows what else. I had also read a book by Morris Sachs called *Musicophilia*,[58] which described a host of unusual musical experiences and talents associated with brain injury. It began with a fascinating tale of a surgeon who was phoning from a telephone booth when the booth was struck by lightening. He subsequently developed a talent for composing music.

My first exposure to the power of music in brain injury was in the 1970s, when I was working as a ward-assistant in Ward A of Plenty Mental Hospital, which was the 'brain damage unit'. (It has since been obliterated from the planet, along with the associated bushland and parkland. All traces have been covered by huge private housing developments, as government mental hospitals, which comprised a massive public estate, were sold off to property speculators by successive Australian state governments.)

I was a 21 year old university drop-out and shocked to the bone by what I saw. There were brain injuries from all kinds of causes there – from alcoholism, from brain tumours, from Huntington's Chorea, from strokes, from undiagnosed diseases, and from autism associated with gluten intolerance. Remembering back I can associate all these causes with particular patients: the nightmarish inherited disease of Huntington's Chorea with Mrs N and Valerie

DeG., inoperable tumour with Vicki, stroke with Mrs Pepper (a nickname), car accident (diffuse axonal?) with Cathie, alcoholism with the six ladies who sat along the wall and were hard to distinguish because of their stereotyped behaviour and similar symptoms and appearances, undiagnosed disease with Mrs Speed (nickname) and Connie, and gluten intolerance-induced autism with Lucille. Vicki and Lucille were the youngest patients there. Mrs Pepper, at just under 60, was the oldest. After they reached the age of 60, women were moved to the geriatric ward.

Vicki had been diagnosed with a brain tumour at eight years old. The tumour was inoperable and had just continued to grow, slowly crowding her brain out of its home. Ultimately it would take her life. Vicki was 21 when I was nursing her. She had black hair, cut simply, and dark eyes, one of which the tumour had caused to point away from the other. Her long swanlike neck was bent forward over her girlishly slender seated body, so that she had to look up sideways to see you if you stood in front of her. She had lost the power of speech, could not walk, and was paralysed down one side.

It was the 1970s and the ward had a television perched high on the wall in a corner of the dayroom. It was turned on at 6p.m. every night. Vicki lived to see Johnny Young on "Young Talent Time." His round face would beam out in black and white and what I found saccharine, kept her sane.

While I was there, Vicki's mother missed her 21st birthday, simply stopped visiting her damaged child. In her life almost totally deprived of love and normal stimulation, Johnny Young and the nursing staff were Vicki's chief source of affection. She and most of the patients settled down when the program, consisting of songs, came on. Most of the wards of that era had a piano, and plenty of patients and staff played, but I don't remember a piano in A Ward.

Lucille, the autistic young patient with glucose intolerance, was perhaps 18 years old, with very attractive features and figure, except when she overate and put on weight. The nurses generally prevented this, as they did with all the patients, whom they dressed a little like dolls and treated like their daughters – even the elderly ones – although expecting them to all remain somewhat passive.

Lucy would stomp around the ward much of the day, and even ventured out into the grounds. She often stood rocking from her front foot to her back foot. You could have rudimentary conversations with her. She gave no eye-contact but would cock her ear to you as if it was her eye, and respond somewhat effortfully in a surprisingly deep voice, although often only echoing what you said. (This was an example of the neurological sign of 'echolalia'.) Her main topic was that she was hungry. The kitchen prepared special gluten-free bread for her, but sometimes she would manage to snatch slices of ordinary bread from the kitchen or from other patients' plates. You would soon notice the change and someone would investigate.

"Uh oh!"a nurse would cry. "Lucy's got into the bread bin!"

Lucy would then become very loud and quite aggressive for a couple of days and need tranquillisers until she settled down. That was her life for years until one day a special program for autistic children opened on the mental hospital grounds. From then on staff would come and pick Lucille up and take her to the school every week day. After a while I noticed a change. She started to improve socially. They must have been teaching her songs at the day-school, or perhaps something they did brought songs she had once known back to her mind. One day she asked me to sing a song. I think I began to sing "Frère Jaques", and was amazed when she joined in. When we got to the end of that song, she said, "Another," in her gruff voice. "You choose," I said. She sang several songs, and smiled to herself and half to me at this activity. She also began opening doors for me as I passed from one room to another. I don't know if she did this for other nurses. Oddly, but not surprisingly, no-one officially commented on or explained her improvement to us low-ranking staff.

Thirty-four years later or so, in 2010, in the psychiatric unit where I was employed in Victoria when James sustained his injury in Queensland, there was little planned use of music and the ward often even lacked a radio. Television dominated the communal areas and often continually broadcast popular music, but with very poor sound. The television noise competed irritatingly with the general noise level, particularly in the areas with hard synthetic surfaces, where it was difficult to follow a dialogue with a patient. The patients themselves often used earphones and ipods, in order to drown out auditory hallucinations, to minimize ambient noise, and simply to combat the boredom and panic at the loss of control that is inherent in hospitals, but worse in psychiatric hospitals, especially modern psychiatric units that have no easily accessible natural environment to escape to.[59]

Compared to my childhood in the 1950s and 1960s, when lots of people played piano and had sing-a-longs, music seems to have become a passive kind of activity and singing is commonly perceived as reserved for professionals. Television and widespread commercially recorded and broadcast music have perhaps made many people self-conscious about their own singing. (It's a bit like political participation, where a wealthy political class seems to have taken over citizen engagement.)

Familiarity with the idea of music as therapy seems to have receded as well. When I spoke to Queensland Workcover and James's psychiatrist, they were pleasant but said I would have to show scientific evidence that music was an effective therapy. Who had time? It is obvious to me that music and art are very enabling media. Perhaps because their artistic side dominates their measurable side and you do not need to have a medical or paramedical degree to use them, their value has been displaced by more obviously technical and scientific, billable, professional activities.

James told Jill and me that he had loved singing as a child but had received emotionally conflicted feedback after his parents' divorce, and had subsequently given up singing. This made me wonder whether there was an untapped musical part of his brain or unused pathways that could perhaps be co-opted now to supplement what the brain injury had left him with.

Fun and sociable as they were, our amateur singing sessions were unreliable and irregular. I decided to look for a professional music therapist in the hope of more intensive and informed input.

Such professionals advertised their services on the internet. There were several in our area, but I could not get any to answer their phone or their email. I found a music therapists' professional association and joined as an associate member in order to get a better idea of what was on offer and to choose from among their members. Mysteriously, I ran into the same problem. No-one answered my requests for tutors for a high functioning brain injured man. After paying quite dearly for membership and waiting for ages, I enthusiastically visited the association's on-line members-only publications. There were exactly two, of very low quality. It seemed that most therapists in the association were invisible and uncontactable and, anyway, only dealt with elderly people with dementia.

From her end, Jill heard about a music therapist agency, and told them about James and me. They told her their email and phone contact details. I left message after message on their answer phone and emailed repeatedly, but never made effective contact.

Singing lessons

One day Jill heard from another member of her choir that he had found a very inspirational choir in Frankston, which a legendary woman called Susan Robinson ran for a group of people who came in off the street. Apparently she had past experience in doing music therapy in a brain injury unit in Queensland. I looked her up on the internet and it seemed that she was also a singing teacher. Her ad described helping people to sing in tune and to sing socially as well as professionally. I was easily able to contact her by phone. She immediately understood what was wanted and sounded interested in the problem. She could conceptualise brain injury as something that impacts comprehensively on the individual and she understood and taught singing and music as something that involved the whole body and the mind. I do not mean to speak mystically here; singing is a very physical and mental process and lessons are quite tiring. They reminded me of physical workouts at a gym.

At James's first lesson, she got him to sing up and down a scale. She was trying to identify his range, but it soon became obvious that he had difficulty actually getting sounds out. Listening carefully to him, she said that she did not think there was any actual obstruction or physical defect and that she felt sure she could get him to find his voice and to sing in tune.

One of the first things she did was to get James to call out "Hey!" to her, loudly. Then she got him to sing "Hey!" on different notes. She also got him to laugh. Gradually he was able to summon more strength into his voice. As I recall, by the end of the session, she had him sing parts of a song. But what really impressed me was that she got James to modify his posture and his pained facial expression. When I compared James to the penguin in Happy Feet, I was not just referring to the sounds he produced, but to the wide eyes and painfully exerted posture, with the throat extended and the arms flapping weakly, like a fledgling trying to fly. Without fuss, she drew James's attention to a tendency to look up at the ceiling with raised eyebrows and creased forehead as he attempted to go up the scale. He soon stopped doing this. Since this had become a slightly eccentric mannerism in speaking as well, I was pleased to see it go. She taught him to use his arms more effectively to control his voice and airflow.

The exchanges between James and Susan were so vivid and physical that they made me laugh with pleasure. It looked as if learning to sing had something in common with learning to be alive.

I experienced the effectiveness of using the arms and posture for myself, since the first lesson looked to be so much fun that I decided to try singing lessons with Susan too. As she pointed out, professional singers do all kinds of things with their arms and hands. This is not just for visual effect; it actually helps to modify their voice production in smoothness, extending notes, hitting higher ones, and altering tone.

Susan used a grand piano, a mirror, sheet music and her own gorgeous voice in the lessons. They took place in the front of a house where a retired greyhound and a cat also attended, albeit silently. There were portraits of Susan at various stages of her musical career on the walls, with long blond-red curls and vivid blue eyes. In one the artist had painted birds soaring around her admiringly. Susan had extensive opera and professional choral experience, and, as I said, a really gorgeous voice. I often preferred getting her to demonstrate by singing for me than listening to my own interpretations. Her interpretations seemed magical, and my own leaden, even though I sometimes surprised myself with her help. Susan's knowledge of musical theory provided many satisfactory and detailed answers to my many questions. I had 'studied' not only singing, but much more extensively, violin, as a child, but I had somehow managed, as with my maths (where failure to ever learn my 'times tables' doomed me) never to learn to read music effectively or enough other technical skills to be able to put my otherwise reputedly considerable skills to use.

James, however, had difficulty doing what I found simple but he engaged strongly with the concept of reading and performing music. When I waited in the next room during his lessons, I would hear Susan telling him how fast he learned. It was true. With each lesson he improved measurably.

But, as I said, singing actually engages you very physically and mentally when you try to learn to do it well. When James began a university unit in information technology at Masters level, he really had no time over to do singing, especially since it required travelling to a suburb forty minutes distant. Preparing for that one day a week at university really took almost all his energy.

Compared to those of a psychologist or a doctor, fees for singing lessons at $70.00 were low, especially in terms of the pleasure and empowerment they gave - and, I would maintain - the improvement in self-awareness, self-presentation and energy. Because they were not reimbursed by Queensland Workcover, and James's income was uncertain, we had to minimize recurrent expenditure.

Our singing at home fell away because we were so busy and Jill was away in France for two months.

James was, however, able to continue to go to the choir that Susan ran in our home suburb. He remembered this commitment even over a Christmas break and despite the intrusion of medical and neuropsychology appointments.

Towards the end of May 2012, he was talking about taking a one-off refresher with Susan. We were in the car and I asked him to sing me a song. He sang one that he had worked on with Susan, "Let it be," by the Beatles. I was surprised at how well he held the notes and that he now sang in quite a pleasant light baritone. There was none of the distressed penguin there anymore.

Chiropractor

When Jill and I began experimenting with music therapy on James, we noticed that he had very restricted neck movement. This added to my growing impression that his posture had greatly deteriorated since the accident. This was not surprising since he had been paralysed down one side and it had been necessary to teach him to walk again. For the observant, posture had been a fairly obvious sign of James's brain injury. He now also moved more slowly. It was only when he ran that one could see, from a somewhat high step, arms flung out for balance, that there was anything unusual. Because of the stigma of brain injury, the more normal his appearance and gait, the fewer social barriers he would encounter.

I took him to a man who initially trained as a myotherapist and later as a chiropractor – Dr Will McLoughlin. Will is also a keen environmentalist.

Workcover Queensland were not willing to consider reimbursing chiropractic treatment. In theory I could have argued that placebo effects are unlikely to skew results in dog and racehorse chiropractics, so what's to prove, but it was easier just to pay the chiropractor.

Will McLoughlin examined James and confirmed huge restrictions in James's neck movement. I thought that James craned his neck forward like an old man, holding himself differently from before the injury, and I could confirm this from photos and films I had shot. James had once said to me that he wondered if he compensated for his damaged ability to balance by looking down at the surface he was walking on, to lessen the possibility of tripping and falling.

On his first examination, Will discovered that James experienced severe pain around the neck area on touching. Neither James nor I had been aware of this before Will had examined him. Will did not do any neck manipulation on the first examination, but he used a little hammer on the neck muscles. James's posture improved instantly. After the session James felt a little less clouded.

Before the manipulation, Will had got James to close his eyes and stand, first on two feet than on one foot, left then right. This is a pretty standard neurological test and James had spent months doing just this repeatedly as part of his rehabilitation exercise regime. He had made quite a big improvement but had run into a plateau. When the rehabilitation ceased, and he had to go to gym without professional supervision, his results on this test deteriorated. After the sessions with Will, his balance improved, but the main visible improvement was in posture and strength and I thought that James improved mentally and emotionally and speeded up somewhat.

Before any real neck manipulations might be possible, we thought we should get a neck X-ray to compare with one taken shortly after the accident. James's GP blanched when we asked for a neck X-ray to take to a chiropractor. He felt James might be running a risk with a chiropractor. He had the idea that James might be holding his neck like that because of his brain injury. My feeling exactly, but shouldn't we try to do something about it?

James would be attending his first Psychiatrist appointment six weeks from then, so the GP suggested we ask the psychiatrist for a neurologist referral before venturing to a chiropractor.

No-one had to date worried about James doing any gymnastics or hard physical exercise, and he had had a long series of X-rays that had revealed nothing unusual. Meanwhile James looked to me as though he might soon start hobbling and stooping, so, with James's agreement, we took the old X-rays, CT and MRI scans from Brisbane Hospital to Chiropractor Will McLoughlin. Will observed that James's neck showed mild changes associated with aging but nothing in the vertebrae that would contraindicate manipulation. Nonetheless, Will worked on loosening James's muscles and increasing his mobility over about three weeks, twice a week, before he 'cracked' James's neck.

We eventually saw Dr K-, a psychiatrist whom I knew from my work at the hospital. Dr K- said it was up to the GP to refer James to a neurologist, and that he had nothing against James going to see a chiropractor. These different perceptions about whose responsibility it is to make referrals etc are common in medicine, unfortunately.

It is impossible to say scientifically whether it was time, singing or chiropractic, or a combination of the three that helped James overall, however there is no doubt in my mind that James's posture was deteriorating again until the chiropractic sessions and that it afterwards improved.

James maintained improved posture until his thyroid condition became obvious, when it again deteriorated along with his concentration and stamina. When he was ultimately diagnosed with Hashimoto's Disease, I remembered that a sign of hypothyroidism was vocal weakness and difficulty swallowing. Futhermore, thyroid dysfunction affects the hearing and memory centres of the brain. I will always wonder if a part of James's suffering in hospital and his diffuse axonal brain injury, particularly around the speech centres of the brain, was due to an untreated hypothyroid episode whilst comatose in the Royal Brisbane and Women's Hospital Neurosurgical Ward. Despite their record of my telling them of his having tested as hypothyroid prior to the injury, and despite my raising this as a possible issue later in treatment, in James's copy of the hospital notes on his treatment, there is no record of the Royal Brisbane even testing his thyroid function.

Update 2012

New MRI and Neurology Reports June-July 2012

James's neurologist (whom he had originally consulted about a condition that arose before the accident – his peripheral neuritis) was eventually engaged by Workcover Queensland to give an opinion on a repeat MRI. This MRI basically confirmed the results of the initial MRI except that they were now more visible than they had been shortly after the injury – presumably because old scar tissue had now formed.

The report from the doctor who interpreted the MRIs was far more detailed than the first MRI report and it makes depressing reading. James's neurologist supplied a report based on that report which confirmed that James would have memory problems and difficulty engaging effectively in the university course that he did a unit in.

Rather than complicate the end of this book with these details, I have put them in an appendix entitled, "Appendix: MRI and Neurology Reports June-July 2012."

Second Neuropsychologist's Report July 2012

In July we received a report from Dr Amelia S-, the neuropsychologist who, called in belatedly, had done her best to help James with his computer course. She wrote that James presented as a warm and friendly man overall. She also described signs of his injury in his personality, facial expressiveness, eye contact, and occasional lack of self-awareness in social situations. None of these things is very obvious to the non-specialist, I find, but doctors who deal in brain injury and neuropsychologists can pick that James's injury is severe, on sight. So can I, because I know what lies behind all the very slight changes in the way James walks, talks and responds. Dr S- added that James could also present as polite and gentle, passionate, reactive and spontaneous. A number of his mental skills had improved on the first neuropsyche testing, some of them going back to the superior and even very superior range.

'James' high premorbid capacity has been an important buffer to the traumatic effects of his brain injury, such that the

deficits he has sustained have meant that skills that were high average to superior —although suffering a major drop — have not gone below average level function. This will however, be experienced by him as a deficit and he will continue to require psychological as well as cognitive adjustment to his changed brain."

The details of this report are to be found in an appendix entitled, "Appendix: Second Neuropsychologist's report July 2012."

James's psychiatrist has recommended that James continue to receive more neuropsychological assistance from Dr S-.

James's self-initiated work on his brain

After the 2 July 2012 session with the chiropractor, we went to the Mt Eliza Milkbar for coffee and cak

es and James bought a book about drawing at the newsagent next door. He again asked if I would lend him my electronic drawing pad because he thought it was probably better than his. I thought he might just not have configured his properly. I kept saying that I would check this but something else always came up.

Ideally James would be on a continuous program of drawing, singing, dancing… trying to enhance his remaining skills and catch up on lost time. The problem is that you never can catch up on lost time.

James also downloaded on 2 July 2012 a lot of songs from you-tube. He began to try to piece together what he sometimes refers to as 'culture' in the aim of finding or reconnecting old memories. For the same reason he buys or borrows dvds of films he thinks he once saw or that may relate to information he once had. Ditto with buying books he read years ago. He has also written letters to a couple of old contacts whom he has had no news from for over a decade, and even to relatives he has not seen in years, in the hope of getting more information about his past.

There is, as I have already suggested, something special about music and memory. At first James could not stand to hear most music and then he could only bear, and sometimes longed for, certain familiar songs. Now he is alive to most music, although retains strong preferences.

In Christine Durham's book, *Doing up buttons*, Penguin, Australia, 1999, I seem to recollect that the author never came to terms with music again. Before her accident she had played the guitar. After the accident she found that the songs brought back memories of what she once had been and these made her unbearably sad.[60]

I am glad that James finds his musical memories comforting and enabling. It came to me recently that fragmented song memories are probably relatively easy to repair, just by listening to the song again. And that songs probably have a lot of connections with other memories and emotions because when we listen to songs they tend to evoke the same thing again and again, but their associations also evolve to take on what is going on around us each time we hear those songs and how we are feeling at the time. Sometimes we listen to a favorite song or set of songs over and over again whilst studying or engaged in some other activity. How much does music anchor in our brains?

I now found myself taking more seriously James's dedication to collecting songs he wants to hear again, and information about singers, groups and their times and politics which he felt he had forgotten. I had disparaged his apparent pursuit of trivia about pop music and identities I did not know much about but which I assumed were associated with teenage fads which, in James's case, were surely outdated.

When he told me that he had finally found a video of AC/DC playing a song on the back of a truck being pursued by fans in Melbourne, the penny suddenly dropped, where previously his interest in this matter had only made me feel impatient. (The song was "It's a long way to the top.")

James was not just pursuing trivia. He was looking for data of all kinds that might be associated with those songs and the stories about the singers of the time. He was looking for himself. How could I not have seen that? Once again, James was demonstrating that he had retained deep intelligence and the ability to work out difficult problems without help from me.

Things like this made me love and respect him. I was pleased to apologise and tell him how I had misunderstood. I encouraged him to do everything in his power to track down those songs and that trivia, as well as to keep up singing.

James also mentioned to me in early July how he still felt dreamlike. With a kind of resignation he said that he had once half hoped for immortality like most young people, but now that his life was half over, (by which he meant that he was over fifty years old) where he once would have hoped to live it on a background of rich memories, now he knew that could never be.

His memory has improved a lot though, so I tried running him through what we had done the day before, when we had gone to an environmental meeting to hear Geoff Mosley, an 80 year old conservationist speak on the progress (or lack of it) of the ecologically sustainable population movement. I hoped that talking about the day before would make James feel a bit more present. He recollected the day with mild animation and said that he did feel a little more present after this. "However, I have lost so much, so many years of memories that I suppose I will never feel whole again."

I asked if I also seemed dreamlike to him and he joked that I was the most impossible creature of all and could only be some kind of dream. But then he said that his memories of me seemed more real than all the others somehow, even the ones before the accident. That makes me feel very lucky and glad that we are still together.

On personality and the wise old neurological ape inside

Are brain-injured people really 'stupid' or is it that intelligence is preserved but hard to access or demonstrate?

I implied in this book that James seemed to have a deep generic sense of what was happening except for when he was obviously delirious. Even then, he was aware on some level, just reacting in a very confused manner. More than two years later, I feel that this deeper sense, this deeper self, remains there and that it is the original James.

No-one really understands the nature and location of the personality, but those readers among you who are aware of their internal dialogue, have you ever wondered who it is you speak to and obtain counsel from? If not some celestial guide, then maybe it is *You*. But who is You? Who is that friend who seems to monitor your every move and support you in distress and in triumph? I have a concept of the intelligent ape within us – as the genetic hardware we are born with. It seems to have its own native intelligence and judgement. The culture and knowledge we acquire from our social environment seem like humble software in comparison. These interface with our physical environment, feeding back information to our wise old neurological ape, who interprets these perceptions according to timeless principles. The particular cultural interface we acquire after birth, through being reared in a specific society – language, beliefs, rules, education – mediates social specifics into ancient categories upon which our wise old ape bases its emotional judgements and its wise counsel. That wise old ape has a personality and a rock-like sanity. It may even be that a limited number of editions of this identity is shared by the species and that the only real difference between people's minds is the way they interface socially, mediated through their body and physical environment.

My perception of Brain Injury

How did I feel about the changes in James? How do I deal with the slowness, some occasional silliness and shallowness, and his inability to follow complex conversations, although he can read complex information? From James's point of view, he has to deal with exactly the same phenomena in me. As I described above, James is still in there. He is not a partial zombie or a lesser edition of himself. It is as though difficult weather conditions prevent him from sending and receiving with absolute clarity in certain situations involving a need for speed, excellent hearing, copious information gathering, or sustained physical effort. On such occasions, I find myself thinking of him as an Antarctic explorer wearing somewhat cumbersome clothing trying to talk to me in a sleet-laden wind. It's a little hard for us to see each other clearly, we have to repeat ourselves, and it is stressful. If we persist, though, because we are both intelligent, we overcome the prevailing conditions and get through to each other. We cannot do the same amount of work together that we used to do because of these weather conditions, but we are still the same people. We can tell you what it's like operating in these conditions and give you useful reports on this landscape. If you operate in a similar environment, we have solidarity. James continues to read and write about politics, to attend meetings and to assert his opinion and it all still makes sense. He retains a commitment to getting his points across and shows a lot of self-control and patience with my characteristic impatience and impulsiveness. It would be good if we could come back and operate together in the more comfortable and streamlined conditions of pre-Antarctic locations, but that isn't possible. Well, some people have to work in difficult locations. You cannot always choose.   At least James does not need expensive and special equipment like people who cannot walk or feed themselves or people so damaged that it is as if they were buried under the ice.[61] So we have settled here and we will keep working on communications and record keeping and we will keep publishing *Candobetter* and go on with our lives.

When we were Human: Back on Tralfamadore

The main title of this book is *When we were human*. It comes from the time that James was still in the Royal Brisbane hospital Neurosurgery Ward, unable to walk by himself and beset by Gilles de la Tourette-like vocalizations, as I was pushing him round and round the ward corridors. He managed to say something like this to me: "When we were human, it wasn't like this, was it? I mean, they treated us differently."   At the time I assumed that he was experiencing a flash of insight into his condition as a disabled patient. Years later, in writing the early part of this book, however, another explanation occurred to me. James used to be visited by Billy Pilgrim, who was kidnapped by aliens and taken to Tralfamadore. On Tralfamadore, Billy had been displayed in a zoo, in an acclimatized compound fitted out with earth-like built habitat. He was a popular exhibit and the Tralfamadorens watched his antics on television. After some time they had the idea of getting him a mate. They kidnapped the famous Earthling pornographic actress, Montana Wildhack, and brought her to his compound. Billy was amazed to have the opportunity to meet, let alone cohabit, with Montana Wildhack and it worked out very well.

Remembering how there was a time when James believed I was a famous person, like Billy Pilgrim, and could not quite work out how I came to be interested in his plight, it occurs to me that perhaps he sometimes thought he was on Tralfamadore when he was in the Royal Brisbane, and maybe he had me mixed up somehow with Montana Wildhack.

When they were acute, James's injuries exposed parts of him that I would only glimpse later. These included a remarkable facility with Latinate, bureaucratic and flowery language, in contrast to his somewhat utilitarian vocabulary in everyday speech. His feeling that I was a famous and wonderful person has scaled back, but he has lost a layer of repression of his feelings, so that he now recognises them more easily. At the same time, he has lost a layer of his personal mask, so that he usually says what is on his mind, where once he almost always hid it. He doesn't foolishly tell anyone anything, but that he tells *me* several times a day that he loves me, thinks I am lovely, and so many other compliments. At first I thought that he was laying on flattery to a degree that one usually associates with insincerity. My natural response was to back away, to become suspicious at what he wanted, or what the punch line would be. Now I realize that he is simply saying what is on his mind and that this was on his mind before the accident. He thought all those things about me before the accident, but rarely said them. That doesn't surprise me because I thought similar things about him, but was often unaware of them and rarely said them.

Computers and computer skills

Although it should be obvious that brain injury will impact on programming abilities, no-one to date, as far as I am aware, has documented how this panned out in a particular case. In James's case, I think that there were two prominent factors. These were fluctuating consciousness and the associated difficulty in laying memories. Fluctuating consciousness affected his attention in the short term, as he tried to think problems through and write and review code. Lack of continuous long term memory meant that he was unable to build on what he had done a day or a week before because he could not retrieve the information.

Test-bed for Candobetter

In early July 2012 James explained to me that before the accident he had so many ideas for expanding and enhancing the capabilities of the *Candobetter* site that he had barely articulated them to himself, let alone to other people. After the accident he still remembered these ideas and wanted the chance to retrieve and carry them out. "At the moment," he said, "*Candobetter* is still a compromise."

Although *Candobetter* had been saved by LVPS Hosting, the web-technicians there weren't Drupal specialists. They migrated the site from a Debian Linux operating system to a CENTOS Linux operating system, without updating its Drupal, MySql or Apache.

This left James still trying to create a working test bed where he could update Drupal, MySql, PHP and Apache and bring *Candobetter* back under his control.

Attendance at his hospital out-patient VicRehab program finished in late February 2011, but, as mentioned earlier, Workcover Queensland arranged for him to attend a commercial gym three times weekly, and he continued weekly appointments with a psychologist, monthly visits to a GP, and was having driving lessons three times a week in order to try to recover his driving licence despite his disability.

Much of the rest of his time he spent on his 'test-bed' trying to reproduce on his own computer what LVPS hosting had done on the Canadian site, but using Ubuntu Linux, Linux Mint and other variants of Linux. He ran into constant problems and was always on the lookout for solutions. He began spending quite a lot of money on expensive Linux magazines, as well as taking others out on loan from the library. At least he was stimulating his mind, even if he did not remember much of what he read.

In one case, his professional judgment seemed obviously impaired. He read in one of the Linux magazines about what he thought was a cheap new virtual server for home use. Called a BC3, it was made in Sweden, and programmed with a Debian Linux version. Eventually he realized that the BC3 was really only a glorified storage disc into which images, music and documents could be downloaded and made accessible for a household to access with various other devices. It was not meant as a cheap pre-formatted v-server for a large content-managed website like *Candobetter*.

Whilst trying to make the BC3 work in the way he had hoped it might, he did an automatic upgrade, replacing the original software. After this it could no longer even be used for the purpose it was designed for. He had replaced the original software in response to a message suggesting he do an upgrade. For all I know he was responding to a browser message or some other distraction from the internet that was not specifically targeting BC3, in which case this was another sign of poor judgment or poor concentration related to his brain injury.

On my advice, James also bought a laptop which was meant to be programed at his request with an Ubuntu Linux version of his choice. James often later said that my idea of a laptop had caused more problems than it solved, but my reason for urging him to get one was the hope that, by using the laptop as a test-bed, James would be able to avoid interruptions to his computer work when he had to go back to Queensland for Workcover assessments and other reasons. It would also prevent him being isolated from the internet. James could not remember how anxious and at a loose end he had been without a functioning computer the last time he had gone to Brisbane, nor did he imagine how likely it was that anything he started in Melbourne would be lost to his memory by the time he came back. He thought I grossly exaggerated his memory problems. It therefore seemed to him that he had given in to an ill-conceived persuasion on my part.

I had another battle in front of me because James was all for formatting and programming the laptop himself. I was concerned about his competence to do this and convinced him to have it done by a professional. We found one who said they could program Linux but they seemed rather vague to me and were late on delivery, so I convinced James to use Landmark Computers, with whom I already had a relationship.

The Landmark people were adamant that James needed to preserve a Windows format on part of the computer for technical and legal servicing reasons. James, in his turn, had very specific requirements of them – notably about which Linux operating system to install on the laptop. They ultimately installed a version he did not want and he took the computer only because he was leaving for Queensland the next day.

Within days James was blaming me for his having purchased a laptop with a defective keyboard as well as the wrong Linux version. I supposed that he was running into fingering problems due to his brain injury. I disabled some preprogramed keyboard shortcuts, but nothing made any difference to his experience. Subsequently, however, I bought a similar Asus laptop and ran into exactly the same problems. I also ran into them on a small Asus netbook called an Eee PC. However I used those computers so rarely that it actually took me a year or two to realise that there was something seriously wrong.

Eventually James bought a plug-in keyboard for his dud laptop.

Infuriated by persistent problems, after he returned from Queensland, James downloaded a K-buntu desktop program on top of the Ubuntu desktop and was unable to configure K-buntu to connect his laptop to the internet. This reduced him to having to use the despised windows compartment of this laptop to connect. I probably said something insensitive like, 'I told you so.'

Mind you, he hadn't actually wrecked his computer this time. It was simply that he didn't know how to configure an internet connection with the new Linux operating system he had downloaded.

Prior to the accident, James had been performing extremely sophisticated program design and testing. Now it seemed to me that he was foiled by simple Linux-based programs which anyone can download over the internet.

James's view at the time was that the programs were full of bugs and had poor instructions. There was probably some truth in that as well, however the poor self-monitoring and the poor attention to detail obviously made things worse. I could see though, that working with me there did help. So the situation wasn't beyond benefiting from help. However my assistance lacked the appropriate programming skills. I was myself confused about the difference between an operating system, a language like PHP, a database like MySql and a content management system like Drupal. I could see that operating these programs together was a significant challenge requiring experience and skill in a non-brain-injured person. No-one knows how much improvement James might have made nor over what period of time, with assistance from a computer professional with expertise in the relevant languages and operating systems.

In the meantime it was a case of the ignorant leading the blinded.

March 2011: Attempted upgrades from Drupal 4.5 to Drupal 6:

James retained little or no awareness of work he had already put in trying to make a test bed over those early months before

Candobetter went off line, and was migrated to LVPS Hosting. In March 2011, I sat with James for a whole day while he attempted again to set up a local test bed using two desktop computers and two screens he had set up in his study.

The point of the two computers was to create a home server where he could tweak a test bed version of *Candobetter* and visualise it (publish it) on the other computer. Of course this situation with two computers and two screens increased the need for concentration and memory.

James initially tried to download a version of *Candobetter* from LVPS Hosting to his local computer. He hoped to upgrade this local version, holus bolus, from Drupal 4.5 to Drupal 6.

He was not sure whether the problems he encountered were because the leap from 4.5 to 6 was too big or because the site was too large or because his brain injury prevented him from overcoming problems which he might have overcome before the accident.

James hoped that the inherent logic of the construction would become clear to him eventually. He realized after a while that a simple upgrade of everything all at once would be impossible. He then tried to transfer *Candobetter* articles (of which there were thousands) into a Drupal 6 architecture on his 'test-bed' site. He had very partial success there, with the web-pages coming out with their columns and layouts distorted.

He then decided to work from small and simple upwards, beginning by designing a very simple site and then adapting an article into it – a bit like rebuilding a chateau a single small room at a time. Unfortunately he had problems even making the page visible on his local site.

I spent some time monitoring what he did as he went through the problem to show me. Having me there looking over his shoulder as he explained what he was doing, line by line, as irritating as it might have been, made him focus better on what he was doing. He quickly picked up a couple of mistakes in syntax (phrasing of the program). Just a matter of a missing bracket and an inadvertent $ sign.[62] But that's why attention to detail and self-monitoring are important in programming and why brain injury stuffs up programming. (**In July 2012,** although he did not remember anything else about that day, he surprised me by recalling the exact error and explaining it: "A $ sign prepended to a string in PHP causes unpredictable problems.")

He had found the fault and now the page was visible and interacted on an administrator's level. That was the first of several thousand pages left to do and to coordinate within a more complex architecture before the site was complete. But James was too tired to pursue this work that day and may never have returned to it.

James would tell me that his knowledge of Drupal six years ago when he had first designed candobetter.org, had been seat-of-the-pants and that upgrading a huge internet site like ours could be really very difficult.

"Why don't you use the CENTOS (Linux) operating system?" I asked, "like the people who are currently managing our site and who migrated it for us?" I suggested using CENTOS because a version of *Candobetter* in that operating system environment already existed on LVPS Hosting. He should be able to download it to his test bed.

James then said that he had problems getting CENTOS to work on his computer. He could not explain to me, or I could not understand, what those problems were until July 2012, eleven months later, when he was quite precise:

"I was overwhelmed by the documentation for Linux GRUB (Grand United Bootloader) and did not want to start another partition."

This was the second example of an ability he revealed over time to remember what had gone wrong and it demonstrated to me some ongoing process of reflection and analysis.

Again and again, when he tried to run Drupal and Apache, no matter what the Linux operating system he chose was, they did not talk to each other. Commenting again, nearly a year later, he tells me that he found it hard to give the necessary directives to the Apache webserver that would enable it to make Drupal available to the administrator. (The [system] administrator in a Linux operating system is the person who is responsible for setting up and maintaining the system or server.)

About 9 months later, on 3 July 2012 I noted down a verbal explanation James gave me from his recollection of his problems in late 2011:

"Directives are defined in a confusing way. There's a primary Apache configuration file and a secondary one defined by the primary one. And there is a series of secondary Apache configuration files that are appointed to by the primary one. From memory the primary one, upon which all else is dependent, is /etc/apache2/httpd.conf. That points to the other configuration files which are either within /etc/apache2 or within subdirectories of apache2. And it's confusing to sort of work out which of those very versatile files are the correct ones for any purpose."

He thought that he should be able to do them these eight or so months later, but had not yet tried.

I find it interesting that he was able to recall specifically what those problems were and to analyse them retrospectively. Does this mean that he can somehow not only retrieve some memories from his more acute brain-injured period, but that those memories archive considerably more detail than his mind could handle at the time, but that he can access and use that detail later?

James's explanation of his difficulties make one think that memory problems and difficulty in following directions requiring attending to and keeping subtle details about files and sub-files in mind meant he got mixed up and could not self-correct.

Perhaps if James reads this compressed account of the work he did between 2010 and 2012 that I know of, he might be able to salvage or build something from it. This is, I know, a very faint hope, but now, in 2015, he really does have a cumulative memory. Whether this will be enough to make a difference remains an open question.

The fate of Tibrogargan

Although James spent a lot of time at his computer, in reality he seemed to have difficulty using any operating system.

After his accident, I had ensured the safe transport to Victoria of his old PC, Tibrogargan, with its familiar operating system and interface containing years of James's old files, emails and photos as well as his local version of *Candobetter*. In a way, it was like conserving an undamaged copy of parts of James's brain before the accident.

Unfortunately James had no idea of the precariousness of his grip on this interface and, one day late in 2010, he decided to do a Linux upgrade. The result was that he could no longer find any of his files on Tibrogargan.

His incomprehension of how he had just somehow slammed the door shut on years of electronic memory that recorded much of his personality, hopes, aspirations, photos and experience, a record he could have turned to, perhaps to find his very self, filled me with horror. In fact it is entirely possible that James was upset by what had happened, but he did not have the concentration to remain concerned about it. It was another bizarre problem in a life that was now filling up with them, in a kind of queue, waiting for his attention, one day.

At that time Tibrogargan sat on a desk overlooking the back garden in the part of the house we called the studio, where James and I had a desk each at opposite ends, separated by a partition. Now James abandoned Tibrogargan and tried to continue his work on one of his other computers on the desk in his study/bedroom. He later moved Tibrogargan into storage in a wardrobe so that he could use the desk in the studio to build another computer.

Although I had been trying to keep Tibrogargan in view, I was also going to work in the hospital, trying to write books and edit our website, as well as acting as a chauffeur to my parents, James and the dogs – among other things too numerous to mention. At some point I lost track of Tibrogargan.

Multiple Linux boot-ups

It is possible to partition a computer so that you can run more than one operating system, as many Linux users do with Microsoft or Macintosh computers. When you turn your computer on it gives you a choice of systems to 'boot up' from. The choice is usually between a proprietary system, e.g. Microsoft or an open source e.g. Linux operating system. Due to his memory problems, however, James had on occasions repartitioned his computers and there were more than two systems on them. At the time he did this because he wanted to save whatever data might be on the earlier partition, but he was unaware of creating an archaeology of several layers of old systems.

On 16 October 2011 he wanted to simplify the number of partitions on his desktop computer, and to save any important data. I asked if I could be with him to observe and help and we decided to film the entire session.

He went about his task in a methodical way, writing the location of the partitions and describing their contents in a notebook in longhand. The only thing wrong with this methodical approach was that he also had multiple notebooks and I knew that the chances of remembering what he had done and referring to the appropriate notebook for details were small. That was another reason for filming the session.

To his amazement, James found that he had eleven different Linux boot-ups or operating systems, with only a couple containing significant data. (For Microsoft users, that's like having several copies of different versions of Windows on the one computer.)

With regard to the 11 or so partitions we found inside James's PC, James was surprised, but I could not tell just how surprised. I knew that James did not have any real idea of the extent of his inability to lay down memory and how this affected his ability to work. I hoped that the view of old partitions might make him think about how these came about. The key to helping him work things out was not to spend time going down historic highways and byways. Still I did try to get him to understand that so many partitions which he had mostly forgotten were a sign to me that he had spent months working on stuff in a semi-conscious state in which mistakes were inevitable. I wanted to use this opportunity to reinforce the need to identify achievable aims, to plan things in steps and to record progress, instead of trying to rely on diminished innate executive function. In part the videos show two people politely and onerously trying to explain parallax views of a shared event. You can see the tension, but no-one explodes. I also thought that, if James continued to deny his memory problems, I might be able to show him the videos to help him understand. In 2015 he has still not seen them. Four hours of videos – who has the time? But by now he has some understanding of the extent of his amnesia.

There were two hard disks in his desktop computer. Both contained Linux programs and partitions and one or both had been scavenged from other computers. Mysteriously, when you booted up from the first hard disk and went looking for Linux programs you would get a different selection over both hard disks than when you booted up from the second hard disk.

James took notes on the number, kind and location of the various Linuxes. We then went through them to see if they functioned. Only two functioned: a Linux Mint 11 and a Ubuntu version.

While James was making coffee in the kitchen, I explored and found an index to the operating systems and was overjoyed to suddenly recognise some old Tibrogargan folders. After some investigation it seemed to me that I had actually found Tibrogargan! Her hard disk had been scavenged and reinstalled in this computer. I had found a mode or a window or a level whereby I could visualise these files in the operating system that James had found too opaque to work with when he had given up on Tibrogargan. The files there might be quite historic, going back to well before his accident, records of his former life no longer accessible to his memory.

When he came back into the study, I exclaimed at my discovery, but found that James had no recollection of a computer called Tibrogargan. I carried on about how much she had meant to him and could still mean, but it meant little to James. As tired as he was, I insisted on James making a copy of these files on a new internal hard drive, which I labelled 'Tibrogargan'.

When we got back from Queensland, a month or so later, however, I asked him where the drive labelled Tibrogargan was and he said, "Oh, that? I took it out and brought it to Queensland to work on and I left it there." Tibrogargan no longer meant anything to him and I have no idea where she is now, on what internal or external drive. However her files were on two different drives, and one may still be here in Melbourne. Her old case is in the garden shed, but all the hard disks have been removed. Beside her is Kazak, two other old PCs and a new one, half assembled, in its delivery box. They are covered in dust and insect activity.

Eight months later, half way through 2012 the people who had sold James his laptop found that he had a large number of different Linux operating systems versions to boot up to on different partitions on his laptop. They said that some of those programs would prevent the others from functioning.

At the time, James still could not say why he had so many different partitions and Linux versions. He could not remember most of them. That was, of course, part of the problem. He would forget what he had done and start again, and starting again meant a new partition and a new version Linux operating system. The computers with their multiple partitions and operating systems and no functioning index to the partitions seemed like a reflection of the state of James's brain.

Contact from an old computer professor has many meanings

At the end of this strange day back in October 2011, attempting to map James's computers, James found out (through an internet search for another document) about two papers co-published with his name by his old Honors professor, Peiyi Tang, in 2004. These papers post-dated the end of his ill-fated employment at the ANU, (which went from Feb 02 to Feb 04). James contacted the professor, who wrote back suggesting they might do some work together. [63] Peiyi Tang was now at the University of Arkansas, but internet allows this kind of collaboration. I have included their correspondence as an endnote.[64]

I wrote to James's solicitor about this on 19 October 2011. The suggestion of future collaboration by a university professor, now gave hope that we might build on this in some way to show that James definitely had pre-accident potential to get highly-paid jobs in programming research and design rather than continue to treat him as a man whose low-paid prospects in cleaning had been interrupted. The difference in earning potential stood to be reflected in any compensation James might ultimately receive for loss of income.

At home, more poignantly, the problem was, could he still perform? With regard to his future performance, I personally would have thought that some of the results in his neuropsyche assessment - memory and attention to detail - made this unlikely but I had not entirely lost hope and thought there might be a way around this.

On the other hand, despite the test-bed creation failures I had described to date, I could also record that James had tried to identify and master problems in an intelligent way. A Melbourne-based psychologist, Amy S-, when she saw him a few months later, felt that he might still have the capacity to program, but that his major problem was a 'disorganised approach' which had pre-injury been an 'affordable' sign of creativity. She meant that when he had had more energy and memory, he could afford to go off on creative frolics, but in his current depleted state, he tended not to return to his main goal.

Building a new computer

Quite apart from his intellectual capacity to program, James felt that he was not then yet in a computer *hardware* ready position to try to work with Peiyi Tang but *might* become so in week(s) or months. He decided that he needed to build a really good computer because lack of power might be a problem, particularly if he did some work with his old professor.[65]

The next day James went to MSY, which is a store franchise that more or less wholesales computer parts to the public. He came home having spent about $1000 on a high end motherboard and power supply and two monitors. Since you could then build an ordinary computer out of parts for under $700.00, this was fairly expensive but the aim seemed reasonable. James, however, said as he told me about his purchase, that he had just realised that he could have saved time and money by simply purchasing a pre-built computer on E-Bay, which is something he did successfully when in Brisbane a few months earlier,[66] so that he would have a computer available any time he went to Brisbane.[67]

Now he had purchased the computer parts, he had to build the computer. This was the third computer he had attempted to build (to my recollection) since his injury. The first one did not work, but was made up of old parts, cobbled together. The reason it probably didn't work was because a fundamental component could not be replaced, rather than James's incompetence. Building the second one seemed to go well but, as I may have mentioned earlier, it failed to boot up. At the time, on that score, we travelled up to see our Microsoft computer assembling friend, Ilan, where he lives in the country, for a diagnosis and help. We found that James may have forgotten to remove a circuit interrupter of some kind after he placed it in the computer according to instructions at an early stage of building. Failure to remove that circuit interrupter could blow the motherboard.[68]

Writing this I realize that our acquisition of computers must seem quite profligate, a sign of the disrespectful manner in which we humans pillage the planet for raw materials and rare earths used for electronic components and then mass-produce devices so cheaply that people accumulate far more than they need at great cost to the environment. We must sound as if we had more money than sense, to put it mildly. So I should explain that James originally intended to build computers and resell them as a way of making an income. He still had this in mind when we had packed to come from Queensland to Victoria and he continued to entertain the idea as a possible occupation for a year or so into recovery, after which it vanished he forgot about it completely. Now the proliferation of James's computers was also perhaps affected by James's unreliable memory and judgment in thinking that a new computer might solve his programming problems.

Anyway, back to the project of the expensive new powerful computer: James spent a couple of days building it then flipped the on-switch and ... nothing happened.

Instead of driving one and a half hours to Ilan's place in the country, I had the brilliant idea, I thought, of getting James to contact an officer of Linux Users Victoria (LUV), D-. D- had seemed to me very helpful to James when we had both attended the 2011 Opensource and LUV open day. I suggested that he pay D- for his help. I thought that James showed improved judgment in agreeing to this quite quickly.

D- and James spent a couple of hours on the computer. D- suggested that they remove all but the vital basics and see what worked and what didn't. So they did this. James had stuffed a lot of extras in before attempting to turn the computer on, which struck me as poor planning, since more parts would complicate diagnosis of any problems.

The computer had two huge hard disks, each one 2 terrabytes in size. D- and James found some incompatibilities, notably two different kinds of cables, and a defective or incompatible second disk, which they removed. The computer would turn on, but was not ready to be programmed, I think, when D- left.

D- returned a week later. They tweaked the computer a bit more and then they discussed what Linux program to put in it. James had 17 different Linux program disks but he wanted to use one he had not previously tried – Mandriva -which was supported by the French and Brazilian governments. I agreed that this was a good idea. D- felt that the French government was only supporting that Mandriva because the person who had written it was French. I thought that didn't matter; French technology is good and the support is likely to be good as well.

D- convinced James to install Mint instead because the DVD that Mandriva had come on in a Linux magazine was not actually the full working version. James wanted to download the full working one. D- said that while he was there they might as well make some partitions that James wanted and load up at least one Linux version. But D- also said, "Why do you want to make partitions? Why not make virtual boxes? Partitions can bleed into each other because they are all using the same basic set up whereas [in his opinion] virtual boxes were close to 100% independent and failsafe in this respect.

I thought of the problems James had had with those many different Linuxes he had loaded on various computers. Maybe they had been interfering with each other. I took Daniel's side, and urged James to use virtual boxes.

When Daniel left, James reformatted all the work they had done together because it wasn't what he wanted. Boxes would not do. James thought that D- was a one-trick pony who pushed boxes because that was what he knew and because he did not have enough knowledge to help James do what James would normally have been able to do by himself.

James was quite cross with both of us, D- and me. I was disappointed not to have found someone who could work with James to help him achieve what he wanted and at the same time give me an idea of what problems were due to James's brain injury.

Computers, Internet Technology and Programming

When James was in Intensive Care, the main issue was whether he would live. When he lived the issue became whether he would ever live independently. When he lived independently (albeit with support from me and others), the issue became, how much better could he get?

After that, we began to bargain with fate about the recovery of his intelligence and quality of life: could he recover reliable abilities with information programming and design? The one after that would be, could he ever again run for political office? In the meantime, since no lay person could determine the extent he had retained his computing skills, a psychiatrist decided to test this by sending him back to university where he would be formally examined as a participant in a post graduate programming unit.

Two Years post injury: Confronting computer programming problems

James felt that he could still program as required. To me it seemed a very tall order. The reason that it seemed a tall order for me was because of those many months now that James had been trying unsuccessfully to get up to speed with his coding in order to upgrade the original Drupal "content management" for *Candobetter*. He was well past the first 18 months of recovery, with relatively rapid recovery statistically limited to two years at most.

It took James days to recover from any sustained effort at coding and checking. Rest was itself an interruption to memory; James could never be sure what he would remember and for how long after a nap, recreation or attending to some other task.

Reviewing what I have written here, it occurs to me that, whilst describing the failures, I nonetheless record James trying to identify and master problems in an intelligent way.

All interruptions were catastrophic and occurred constantly, beginning with James's regular rehabilitation commitments from the time he was released from the Mater Private. At first his memory had been so poor that there could be no continuity beyond a few hours in a day, after which everything began again, clean slate. Gradually parts of the slate started to be preserved, but it was like a blackboard with partially erased and blurred work still on it, making the record hard to read. As James's memory improved his sense of not making up ground with his computers ate away at his resolve to get on top of the problem, making him vulnerable to interruptions in another way.

Recovering memories and skills took time and effort on James's part, but living in the present also required time and effort. James had irrevocably lost time and he had less energy available for effort.

Running counter to all these obstacles to skill recovery were the positive forces still present in his personality. James retained a strong internal judgement and conviction that he would ultimately overcome the problems. He was aware – at least superficially - that his potential had been irrevocably damaged, but he believed and I believed that there was a lot he could still do. I was inclined to trust his intuition, but knew that his insight was still poor and was affected by his memory.

James also needed to find extra hardware to assist with keyboard dexterity problems, a new chair, and new screens.

Although I had done some research into these areas, we had never had time to discuss what I had found out. James did his own research on what might help, then went out and bought it.

Psychiatrist prescribes university

James's contact with his old lecturer, Peiyi Tang, was probably a major factor in getting Workcover Queensland to seriously appraise James's computer skills. Prior to this they had had no obligation to think of him as anything more than a cleaner who dabbled in computers, the internet and politics as a hobby. Simply because a university professor had recently corresponded with him and mentioned working together on the basis of work James had done at university level with the professor, this had changed and the Workcover doctor had recommended that James do a computing unit at university to see what would happen, subject to review by a psychiatrist. Dr K-, James's psychiatrist had the very difficult job of assessing the reality of James's belief in his retained skills, but was unable to assess these on the sole basis of interview. He felt that only at university level could James's skills be adequately assessed. He backed up James's choice of a Masters level unit in Components of Internet Technology and recommended that he receive neuropsychological coaching whilst he studied. Most unfortunately, this specialist coaching was badly delayed.

Choosing a course and a university

I was impressed by how efficiently James chose a course from a huge variety on offer. He chose the university with some help from the Commonwealth Rehabilitation vocational psychologist, Clare, with whom Workcover Queensland had allocated him only six hours. She seemed to give a lot more than six hours of her time. The contents of the course appeared to be appropriate.

James had managed to convince me, the Workcover doctor, his psychiatrist and a Commonwealth Rehabilitation psychologist that he was in there with a chance. The psychiatrist prescribed the unit in order to evaluate James's ability, not necessarily in the expectation that he would re-engage professionally in the area.

James would be about one month off his two year anniversary when he started the university unit.

C# and Microsoft Visual Studio

It did not occur to James and was perhaps not obvious from the course description in the prospectuses that the university course would assume familiarity with Microsoft operating systems – (i.e. Microsoft Windows). Shortly after starting it became obvious that James would be expected to learn a new coding language called C# and, using that language, to operate a complicated and expensive[69] piece of Microsoft software called Visual Studio 2010. According to James, Visual Studio was very similar to Java and Netbeans, but it cost around $4000, whereas NetBeans was free. Visual Studio was a Microsoft programming environment with architecture for people to build interactive websites. It was supported by a structured internet virtual environment called Windows Azure (http://windowsazure.com) Windows Azure existed in a variety of upgrades and transitioning between them could be problematic.

James had never become dependent on graphical interface objects (i.e. 'windows' and 'icons') the way the majority of computer users are. He was about 27 when home computers were just becoming available to the masses. People only began to use 'mice' and 'click' highlights and move around between 'windows' after 1983, with Apple's *Lisa* and then Microsoft's *Windows* in 1985. Before these graphic systems took over, people keyboarded in codes equivalent to the mouse clicks on icons and boxes of today that let them create and save text or calculations. James became used to typing command lines into a small hobby computer a few years before he acquired his first home computer, a Dick Smith IBM PC clone, around 1986. He did not really need those clickable little yellow boxes called 'folders', other 'icons' and 'windows' that are so convenient but which hide a lot of processes that enable home software engineering and consumer independence).

Later James went on to programming mainframe computers for a large Australian transport company and he also managed an early internet provider in rural Queensland. Subsequently he did computer science at university, which still involved original coding (as opposed to clicking boxes and entering prescribed fields), then went into a software engineering position that would equip him for a doctorate, and which also involved complex code-writing and software engineering, using open source systems like Linux, rather than manipulating 'graphical objects' in commercial programs.

When I met James around 2004, graphical interfaces were well-established in the open source programs he used on his personal computers, but James still tended to use command-line codes which took up less space and time than the graphical interfaces. Even when he surfed the internet, James often limited his browser to its 'text' function. Cartoons and other digital art on *Candobetter* raised the site visibility and popularity and it was James who first suggested that I should add illustrations, but he was often unaware of these unless I drew his attention to them, because he browsed in text mode.

If he had not sustained a brain injury, he would not have had any problems at all familiarizing himself with Visual Studio, the Microsoft Azure virtual platform and its C# code, because, after all, he did know how to use a window-type format and he was used to learning new coding languages. It seemed to me, however, that when his brain was trying to save itself after the accident, it went through some kind of triage[70] operation where it decided what to ditch and what to keep. No doubt there were constraints in what it could choose from, but it appeared to retain as much code as possible, but not to bother with the idea and use of graphical interface much at all. Because the Visual Studio interface was very graphics-dependent, even for coders, it actually worked as an extra obstacle for James. Learning a new computer language would usually be beyond someone with his extent of brain injury, but when James was confronted with a new language interface that was mostly concerned with building interactive boxes and windows, it was even harder for him to deal with.

Difficulty understanding lectures

James's study unit only required him to attend one day a week, on Mondays. There were two hours of lectures followed by two hours of tutorials. When I realized this I felt sorry for him. I was not surprised, but James was, when he returned from his first day in a shaken state. He does not have much awareness of his auditory processing handicap as his mind seems to normalize what he does hear. In this case the problem was too dramatic for him not to notice. Of course he had understood just about nothing of the hours of continuous lectures or the tutorials, because of his inability to take in much verbal information. There had been overheads, but they had gone so fast and they involved a lot of rapid cycling through graphical interfaces. It was a brain-injured person's nightmare.

He hoped to overcome these problems by downloading the digital presentation supplement to the lectures, in the future. They were supposed to be available a couple of days ahead of any lecture, but for the first two or three, there was a delay due to the crash of a staff member's computer. He later discovered that the digital presentations could be downloaded with a sound recording. He had to learn how to find his way around the university intranet, to recognise and to access these services. He lacked the skills to quickly recognize the graphical cues and use them semi-intuitively, the way a person without his handicaps would be able to. He did become better at this.

Difficulties with unreliable versions and over-securitized updates

When he tried out the Visual Studio software, as well as the problems understanding and learning its graphical interface, he ran into all kinds of other problems with licencing, downloading, passwords and availability of versions. These problems were resolved after about three weeks, but in that time James had got very very far behind.

As he tried to do work, the difficulties of his brain injury were also compounded by a loss of confidence in the Microsoft Visual Studio software. We wasted hours trying to download new versions and check on-line to verify whether the problem was in his comprehension or the software, but could not afford the time to persevere alone. Ultimately he had to go back several times to his university administration to obtain the correct versions. A university vacation meant that several days were lost again because there were no staff to consult.

Standardizing the software environment, reducing distractions and looking carefully

On 28 May, following some ideas from James's neuropsychologist, Amy S., James decided to go and work in the laboratories at the university in order to be sure that he was using decent software and to minimize the distractions of internet surfing, posting comments and writing articles, all which naturally grew in proportion to the difficulty of his assigned tasks. I decided to accompany him and try to carefully monitor and understand his work difficulties.

The result was that my optimism increased a little and I approached the Disability Unit and James's lecturer, Matthew K., by email requesting more time. [71] (Perhaps I should explain why I took this role on. It was not because James was not capable of approaching them himself, it was to save him the time and effort because he needed to devote so much energy to the basics of the course itself.)

In my email, I said that I was aware of James's difficulties in the Components of Internet Technology Course, but that he had made some progress that day, which I related to changes in his working environment, among other things.

I described how I had emphasized to James that he needed to very carefully examine all graphical interfaces (windows, boxes, small fonts and big fonts etcetera) and read all instructions. His neuropsychiatrist had suggested to me that it might even be necessary to remind James to explore the screen by moving his head. He could also use assistance to spot typos. He needed to remember to work as 'Administrator' in Visual Studio. It is very helpful to have one *Windows* computer with *Visual Studio* to read instructions from whilst working on a second one. The light was very good in the laboratories and this might also have been important. The lack of distractions in the laboratories was probably also helpful.

We were at the Lab computers for a couple of hours. During this time James had become so tired that he forgot where he had started and where he was going on a programming task. He might have given up if I had not been there to prod him and remind him of his objective. He then, however, picked up where he had left off. In fact, his ability to work, concentrate and remember had improved since the beginning of the course, so I hoped that it might continue to do so during the rest of the course.

At the outset, with the help of the University Disability Liaison Unit, we obtained consent to ask for up to 50 per cent more time for James to carry out work and examinations. (I later learned from brain injury discussion sites that you should really ask for as much time as there is, subject to the life-time of the course.) In response to my emails, the Disability Unit helped us to liaise with his lecturer in a request for an extension on his second assignment.

Microsoft Visual Studio and a pertinacious lecturer

The problem for his lecturer was that all the assignments were starting to back up and his lecturer did not want to compromise James's ability to do the last two assignments by running beyond the total time allocated to the course. James was given a bit more time, but his problems were very obvious to me. When he told his lecturer and the disability unit that he felt he would have to withdraw from the course, his wiry overworked young lecturer, Matthew K, stepped heroically into the breach, a finely drawn Japanese anime figure, pale faced, dressed in black.

Matthew spent hours, several days a week, that he did not have to spend, with James, during a time when he would have seen no other students, over 'swot-vac' when the university is empty, then in between lectures and tutorials, in day time and in the evening, sometimes twice in one day.

James's body and mind, however, could only deal with so much at a time and sometimes fatigue would overwhelm memory and the ability to execute what he had learned. On the other hand, pushing James also probably stimulated new connections. Brain injury can reduce the intensity and longevity of emotions and so reduces the associated strength of curiosity and learning. Stress in itself probably could temporarily enhance emotional tone and learning. It was hit or miss, swings and roundabouts.

James respected his lecturer, was inspired by him, and had confidence in his knowledge, so James was prepared to follow his instructions. It turned out that James had been trying mistakenly to develop an application using an XML file to express code instead of the components that were intrinsic to Windows Azure and Visual Studio. Visual Studio was quite hard and anyone trying to use it for the first time was bound to run into some problems. When James ran into problems he thought he could get round them by directly editing the XML code. He was using old knowledge. He had some prior familiarity in using XML files to write the code to construct webpages, so this was a partly logical fall-back, but it seems to me that he would not have found it logical if he had appreciated the integrated approach of Visual Studio, which meant that Visual Studio could not read XML files, or at least could not import them from outside itself.

How much of this closed-system approach was defendable on programming system terms and how much was an adaptation to Microsoft's need to keep people dependent on its products and derive financial benefit from this, was unclear to me. Were similar Open Source programs and systems, like NetBeans, more 'open' to imported elements in this case? Was there a certain amount of cowboy to James's programming which had worked before but which now added to the already overwhelming unpredictability of results?

James came back from his many sessions with his lecturer Matthew intending to put into practice everything he had learned, but I can only guess how difficult this was for him. An initial obstacle was that he rarely could do that evening what he had hoped enthusiastically to do on his way home. His brain usually was too tired to focus.

By the next day a lot of the memory and emotional drive for a particular task had dissipated. The more he took in at any session, the more it took out of him. The longer the break he took to recover, the worse the situation got. We must add to this the impact on him emotionally of not meeting his own intuitive expectations.

On the other hand I knew that, even when he was taking time out, his brain was working on the project, just so much more slowly, and under such great handicaps of fatigue and loss of memory/emotion/drive that he could not count on the result. (Yet, sometimes he succeeded.)

I had a feeling that Matthew was able to break down the tasks and simplify James's approach for him, so that, if he could possibly follow Matthew's instructions without break or fatigue, he might get real results. Five days before the last deadline in James's course, I accompanied him to see Matthew and observe them together.

Matthew's work, despite the lecturer's overall kindness, could not meet James needs in one way, however. Like 99.9% of the population, Matthew seemed unable to understand the severity of James's problems with auditory processing. Matthew therefore did not speak slowly enough to give James time between spoken and visual information to take things in. Not only did Matthew fail to slow down, but he characteristically spoke unusually fast. It was in his nature to rocket along. In fact, his speech had a mild buzzing quality because he actually partly suppressed consonants that would otherwise have impeded his race to the end of each idea. Even I found it hard to keep up. This observation should not take away from Matthew's major contribution to James's rehabilitation, it should instead demonstrate how very hard it is for society to understand and then to deal with slowed auditory processing. I also tend to speak extremely fast and often fail to slow down for James, even though I know it costs me eventually.

For James, the time and fatigue factor probably also meant that he lost that sharpness of approach overnight. Where most people would have taken days, James might take weeks or months. His memory had improved quite a lot since the course began, just in the normal curve of healing, which means he now had more to work with than he did at the outset. It was indeed unfortunate that there were some initial software problems and that we had a huge delay in getting him a neuropsychologist.

I liked to think that with more time - well after the course would be finished and forgotten - at least some of these things would come together. At the very least, I hoped that Matthew's work had sharpened James's own perceptions of where he was likely to make mistakes so that he could at least use that understanding to work more effectively with his Linux-based applications that were more familiar. Nowhere else could he get the kind of assistance with insight into professional skills and his injury that had been given him so generously by Matthew K.

I was amazed that James got credits for his first and second assignments. When I expressed my surprise, Matthew said that he had marked them easily for all students, but that it was the third assignment – a 3000 word essay – where he feared James might not make the grade.

The last day of James's extension in time was a Tuesday. It was the Friday before that I went in and observed James with Matthew. They worked together on an assignment where James was to design a basic application in Visual Studio using C#. The application was a virtual machine for processing micropayments for website articles. We got there at about 3p.m. and James left at about 6p.m.

During the session many software problems arose that were actually generated by Visual Studio and which would have baffled most students. If Matthew had not been there, James would not have got past them. Matthew had difficulties himself that took a half hour to solve on two occasions. James spent some time working when Matthew left to do something else and ran into a problem that Matthew had already predicted but which James had not taken in due to the rapidity of Matthew's commentary. Writing at furious speed, I had managed to take down what Matthew had said, however, which then informed James. When Matthew came back, he told James that he had done enough to pass that assignment and advised him to spend all the rest of his time on the essay question.

The essay question was the comparison of different electronic payment systems available on the internet. James was most interested in a Swedish system called Flattr. He wanted to work on the essay over the weekend, but was completely unable to bring himself to do so before about 9p.m. on the Sunday. He was due to go into the university on the Monday and consult with Matthew again. He had produced 800 words and had forgotten that the required word length was 3000.

It had been frustrating to observe his involuntary procrastination. Too much rode on it for me not get emotionally involved.

"There is no way you are going to pass with that," said I, somewhat brutally.

James decided that he was going to hand it in as it was. He had had enough.

Because he had stayed up until about 3am I did not trust him to drive in to the university, so I drove him in. I would visit our friend Jill nearby.

As we pulled up near University Building H, James pleaded with me to come in with him.

"I just know he will try to make me stay and do some more work on it!" he moaned. "And I won't be able to resist."

I thought James was probably right but I did not want to get involved because I knew that I would not be able to resist increasing James's chances of passing. From reading the course notes, talking to James, and googling the subject, I probably knew enough about his project (even though I could not have performed it) to know what he could add to his essay to give him a good chance of passing. I did not want him or his insurers to come out of this unusual rehabilitation experiment with a false idea of his capabilities.

I made a hasty getaway. Later James rang me at Jill's place. He would be a bit later. Jill and I went out to dinner with friends. I got home at 10p.m. and James got home some time after midnight.

He had agreed to spend some more time on the presentation overnight and then to hand in the essay the next morning.

Ultimately he handed it in the next afternoon, because Matthew eked out a little more time for him. When he handed it in it was about 2,800 words and read quite well. It still lacked some inclusions he could have made (but I did not open my mouth about this) and I thought there was a distinct lack of technical detail in his comparisons, which detracted from his conclusions. It seemed to me that, on top of the two credits he had achieved in the first two assignments, he just might pass. I was ambivalent about this. One the one hand, passing was so important for his self-esteem, especially since before the accident he was used to getting high distinctions if he worked hard on a project. On the other hand, what if his passing served as an excuse for insurers to pretend that his brain injury was less than severe and therefore leave him financially undercompensated? Could they fail to take into account all the extra help he had needed? But, as his neurologist said, James had to devote all his time to one unit of a course where other students will have done several units at a time, often whilst holding down jobs.

As I said at the outside of this section on computers, James's injury was so very severe that it was only because he was really extremely gifted that he was able to contemplate doing anything at all with computers. Reading and writing again were initially thought to be big goals. Walking was at one stage. Learning C# and Visual Studio ... extremely unlikely.

Despite all this, he passed the course! (Just.)[72]

Need for information technology skills in the rehabilitation field

Later, when he described his progress to his psychiatrist, it was clear to me that James was falling back on his pre-injury competency. He estimated that he might be able to work as an info-tech tutor in a university, even though he had barely passed this most recent course. In fact the whole semester had passed in such a complex blur of pain and external disorganising factors that James was unable to process his experience.

To me this was a terribly wasted opportunity. It was, of course, fantastic that an effort was finally made with the help of some external consultants, to give James any opportunity at all to test his information technology skills. In a sane university system, however, it would have been possible to have sat down with the faculty dean or someone similar and review the initial problems within the first few days of the course. Then perhaps another course using Linux and open source software, at that university or another, might have been possible. However, universities have become so commercialised in Australia, with so much responsibility devolved to insecure tutors and PhDs, but authority retained by administrators most concerned with getting and retaining fees, that this was not possible. The lectures, although multimedia and recorded, required passwords for access on the university system and were often posted very late in the week. Everything was commercial and privileged. Orwellian. The hierarchies between administration and teaching made it difficult to organise humanely. You could be locked out of the system and have your results withheld suddenly for failure to pay fees. This happened to James, who responded to slick encouragement to book in for the next semester early. He subsequently forgot he had done so and was ultimately horrified at the prospect of another semester. It took legal consultation to convince the university that he was brain-injured and was not meant to do more than one semester. This despite the fact that their own disability officers had been involved from the start and were willing to explain the situation. In the meantime the university withheld permission to write a report for James's insurers from his tutor, but fortunately, before this edict, he had provided a draft report we were able to use.

I still think that if the rehabilitation system had provided James with skilled assistance to plug injury holes in his basic Linux skills early – which I was not able to do - these would have helped him progress to later more complex skill recovery, including regaining conscious awareness and thus memory of steps taken by his fingers in programming.

There must be more and more programmers and software designers who use programming landing in neurosurgical units every day. They form an important subpopulation of our society. I still don't understand why neuro-rehabilitation providers only supply basic coaching in reading and writing. They hardly even deal with arithmetic. They behave as if computer programming were an unearthly skill. Hospitals won't provide ordinary internet connections for their patients when these might save them from profound isolation and give them a platform to begin their own rehabilitation, referring to their email and other documents. It is not enough to rely on employers to retrain basic programming skills; they intervene too late, often throwing employees in the deep end to sink. And not every programmer is employed as a programmer at the time they sustain a brain injury. It might not get someone back to work in the field, but basic coding is such a major part of a programmer's function and memory that it should not be left untended. I was even unable to find a single neuropsychologist with programming skills. This amazes me in light of the potential usefulness of computer-modelling and training in rehabilitation. The process could even be automated into voice and screen programs and should not be left to Microsoft or Apple. Of course no rehabilitation should be left entirely to machines. The human contact is vital.

UPDATE 2014: Programming

On the 4th of October 2014, I approached James again about the need to have someone local to help manage *Candobetter*. This time I showed him the following advertisement I was considering placing online.

> *"I have an unusual casual job to fill in Frankston that could become a weekly thing for about 3 hrs week. It would interest someone who enjoys teaching or coaching and has good knowledge of Drupal, php, msql, but isn't looking for full-time work.*

Initial consult re management options non-commercial drupal website http://candobetter.net

The person who built the site in 2006 has had a head injury and has forgotten how to manage the site. We would like someone to

(a) try to reteach him his computer skills from ground up or wherever he is at

(b) include his partner in these lessons

(c) help him and or me to manage the website

The injured developer is very pleasant but has memory and hearing problems. You would be meeting us in our home.

We would like to take it session by session, say, once weekly, and review after each session, initially."

James looked at the ad and said it looked okay. He even seemed a little pleased. After I had posted it, however, he said that he really didn't think it was necessary and added that he did not think we would find anyone qualified to do the job without paying them more money than we could afford.

A bit later I went into his study to explain that I had made the decision to advertise because I was becoming really anxious not having an independent back up. Just as something had happened to James, just as everyone gets old and businesses fold, something could happen to the site that the hosts at LVPS hosting would not be able to fix or even that the webhosting site might go broke and ask us to find another site suddenly, which we would not be able to do.

"We need to be able to have a copy on a local computer at home in case something goes wrong, and you are not able to do that."

James disagreed with me, saying that of course he could download a copy to his home computer. He just hadn't done it because he had his mind on other things. I bluntly said that he had tried to do this many times in the past and had not succeeded. This was obviously a very confronting thing to hear and James blustered and became indignant, insisting he would do the task right now.

After he had calmed down a few minutes later, I asked him if he could imagine why I lacked confidence in his ability to do this. He said he didn't know. I told him that he had tried to do the task before several times, but not recently. I said that the matter needed to be attended to and he was welcome to try but it was not unreasonable for me to lack confidence and it was for that reason that I was trying to find a coach and defacto manager.

We received only one reply from our ad for a teacher and this was from someone in another country who thought perhaps we could do it via Skype. This was impractical because of the need for personal presence in teaching, especially of someone whose auditory processing and general responsiveness was already complicated by injury.

I located someone at VG computers who was willing, in the meantime to install and configure Drupal and migrate *Candobetter* to a personal computer for James, to secure the site in the meantime. We did not want a server, just a home computer with the website, but the quote was for a server, and in the region of $2,500.

Before we could resolve this confusion, something else came up as it so often does, and the matter of expediting a home copy of *Candobetter* receded again.

A week ago my internal scheduler suddenly woke me up in the night and caused me to think about the problem again.

The next day, on Sunday 30 November 2014, at a café with Jill and James, I raised the subject again. James was adamant that he could do the job himself. With Jill as his witness he said he would purchase a new computer the next day and start work. He estimated that he would finish the task within two weeks. This would be necessary because he was due to fly to spend Christmas in his house in Brisbane.

He was unable to buy the computer until Tuesday due to circumstances beyond his control. On the Thursday he did not appear to have started work yet.

He had to abandon his project to review my second book in the Demography Territory Law series, in order to attend to the computer. I know that with an ordinary person the memory of a book fades quickly enough; so the chances of a review were dwindling.

On the 11th of December 2014, James ran into a problem with the new computer that he was trying to convert to dual boot with Linux. He found that it simply would not boot off a 'gParted' cd. Further investigation revealed news that the new Windows 8 operating system would not permit any other system except a couple of Linux versions which had paid Windows for the privilege. That made those systems untrustworthy because it meant that they had lost the function of independence which characterises Linux.

We found this out by searching the problem on the internet.

It looked like Microsoft had almost completely snookered the once free computer development system for its users. Where people could once easily be completely independent of Microsoft, they could no longer be, we read. Windows 8 apparently required internet connection before it would complete boot-up. After that it seemed it demanded almost continuous connection to the internet anyway, greatly increasing owners' vulnerability to surveillance and compromising their general network independence. [73] Rumour had it that China has banned Windows 8 for this reason. Apparently it was still possible to boot from a cd using a Macintosh computer, but James had bought a Microsoft computer. Since he had thrown away the packaging, he was unable to take it back on the grounds that it was unfit for the use he wanted to make of it.

Eight months later he would review the problem and find an answer.

Update on dogs

Lucky

When last we visited this subject in any detail, poor Nubi had been euthanised and Lucky was on the rise. As soon as Nubi had been removed from the scene, Lucky stopped spending much of his time safely in the middle of the double bed. He soon learned to walk off the leash and we went for many delightful rambles. His deafness meant that one needed to keep a strict eye on him. On my 60th birthday in 2012 Jill, James and I went exploring the coast around Jam Jerrup, South East Gippsland. We found a track leading through a semi-deserted cabin park of some considerable age, with lawn, forest and a long beach. Lucky sped around exploring every aspect, disappearing for minutes at a time, then coming back to us. We would catch his attention when he appeared to be looking for us by waving our arms. After an hour and a half exploring, we turned back towards the car. Approaching the car I looked around for Lucky. No-one could see him. I walked all the way back to the cabin park with my heart in my mouth. I was asking a group of people with fishing-rods sitting around a cabin and a trailer if they had seen a large red fluffy Pomeranian when I spied said dog under their cabin. Lucky was peering out short-sightedly. There was a great reunion and we walked back to the car together. When Lucky saw the car itself, he did a dance of joy. It is hard to imagine what was going on in his head but it seemed that maybe he had thought that the car had left without both of us.

Lucky was perhaps the most perfect of dogs I have encountered. He was increasingly responsive and affectionate. He would stand with his front paws clasped around my leg for minutes on end and he loved being picked up. He also loved being brushed. Being a dog, he loved hunting and he had a particular style. He would run round and round a bush or in a circle in long grass, attempting to flush out small animals. He also noticed birds more than most dogs. And he could chase cattle, we discovered.

One day James, Jill, Lucky and I went for a visit to our friend Ilan's farm, in South East Gippsland. A regular task there was moving a small herd of heifers between paddocks in order to vary their grazing impact. On this day we needed to get them from one end to the other of the farm without their escaping down the side into the large sloping paddock with an orchard at the top and a stream at the bottom, which ran the length of the property. To my surprise and enchantment, Lucky, despite his diminutive stature, turned out to be an agile and enthusiastic cattle-herder.

When we had Lucky I used to drive to Saint Kilda three days a week in a part-time job as a credentialed mental health nurse. The days when I was absent were difficult for James, who I found spent quite a bit of time lying in bed in a state of apathy. He would only spark up in anticipation of my return. He knew that Lucky also looked forward to this. One day James opened the gates so that I would be able to drive straight into the garage and Lucky would be able to greet me. Unfortunately, James was still in a mode where he tended to assume that dogs thought in much the same way as people do and he simply expected Lucky to wait with him for the car to arrive. I was, meanwhile, driving home along the Nepean

Highway when I suddenly thought of my old dog, Bianca (deceased several years ago). I had such a clear picture in my mind's eye of her sitting up in the passenger seat that I reached out as if I could touch her, in a kind of experiment to see how strong my recollection was. To my surprise and great pleasure, it was almost as if I could feel her fur and muzzle in my hand. I enjoyed this dreamy sensation for several minutes and only took my hand away from my imagined image as I turned into Frankston itself. Soon I was driving up the hill to our house. Outside the house James was standing in close consultation with a woman. They were huddled together. The gates were open. I knew something was wrong and, as I drove in, I could see they were bent over Lucky in James's arms. Lucky had run across the road to join another dog on the other side and a car had hit and killed him instantly.

I jumped out of the car and took Lucky inside the house and layed him gently on the bed. There was a solitary drop of blood oozing from his mouth and he was not breathing. But his eyes were still as bright as they had been in life, gleaming with apparent health and he seemed to be conscious and looking at me. I could not bring myself to believe he was really dead. I held his paw and stroked him and talked to him and told him how much I loved him and would miss him, over and over again, while James quaked in the background.

I will never forget that dog.

Harvey and Susan

That evening, although I sensed it was wildly impulsive, I went looking for another Pomeranian on the internet in an unrealistic effort to take my mind off Lucky by finding another Pomeranian. I found one named "Harley" on an animal rehoming site. On getting in contact with the person who was looking to rehome him, I discovered that there was also a Tenterfield terrier called "Suzi", but the two dogs – both twelve years old - were being advertised separately because it was thought they had more chance of 'adoption' singly. To the contrary, two dogs struck me as ideal. The Rehoming Service was initially sensibly suspicious of my fitness to look after a dog. If the last one had been killed because a member of the household had left the gate open, how could I be sure the same thing would not happen to the next? My answer was that the lesson had been so painfully learned that it probably would not happen again. We also had completely new fencing (involving a regrettable argument with new neighbours). The questions I was being asked were however very sensible and it was only when I said that we were prepared to take both Harley and Suzi that the rehoming service person gave in. This was because Harley and Suzi's owner, Bob, was about to leave the country and no-one had shown any interest in them singly or together. I made a time for Bob to come round with both dogs. He was around 70 years of age, recently widowed after his wife's long illness, and intending to emigrate to Borneo to help save orang-utans and live with a Malaysian family.

I was shocked at the size of Harley and Suzy. Having only seen their photos on the internet I had assumed they were normal size animals, but they were, in fact, bonsai-ed. They were so small that it required effort to think of them as real dogs rather than stuffed toys, especially since they made not a sound at our first encounter.

Bob told me how little they ate as if it might make me more inclined to take them on. He said they ate once a day at six p.m. a very small amount that could be enclosed in his fist. He said they would eat, 'literally anything'. He described Harley with blokey affection as a 'pig' and said he had to be stopped from eating Suzi's food. Both dogs had been conditioned to lie flat on their stomachs and await the command to eat, which was "Okay!" Both slept outside in a kennel. It was winter at the time of Bob's visit to our house and, although Harley was too furry for me to see his ribs, I could count every one of Suzi's and every vertebra along her back as well. She moved slowly and jerkily, like a clockwork toy or a chameleon and, although she and Harley were both twelve years old, there was no fat at all around her neck. I could not imagine how these dogs made it through the winter nights outside. I could not imagine Suzi, particularly, making it through another week by the look of her. Her eyes were dull wells of despair. She was obviously starving to death. Bob mentioned casually that there had been 'another one' who had one day just 'keeled over and died'. Obviously we had to rescue these dogs. Before Bob left I placed a variety of foods in Lucky's old plates in the front hallway. We arranged to contact the Rehoming Service and Bob said he would bring the dogs back with their beds and papers the next day. On their way out the dogs scrabbled to seize morsels of food from the hallway.

We renamed Harley, 'Harvey' and Suzi, 'Susan', because we felt the motorbike branding was a little too heavy for such tiny creatures. I do not think that Bob meant to be cruel in the way he fed and sheltered Harvey and Susan. After all he was keen on saving orang-utans – although there was also a Malaysian woman involved. It seemed to me that Harvey and Susan had probably led fairly comfortable lives inside while Bob's wife had been alive, but that Bob had imposed a manly concept of discipline and a 'dog's place on them after she had died. He was possibly forgetful and did not regularly feed them even the meagre amount he claimed to. Perhaps when his wife had been alive they had frequently fielded tidbits and snacks, although neither dared to beg.

Bob returned the next afternoon with the dogs' beds, bags of dried 'kibble', assorted 'snacks' and a stout cord with a lock that was meant to attach Susan to a post if she were left untended.

"Regrettably," said Bob, "she has a tendency to *wander*. She goes into other peoples' back yards. And she can climb, so we had to resort to this awful tethering." I supposed she was looking for food. She shows no sign of wanting to run away from her home here.

As soon as Bob was safely off the premises, I served Harvey and Susan a generous prepared smorgasbrod of roast chicken, grilled lamb chop, cheese and sliced corned beef. The dogs were predictably amazed at the spread and hoed in repeatedly. The real food had an initial effect of some diarrhoea and throwing up, but mostly they kept it down and ate as much as they could. After a couple of months their feeding habits declined to what I recognise as normal in dogs – mostly only eating every second day, hiding extras in the yard in case of famine induced by forgetful owners, and snacking occasionally. There is always food on their plates. I discovered that Susan loves cashew nuts and both of them are fond of most nuts. They no longer eat 'literally anything' but they enjoy a variety of foods and Harvey really likes raw carrot slices. They have, unfortunately learned to beg, and we now serve them a little of what we eat each night in order to make this unnecessary. Last Christmas we put a selection of Christmas meats down on plates in the dining-room and there was absolutely no begging. However, they do beg if we don't share.

It has taken these dogs a while to become 'real dogs'. They had never been walked off leash. They had never been out to the bush.

James Sinnamon with two rehomed friends, Susan and Harvey, Dec.2013

Bob had said to us, looking around our rambling house and gardens, "You know, I don't think they will miss me all that much. They've never had anything like this."

He said they would probably love going bush and they do. They go to the same places as Nubi and Lucky did around our area – down to Sweetwater Creek, to the Mount Eliza dog park, to the huge leash-free forested areas of the Mount Eliza Regional Park near the Moorooduc Railway and, best of all, the several hectares of fenced in and thus anxiety-free forest dedicated to leash-free dogs and their owners called the Community Forest, next to the Briars on Nepean Highway in Mount Martha. Harvey can do several laps. Susan manages to go round easily, although much more slowly, sniffing landmarks all the way. When Susan first moved in with us there were four more holes to go on her collar. Today she is on the last hole. Although she eats as much as she wants and whenever she wants, there is a little fat on her haunches and neck but no extra and she is visibly muscled and bright of eye.

Change of diet and budgeting

Since his organized physical rehabilitation had ceased and he had dropped out of the local gym due to injury, around 2012, James was spending more time at the computer and less time walking. Even when we went to Brisbane, where we had been forced to walk a lot due to James no longer having a car there, we had reduced our walking, since he had now bought a car for Brisbane use.

Having this second second-hand car to service and repair was what caused James to run up against monthly income limits, making it imperative that he recover his old financial carefulness, which mean that he needed to get used to looking at his accounts every day, instead of just running on automatic.

Apart from these increased costs, we were both gradually getting fatter. We began the 'Fast Diet',[74] doing a modified fast two days a week and felt we were making splendid progress. My ankles stopped swelling up overnight. However three months later we had both actually put *on* weight, my cholesterol had risen and James's glucose and cholesterol had risen.

It seemed to me that we were now abusing enough simple carbohydrate to make up for any fasting. The reason was that the Fast Diet proposed that one could eat anything and everything on non-fast days. This turns out to be preposterous. Anyone can make up those two days simply by hitting the simple carbohydrates with or without the fats – although the two are usually packaged together.

Seeking inspiration on these problems, I found videos featuring Dr Lustig of the University of California Television, who lectures about the role of fructose in the epidemic of diabetes and obesity. He points especially to corn-syrup (which goes by many, many confusing trade-names) as in a class of its own, however he implicates all forms of simple carbohydrate, notably sugars and starches. Wondering how I might actually do without these things I had become hooked on, I surfed youtube on sugar addiction and found a video where a woman said, "You go to bed and sleep. While you sleep, you burn fat. Then you get up in the morning, have a piece of toast, and start the insulin cycle all over again, making more fat." This sounded just like me and inspired me to stop eating toast in the morning and then to cut out potatoes, rice and pasta. I found that, if I didn't begin with the simple carbohydrate – bread – I didn't develop my usual unsatisfiable appetite. I convinced James to do the same. We had got used to fasting as a way of avoiding turning on our appetites, but eschewing toast in the morning had the same effect of not turning on our appetites. We were glad that eggs were no longer considered dangerous. Hard boiled eggs and omelettes now took the place of breakfast cereals and bread. We ate plenty of other vegetables and fatty protein however, but cut down on fruit due to the evolutionary theory that it is a high fructose pre-winter food designed to increase fat storage.[75] I very quickly lost four kilos and James lost about nine. Both our blood sugars improved. James's cholesterol improved quite a lot although my cholesterol reduced infinitesimally one point. Unexpectedly my chronic complaint of neck and arm ache seemed to fade away and I developing renewed strength in my arms. We hope that some similar healing process might benefit James's brain. Certainly, since he was on the brink of diabetes, which wrecks muscles and brains, it would help stop that process. His endocrinologist and his Brisbane GP identified our diet as 'the Caveman Diet' and approved of it.

Maintaining political and contemporary awareness

James frequently tunes in to news and discussion programs on the radio. He repeatedly scans the independent and non-anglocentric news and analysis on the internet: Global Research, Voltaire.net.org/en; Landdestroyer.com; Paul Craig Roberts; rt.com; Presstv.ir and com; sputniknews.com and a number of Syrian specific websites, including the Syrian Arab Newsagency: sana.org.en; Syria news.cc and Syrianfreepressblogspot.com. These are his main alternative news resources but he also goes outside them. We are both on an international informal list for exchange of views and reports on Syria, Ukraine and similar NATO targets. We also attend and report on meetings and events on issues like peace activism, wildlife issues, forest issues, bushfire issues, population pressure, and planning and democracy alerts – especially where these involve local or state activists, rather than long-time registered NGOs.

James continues to read hard-copy newspapers, despite my horror at the way they then pile up awaiting an unreliable process of making cuttings and filing. James also watches the news on SBS, ABC and the *7.30 Report* to see how they report on the issues he follows, which, along with domestic politics of privatization, population numbers, foreign property ownership, include involvement in what are arguably illegal resource wars in the East and Middle East.

I am impressed by James's ability to get in and out of the tv room, to watch an item on what he calls, the 'Disinformation' then to come back twenty minutes later to catch a few more minutes of another news program. He follows specific topics and is looking for presentation bias, which he is able to show over time due to his notes. These notes are particularly revealing for Australian reporting on foreign affairs and our involvement with NATO in wars near Russia and in the Middle East. Knowing something about the history of oil exploration in that area, colonization and war, as James and I do, is very helpful for critical reporting.

Although I get most of my 'news' by email, from the internet and by covering events that I deem newsworthy, if I do come in to see what he is watching out of curiosity, James hates the way I get transfixed and object to his channel-surfing, once I'm following a 'story', however puerile. James says he prefers the ABC and SBS out of habit. He thinks there was a time when their news programs seemed more reliable than those of the commercial stations, but that there is probably little difference these days. Although the ads on the commercial stations remain a strong deterrent, SBS now has ads all the time too. The ABC has frequent similar interruptions in the form of program and station identity promotions, perhaps so as to diminish the competition it poses to the ad-bearing stations.

Since his accident James has purchased and borrowed many books. Their subject matter generally relates to current wars or those of modern history, citizen privacy issues, outstanding leadership in politics, and environmental issues. He is particularly interested in deeply understanding: Greek politics and conflicts starting from the Italian invasion in 1940; the conflicts leading up to the breakdown of Yugoslavia; the history and current conflicts in Eastern Europe and the Middle East; and the role of some 'socialist' groups in derailing citizen engagement in grass roots politics in Australia. His interests are however very eclectic and I have the impression that he selects books on a similarly wide range of topics that he did before his injury, and dips into them much as he would have before, to compare accounts and opinions. He gleans information but finds it much more difficult to marshal and restate in written articles. It seems that he has intact knowledge of history and theories which he is still able to modify critically, but it seems hard for him to consolidate new learning in less well established knowledge and experience.

I have not been able to keep track of what books he has actually read substantially. I do know that he has read in their entirety *The Snowden Files*; *I Choose Peace*, a book by left-wing wartime British politician, Konni Zilliacus; Max Hasting's, *All Hell Let Loose: The world at war 1939-1945* and Sheila Newman, *Demography Territory Law: Land-tenure and the origins of capitalism in Britain*. He bought two or three books that challenged the views of his pre-accident read, *Armed and Ready*, examining the defense capabilities of Australia in 1914. I know he has several other books that he started reading, but he cannot remember which ones he actually finished in 2010, 2011, 2012, 2013 and the early part of 2014. What he reads now, he seems to remember but, frankly, it would be very difficult to compare his book remembering abilities before the accident with those of today. I can only test him with my own books, but those books are very difficult anyway. The whole nature of remembering book content (as in the exposition and defense of different theories) is difficult. Many people do not actually take in theories and therefore cannot retell them. Most of us don't take in theories on subjects we are not very interested in. Most of us can easily take in opinion and some theory on familiar subjects.

Candobetter.net

Any reader who would like to see a record of James's writing before the accident and then its evolution after the accident can go to http://www.candobetter.net/JamesSinnamon. The articles here start in 2006, but initially consist only of republished material or copies of letters sent or comments received. James evolved as a journalist and brought another dimension to *Candobetter* via his election reporting where he ran as a candidate. Note that whilst this could be called a kind of immersion journalism, his candidatures were sincere and he was learning through experience. James was also owner of citizensagainstsellingtelstra.org and this association boosted readership initially. This site and a later one, citizensagainstsellingtelstra.net, fell prey to cybersquatters when the rent became overdue after James's accident. James was able to buy back citizensagainstsellingtelstra.net from Crazydomains.com.au early this year when squatters gave up their speculative occupation of the site and the price went back to something reasonable. (A number of early links to the old citizensagainstsellingtelstra site still need to be fixed on *Candobetter*.) However, the loss of connection with the citizensagainstsellingtelstra websites has been another factor post-accident that impacted on readership of candobetter.net. James's ambition for *Candobetter* (thought up at Woodford Folk Festival in late 2003 after discussions in their Green Room with me and others) was to start an open political discussion site to replace local indymedia.org sites which, tragically, had been taken over and compromised by particular political groups and goons. He also wanted to supplement the few other independently owned Australian political discussion sites that were around, these being Margot Kingston's webdiary (an anti-PM John Howard site now archived by the National Library of Australia at http://pandora.nla.gov.au/tep/21852), economist John Quiggin's site: http://johnquiggin.com/, http://www.onlineopinion.com.au/ and Larvatus prodeo: http://larvatusprodeo.net/. (Kingston's webdiary was abandoned some time after John Howard lost government. Lavatus prodeo also discontinued itself in 2012, started up again in 2013, then stopped in 2014.)

James had some run-ins with Webdiary and Larvatus Prodeo because of his commitment to the 9/11 controversy. Looking back at how perseverative he became on this matter on several occasions, notably on Lavatus Prodeo, he now thinks that his unpopularity for some associated posts might have been reasonable. He wonders if he was ill at the time and I wonder if he was not then suffering from a Hashimoto's disease attack, which can affect judgement. He and others had run-ins over 9-11 with the owner of the Australian Running on Empty peak oil list.

He can no longer remember what it was that converted him from relative indifference to such certainty of a conspiracy in the 9-11 incident. Neither can I. Because I knew that focusing on 9-11 would alienate a lot of our readers and affect perception of my social science work, I asked James to categorise it under the tag of 'heresies' on *Candobetter*, where it remains. I did try to understand the matter myself and found the mystery of the free-fall of the World Trades Center third building very persuasive. I agree that there are many circumstantial indications of US Government involvement in the deaths at 9-11, that the official inquiry was inadequate, and that the US used the tragedy as an excuse to invade Afghanistan and Iraq, leading to Syria and Ukraine interventions, which were politically strategic to petroleum assets. I was familiar with the history of colonization and wars with regard to petroleum from early times, having specialized in this area.[76] Without a trial of specific accused persons and requisitioning of all possible evidence, I don't believe it will be possible to prove for or against the US Government's role in 9-11. Such a trial would be very difficult in the US because of the rank of those likely to be accused. There is an argument that the US involvement in illegal wars is disgrace enough, but it is true that, if the US actually did mount 9-11 as a false-flag attack, this might bring about sufficient public curiosity and alarm to make major political changes inside and outside the United States. It would probably also require international criminal trials to actually register on the public at large. In general terms of probability of such a false-flag attack, I admit that it is not beyond the bounds of possibility and that there have indeed been many precedents of staggering criminal conspiracies by governments in world history. Despite the aphorism that you should suspect a stuff-up rather than a conspiracy, history tells us that conspiracies are more the rule than exceptions in politics involving the rich and powerful whenever something that would not get public approval seems necessary, although stuff-ups are also common. Spying, which seems to be conducted by all governments, requires conspiracy. Spying on peoples' emails by the United States National Security Agency (NSA) is an example of a new level of conspiracy. Notably, almost all wars have required calculated conspiracies at high levels to get ordinary people to fight. Political assassinations by governments also require conspiracies or declarations of war. Not all these conspiracies were criminal in the strictly legal sense. If you would like to read more on James's views on this matter go to http://candobetter.net/911truth

You might think that brain injury would carry such a fog of its own that no brain-injured person could develop insight about pre-accident behavior, but you would be wrong. James says, of his period of hammering 9-11, "It was obvious that I was not thinking clearly. I just kept the argument going on and on, way beyond what was necessary. It reached the point where it was obvious that my points had not been answered and I could have stated this, but instead I effectively badgered those who disagreed with me again and again until they gave up. By that time I had made myself look like a ridiculous obsessive compulsive."

Although we already had a JFK section, as well as a Martin Luther King (MLK) section on *Candobetter*, since his accident James has also written or published articles on Kennedy's birthday since December 2010, which are to be found under this tag: http://www.candobetter.net/node/3553. I probably corrected typos and repetitions, but these have greatly reduced over time. It is easy for James to remember Kennedy's birthday because it is the same as his own. The interest in Kennedy and also Martin Luther King, were there well before the accident, but I think that James feels particularly close to these figures because they died before they could complete their work and James himself almost died.

Candobetter acquired its defining perspective as a website for reform in democracy, environment, population, land use planning and energy policy from James's interaction with my 2002 environmental sociology thesis, the Growth Lobby in Australia and its Absence in France (See this page for links:

http://candobetter.net/node/1882). The methodology of focused beneficiaries and diffuse costs remains useful for understanding how unpopular and costly policies persist in democracies. The intersection of these ideas and methodology[77] brought about a unique historical perspective and a way of interpreting current events for *Candobetter*. But the accident knocked this knowledge out of James's memory and it has been hard for him to refocus on these fundamental intersecting areas and the methodology. Here is what he wrote in 28 January 2008, updated on 24 January 2009, well before his accident. The original (with footnotes and links is at http://candobetter.net/node/1002:

"How the growth lobby threatens Australia's future

Why is it that the Australian government, and other governments, principally in the Anglophone world, deliberately encourage population growth when common sense and intuition, not to mention the hard evidence, tell us that a larger population cannot possibly be in the interests of the current inhabitants of this country or of the rest of the planet?

We are long past the point where adding extra numbers in any way increases the synergy of the inhabitants of this country. Consequently any additional population growth must necessarily make each and every one of us poorer on average as the per capita access to natural resources necessarily decreases in proportion to the increase of the numbers of people.

However, it gets even worse than that, because of the dis-economies of scale inherent in large populations. An obvious example is that Australians are paying extra water rates to finance costly water desalination and sewage recycling plants required to provide water for additional people.

Had we stabilised our population, this would have been totally unnecessary. We could all have been adequately supplied by our existing less unnatural and less technologically complex water infrastructure.

Similar points could be made about transport, electricity, health, education and other services.

To cope with increasing numbers, it is necessary to destroy ever greater tracts of native bushland, to abuse our topsoil and waterways, and to unsustainably dig up ever more of our finite endowment of mineral wealth.

Three and a half decades of extreme 'free market' economic policies have further compounded these problems. These policies hinder governments from making use of what economies of scale are possible. They prevent effective planning in the interests of all members of society. Obvious examples include the huge inefficiencies of the private property market and the shambolic state of Australian urban planning, a result of the dismantling of Whitlam's Department of Urban and Regional Development (DURD) by Prime Minister Malcolm Fraser in the late 1970's.

In some ways it may be the case that immigration does indeed enable the transfer of wealth into, as well as out, of Australia:

The purchase of a home by wealthy or middle class immigrants as a means of buying Australian citizenship, which is effectively a transfer of wealth from the source country into Australia;

The poaching of skilled workers, often trained at the expense of taxpayers of other countries, including of poor third world countries - a practice, for which the Queensland Bligh Government has become infamous;

The selling of Australian university degrees and vocational training, which has notoriously become yet another means of purchasing Australian citizenship.

Little, if any of this wealth trickles down to ordinary Australians and whatever benefit they do gain is more than negated by the loss of previously available educational, training and employment opportunities, and consequent housing inflation (also discussed below). Even if it can be shown that Australia, as a whole, gains, rather than loses wealth through immigration, that wealth will most likely evaporate within this generation.

In any case, on a global scale such wealth transfer is a

zero-sum game, at best.

All things considered, it seems far more likely that we are not only becoming more impoverished, but we are becoming even more impoverished than we might expect to be if we had simply divided the existing wealth amongst larger numbers of people!

In a perverse way, it seems to me that this may have actually made it harder, *rather than easier, to argue the case* against *population growth and immigration.*

Whilst I can't know for certain if this was true for others, I will try to summarise some of the ways that this fed into my own conscious and sub-conscious thought processes and caused me to avoid questioning our high immigration policies for many decades.

My own intuition caused me to conclude that, if, somehow, immigration made us worse *off on the whole, it would surely be* harder, *rather than easier, for any group to* gain *from immigration. Therefore, when faced with the strident assertions from all the seemingly credible authorities that immigration was economically beneficial, I found it easier to deny my own gut instincts and not to make the considerable investment of emotion and time necessary to question this pervasive message. Instead, I just quietly hoped that the advocates of immigration, who promised me a more prosperous, vibrant, interesting and sophisticated society, were right.*

Alternatively, on occasions when the economic arguments did not seem to quite ring true to me, then the only other likely plausible motive would have been an underlying altruism of an enlightened elite more willing, than the ordinary, backward, redneck, xenophobic masses, to share the wealth of this country with others less fortunate than ourselves.

However, the conclusive evidence, after many decades of this social engineering, is that Australia has, instead, become a poorer and more dependant country as consequence and this has not been brought about because Australia's elites are self-sacrificing and altruistic.

Contrary to what I expected, a small group has,

paradoxically, not lost, but rather gained from this chaos and suffering, at our expense. That group is the growth lobby. *First identified and described in detail in Australia by Sheila Newman in the Growth lobby and its absence (2002) [1], it is really a group of land speculators and landlords operating in an organised way on a corporate level.*

Land speculators and landlords openly welcome the way that growing demand increases the price of vital resources over which they have acquired a monopoly. They profit from commodifying and then controlling access to resources and services which include water, land, power, housing, roads, food production and transport, which each one of us needs in order to live a dignified life, or even simply to live.

As one consequence, in Brisbane at the start of 2009, even previously well-off professionals are being impoverished by insatiably greedy landlords, who exploit these circumstances to increase rents at every possible opportunity. A surveyor, who lives near me (who acknowledges that his own work entails the destruction of bushland to build new housing developments to cope with population growth), told me how he was unable to travel back to Germany this year for his holidays, because of being personally affected by recent rent increases.

The growth lobby *also includes property developers, financiers, building companies and suppliers of building materials. There are also others that gain from population growth through high immigration, such as immigration lawyers, employment agencies and cheapskate employers.*

Whilst these activities may provide a facade of economic prosperity, none are capable of increasing the underlying ability of this society to provide for its own needs.

Queensland Premier Anna Bligh in April 2006, then Deputy Premier, ludicrously defended population growth on the grounds that it was necessary to keep people in the construction industry employed.

How could Premier Bligh, supposedly an intelligent person, have failed to ask herself the obvious question: How are those additional people then to be employed? Must we build yet more

houses to keep them employed and import yet more people to
Australia in order to provide a demand for those houses?

At some point such 'growth' has to end and Queenslanders
must be able to find gainful employment by meeting the needs
of other Queenslanders instead of future inhabitants.

The longer Australians put off stabilising our population
and establishing a steady state economy, the worse will be our
circumstances.

But the growth lobby *wants this situation to continue*
indefinitely. To ensure that the forced march to dystopia
continues, the growth lobby *pours funds into the coffers of*
Australia's major political parties, including Anna Bligh's
Labor Party. It creates obligation and dependency in our
political parties and governments. In turn, our governments
endlessly facilitate the real-estate economy, in the face of every
democratic objection, merely to keep themselves in Government.

If we are to hope for any kind of a decent future for
ourselves and our children, these corrupt arrangements must be
brought to an end and the power of the growth lobby *must*
be broken."[78]

I think that it was in the second half of 2014 that James's
memory finally became fairly reliable. I feel that his writing skills
have improved and will improve because of this.

On 14 October 2014, after months of work and a few years
threatening to do so before each election that occurred after his
injury, he published "Issues that should be decided at the 29
November Victorian State elections" at
http://candobetter.net/MakeYourVoteCount/. The election was
held on 29 November 2014. He did not remember that he had
attempted to do the same thing for the 27 November 2010
Victorian election or for the September 2013 Federal one since his
accident, but had never achieved this aim. Each time he had also
planned to devise electronic questionnaires to monitor the policies
of parties and independents, but each time he underestimated the
amount of planning, concentration and programing such a project
would entail, especially in light of his decreased stamina. Leading
up to his Victorian election policy article in 2014, I showed him
where he had published such a poll on 18 March 2009 for the

Queensland elections of the time, in which he also presented as a candidate. See http://candobetter.net/QldElections/survey. He seemed a little shocked to see he had already done such a thing and he began to see or remember how complex it was, involving a lot of work, even for a person of normal health. I then suggested that events had overtaken him anyhow, because a number of websites were now providing policy comparisons between the parties and allowing people to ask about candidates' attitudes to matters that concerned them. James was then about to abandon what he had already done, but when I looked at it, I could see that he had actually come up with a variety of really interesting policies, after talking to various *Candobetter* writers and to Victorian activists. He also arranged these policies in groups and then in a hierarchy that reflected their likely appeal to the electorate. So, where he might personally have preferred his policies under "Protection of civil liberties, freedom of speech", he had placed them last in light of lack of public awareness of the basic premises and also, possibly, the federal and international nature. Nonetheless I was impressed by the originality of two of the policies here, which suggested a debt of gratitude owed by the state of Victoria to Australian-born and long-time Victorian resident, Julian Assange and also drew in Edward Snowden, thus putting an emotional aspect to something often misconstrued as dry – telecommunications surveillance.

On March 4, 2015, the Construction, Forest, Mining and Energy Union (CFMEU) organised a protest in all states of Australia against the Abbott Government's proposed workplace legislation changes. Since Australia has forbidden secondary boycotts and thus made national strikes illegal in Australia, this seemed the next best thing. Although both James and I had good intentions of distributing information there, neither of us prepared early enough. I simply wanted to use it as an opportunity to advertise the importance of the alternative press and the existence of *Candobetter*, but James wanted to write something about changes to surveillance laws. He stayed up until about 4am trying to write a pamphlet, and made some good changes to signs I had designed to be printed and laminated. Unsurprisingly, when the alarm went off at 7am, he was too exhausted to go.

The next day, however, I found that he had posted his proposed pamphlet on *Candobetter* in the form of an unpublished article. It seemed to me that he had pulled together a number of things very well and used a good analogy in his good cop bad cop analysis. Something really good had come out of his efforts after all. I illustrated it and published it on the front page as "Union-bashing and War for United States global hegemony – two sides of the same coin?" at http://candobetter.net/node/4321.

Union-bashing and War for United States global hegemony – two sides of the same coin?

"This article gives a 'good cop bad cop' analysis of Australia's two party history since the fall of the Whitlam Government. It looks at the evolution of economic theory and foreign involvement in wars over this period.

As United States whistleblower Christopher Boyce (aka 'the Falcon') revealed on SBS Dateline on 18 February 2014, after the CIA helped to destroy the Whitlam Labor government in 1975 it attempted to consolidate its victory by buying influence amongst newer leaders of the Labor Party.

Consequently, when Labor finally won office again on 5 March 1983, the reforms of the Whitlam government were abandoned. Instead the Labor government of former ACTU President Bob Hawke helped pioneer in Australia the adoption of neo-liberal 'free market' economic policies that would later be adopted around much of the rest of the world. Australian finance and industry was exposed to globalised international competition, including the removal of government control over the Australian dollar's exchange rate (i.e. the "floating of the dollar") by then Treasurer Paul Keating in December 1983. Government services and the public servant numbers were reduced. Also publicly owned assets were privatised with no electoral mandate – QANTAS, TAA (the domestic air service), the Australian National Line shipping company and the Commonwealth Bank.

Since 1983, Labor governments have generally played the role of 'good cop' whilst Liberal/National Coalition governments have generally played the role of 'bad cop'. Prior to his defeat in March 1983, 'bad cop' Malcolm Fraser imposed a 12 month wages freeze during which the value of real wages fell 9.1% as a result of cost of living increases not being matched by wage rises. Subsequently the 'good cop' Hawke Labor government with the help of ACTU President Cliff Dolan imposed the Prices and Incomes Accord on the Trade union movement. The accord allowed restricted wage rises and promised increases in social spending.

Labor has also backtracked away from Gough Whitlam's opposition to war. In 1991 the Hawke government sent troops to fight in Operation Desert Storm after Iraqi dictator Saddam Hussein had been tricked into believing that the United States would not retaliate should Iraq take action against neighbouring Kuwait for slant-drilling into Iraq's oil fields. Sanctions imposed against Iraq after its 1990 invasion continued until after Saddam Hussein was overthrown after the 2003 invasion. This continued even after bleeding heart' Paul Keating became PM in December 1991. Many hundreds of thousands of Iraqis died as a result. One estimate puts the death toll at more than 1.5 million including 750,000 children under the age of 5. Other wars, which gained bipartisan support from the Labor Party include the invasion of Afghanistan in 2001 and the invasion of Libya in 2011.

Both sides of Australian politics, with a only a few honourable exceptions, are shamefully complicit in the United States' proxy terrorist war against Syria which started in March 2011. In that war, 210,000 Syrians have so far died at the hands of jihadist invaders from nearly every corner of the Muslim world and a number of other countries . Australia expelled the Syrian ambassador under the pretext that the Syrian government had allegedly massacred 108 of its citizens, including 34 women and 49 children at Houla on 25 May. This allegation that the Syrian government had massacred Syrians in a region particularly renowned for its strong support for the Syrian government is contrary to common sense and all credible eyewitness and forensic evidence.

Both sides of parliament have also sided with the neo-Nazi government of Ukraine which came to power in February

2014 as a result of a CIA-orchestrated coup. Both sides have also unquestioningly accepted the lying mainstream newsmedia narrative that the 298 passengers, including 27 Australians, aboard Malaysian Airways Flight MH17 were murdered by East Ukrainian self-defence forces with a BUK surface to air missile on 17 March 2014. This is contrary to all eyewitness and forensic evidence, including the photos of the pilot cockpits apparently riddled with cannon shells on both sides which indicate that MH17 was shot down by 2 Ukrainian Sukhoi 17 fighters.

Both Labor and the Coalition have voted in Parliament to support dragnet collection of all our Internet and telephone communication meta-data by our spy agencies. This is in spite of whistleblower Edward Snowden's warning that this has never prevented even one act of terrorism.

The above are only some examples of how corrupt Australian and global politics has become since 1975. However while there is still a free and open Internet, and a trade union movement with capable leaders we still stand a chance of reversing the damage and re-establish a quality of democracy comparable to what we once enjoyed.

You can help by using the Internet to help tell the truth, and by contributing an to http://candobetter.net and other alternative newsmedia sites and by spreading the word. This article is published at http://candobetter.net/node/4321 ."

Art and Music

James was referred through a local outreach program to an activity centre called IMPACT, which he qualified for due to his diagnosed mild depression, rather than his brain injury. (This diagnostic prioritizing is a reflection of a strange bias for symptoms arising from psychiatric disorders rather than from brain injury. All the sillier when you consider that a large proportion of schizophreniform psychoses amount to slowly fluctuating brain syndromes which over time often become permanent brain damage.) At IMPACT he saw an enthusiastic and skilful counsellor once a fortnight for an hour or two. She focused on collaborating to help him identify and achieve goals. IMPACT actually owned the warehouse building where it had its offices and activity rooms. This gave it stability, identity, space and scope. It thus avoided the expense and logistics of accessing and paying for private or government premises.

IMPACT also provided two art programs a week. One was called "Artwell". These programs provided James with a regular opportunity to focus on his aim of improving his sketching abilities, which his father had encouraged him to do. In these sessions he was able to concentrate enough to actually put pen and paint to paper. I think that the damage to his brain may have been less on the right side and that the damage to the left side may have reduced some inhibitions that the left might otherwise have imposed, because James draws fluently and seems to be able to make an excellent composition from any subject he chooses. (There are quite a few artists and architects in his family.)

His artistic output ceased when IMPACT was unfortunately closed down by the Australian Government, which seemed to shake up all these services and reduce and redistribute the workers in various agencies away from their original locations. I considered this extremely cruel. From personal experience I know that this needless kind of reorganisation is deeply traumatic for workers, let alone highly dependent clients.

A new worker was sent to our house to replace James's IMPACT worker, but she declared James's room (with three tumbling piles of unsorted newspapers and newspaper cuttings) to be a fire-hazard! She would only meet him in places outside our house, which means cafes or the municipal library in the absence of the old IMPACT premises. With regard to the newspapers in James's room, subsequently part of the house was electrically rewired and the electricians needed access to old wires via his wardrobe. Knowing that time was literally lots of money in this case, he moved boxes of papers out into the carport and in one day had cleaned out his room and consigned the papers to archival conditions more closely resembling abandonment.

In the absence of IMPACT James himself fulfilled two other aims. One was joining a choir. Choirs seem to have fragile existence and he was onto his third choir in late 2014. He attended this regularly because he enjoyed singing, even though he almost always found the choice of songs disappointing. Perhaps because of this, he kept losing track of his song books, although he could usually find his paints. He liked me to come to choir with him and I did attend one for a while but I really could not afford to spend the time at that point and I also didn't much like those songs!

Just prior to Christmas 2014 the local council agreed to give space to the Artwell program in Cube 31 which is the local government art gallery premises. So James began to go there again. Unfortunately the art sessions clashed with his latest choir. And then there were the tennis lessons.

At Artwell he entered an art competition with two very well composed paintings of unusual still-life subjects, which he mostly found in the boot of his car and supplemented from his surroundings. My favorite included gleaming black gumboots (mine and his) with a tangle of dog-leads and two old car license plates. He had retained a facility for accurate representation of objects in line drawings and a feeling for colour and balance. The objects also seem to symbolise his relationship with me, the dogs, and our cars. The number plates were not stolen; they came from my car, which had just been sold. The painting is in violet, black and shades of brown, executed in water colour.

James Sinnamon's painting of gumboots and car boot contents

December 2014: Strengths and weaknesses after 4 years and 7 months

I was initially going to end this book in December 2014, four years and seven months after James's traumatic brain injury. I thought that James was unlikely to improve much more, except to slowly consolidate. I wanted some definitive point to end on, and I chose to make it a description of how James survived when he spent ten days away from me in Queensland, as a measure of his independence and recovery. There would be more revelations about his health after that, however, and so the four years and seven months point here merely stands as a part of the onion skin of individual recovery and the questionable efficacy in medicine of assuming all hoof beats are horses.[79]

Since James came to live with me in Victoria, we had almost never spent a day apart. James had regained a lot of self-control, but he remained vulnerable to mild buying sprees, which I would describe as buying household items without consulting me, and coming home with multiples of items we didn't need in large quantities, because they were 'on special'.[80] James's shopping as an emotional distraction tended to happen if I reduced basic attention or support over a period of days. For instance, I might eat by myself at the computer, watch a video by myself (he finds movies taxing on his limited time and energy rather than relaxing) leave him to cook and serve his own food, listen to music all day long, go and see a friend, spend all day editing a movie or writing, working on a painting, ensconced in my basement sewing room, or taking my mother or father out. Even though I would be there most of the time to 'consult', there would not be the same possibility of deep and patient interaction. If this happened over two or three days, James's coping skills were stretched. I also gave him minimal attention when I was trying to discipline myself to complete a project – like editing this book. James would cope for a day or two quite well, but after that, some of these latent habits of overspending or over-accumulating 'things' or overeating or not exercising enough would start to come to the fore.

In counterbalance, as I have mentioned, James had himself cultivated a number of external interests which he maintained independently of me. He continued to go to Artwell, where it had now moved to the local council.

Artwell, as an art-activity at the permanent IMPACT premises, had been well-patronised. Without the supportive program within what amounted to a place people with disabilities could hang out in, it became an isolated activity that people had to remind themselves to attend and then travel to. Mentally disabled people have a hard time organising their daily routines anyway. They tire easily and they don't have much money to travel to and fro for occasional activities. James is much better off than many because he can still drive, can afford a car (just), has my support, does not rely on a public pension, and does not have to pay rent. Predictably, attendance at Artwell dropped right down after it moved. James went there partly to try to keep it available, out of principle, much as he always chose to support local businesses when he lived in Brisbane.

We also shared many activities together, such as meeting with Australians for Mussahala (Reconciliation) in Syria, and attending and filming political/environmental and land-use planning events. James could now be relied on to do some of the filming instead of me. This was handy if I were one of the speakers and, if I wasn't, I could choose to relax instead of film.

Although, in theory, James wanted to edit movies he filmed, allocating the time and energy to learning to edit posed a problem. It seems quite possible that he will never do this because the editing process, once you have learned the software techniques, still requires a lot of concentration and judgement, using skilled visual and auditory attention.

There were definite limits to James's energy and thus to his time. I estimated after four years and seven months that he probably has about an eighth the amount of energy that I had and noted that around 2011 I had estimated that he only had one twentieth. The 'trick' for James was to use that energy wisely. He had a reserve of energy that he could keep ticking over in addition to energy that could be thoughtfully directed. He had a normal need for continual mental stimulus. He achieved this to a large degree by continuously reading, often on the internet, and by republishing, with footnotes, articles from other sites, and, as I have described elsewhere, by attending to the ever increasing volume of SPAM that all website owners and editors have to deal with. He would also take the dogs out by himself, if I was busy. He could now be trusted with them as long as the walks took place in familiar terrain where he was familiar with the likely distractions and could apply learned rules.

Proof-reading skills

In 2011 I had written how James had once been a great proof-reader before the accident due to his very reliable knowledge of measurements, scientific definitions, maths and his superb familiarity with geography and history. He was alert to miscalculations, misspellings, repetitions, contradictions and mistakes of place, person and events. Although he had described himself as a slow reader, prior to the accident, he had actually been a very close and careful reader, rather than a slow one.

After he was injured he really did read slowly. Although he could take in printed information better than verbal information, he could not read much at a time and lost track of what he was reading from day to day. I could not therefore rely on him as a proof-reader anymore, which was a great loss. I noted that, even if he had to some extent retained the ability to pick up mistakes, he did not have the energy to read the volume of information necessary for critically proofreading a book.

I really missed his company and support as well as his expertise, as I concluded *Demography Territory Law: The Rules of Animal and Human Populations* in 2013 and *then Demography Territory Law2: Land-tenure and the Origins of Capitalism in Britain* in 2014. He had, however, possibly as early as the first months of 2011, contributed as co-writer with me to the very much shorter fourth volume in the series, about events leading up to WW1 and the post WW2 era, which remained in draft form. He had also begun reading drafts of this book about his recovery, but had never got more than a third of the way through or so. This was an acid test since he found his own story fascinating – a revelation each time.

On Thursday 27th of November 2014, James asked me for a copy of *Demography Territory Law2: Land-tenure and the Origins of Capitalism in Britain*. I gave it to him but wondered whether he would be capable of understanding it. I knew he had actually finished books that he had bought over the past two years, but these were all or mostly in areas in which he has special expertise and great familiarity – mostly military and political history. As our friend Jill said, my second book in the series was probably more his style because it did contain quite a bit of military and political history, however I was approaching this from a completely new biologically based political theory which was likely to go over his head. Then again, because I started writing this theory before I met James, and talked about it frequently well before his accident, he might have stored some of it somewhere. Against this was his utter failure to recollect the contents of my thesis which had been so important for our website, and – dare I say – our initial relationship.

On the 4th of December he had ten pages to go in *Land-tenure and the Origins of Capitalism in Britain*. He was reading it several hours at a time without being distracted onto other books, whilst still attending to appointments and catching up on the tv news (which he calls 'The Disinformation') and the web. He also had a few long afternoon sleeps and wondered why he was so tired. I thought it might be because of his dedication to the task, although he was also recovering from his badly infected foot, which might have made him tired as well.

On the 2nd of December I, when Jill, James and I were in a café, I tested his comprehension of the contents. I was very persistent in getting him to describe my theory – to the extent that Jill frowned at me and suggested that it might be counterproductive to 'badger' him. She was probably partly right, given the drive of authors to be understood, but that was not the only reason for my intense approach. It is not obvious to people that James still misses parts of sentences, but to communicate effectively with him, you have to realise this. For anything important, I repeat myself and get him to repeat back to me. I gave him quite a lot of time to describe my theory and he took a while to do so. It took him two or three goes over about three minutes to collect his thoughts and re-express what he had understood, but he succeeded in describing what was important.

How long would he remember this? I would keep trying to reinforce the learning because of its importance to me and therefore to our relationship. I would observe to see if he applied his understanding to events around him in the future.

He intended to write a review for *Candobetter* but wrote the short comment below for *Candobetter*.

"*Native Australians suffered no worse than the native English*

The following was posted as a comment in response to The Troubling History of Thanksgiving (28 Nov 2014) by Gary G. Kohlson on Consortium News:

In fact, what native Americans and native Australians suffered at the hands of European settlers is little different to what the native British commoners suffered at the hands of British elites since 1066 and what native Irish commoners suffered at the hands of British and French Huguenot settlers since the time of Elizabeth I.

In Britain, the theft of common land and dispossession of its inhabitants was termed 'enclosure'. Descendents of the dispossessed were sent to Australia in 1788 as convicts.

This is described in the book Demography, territory, law 2: Land-tenure and the origins of capitalism in Britain, New Theory by Sheila Newman, *Countershock Press, 2014.. More about the series here: Demography Territory Law Series by Sheila Newman. I have almost finished reading it and will review it shortly."*

He also wrote a lot of comments in the margins of my book, most of which I have still not read, owing to intervening events. One of his comments was, however, that a busy 17th century map I had used to show the location of the seven united states of the Netherlands did not seem to contain one of those seven states. Since I had not even noticed this myself, I was impressed at his attention to detail.

Unfortunately the review never eventuated, for the same reason that I never read his margin comments. A review would have given a good idea of the extent of improvement in his reading and new learning of complex information.

What I can say now is that his interest as well as his sense of duty to read my work was a big sign of recovery of his previous personality and drive and of one of the important qualities he brought to our relationship.

James and his brother Bill

In 2012 James got to know his brother Bill again, along with Bill's family. We went to visit Bill and his wife, Persephone, and their two small daughters. Bill was now doing an intern year at a local hospital. I had the strangest feelings seeing James and Bill together. They looked and sounded so similar. On the one hand, Bill's evident good health, alertness and confidence, highlighted obvious signs of James's injury, reawakening a profound sense of loss in me. Yet, seeing them together made me strangely elated – because they seemed to fuse in my mind to create an intense illusion of the 'old James' – healthy, confident, strong and smiling.

Since then both brothers have met several times on a friendly and positive basis.

Early in 2015 we went to visit Bill and his family again. This time I was greatly reassured because it was a normal experience for me. The two brothers were now obviously two different people, without one seeming more whole than the other, despite the changes that James has sustained. Time and the treatment to James's thyroid had improved James's posture, speed of processing, memory, confidence, alertness and expressiveness – even though he would never hear at the same speed as people speak and he would never regain the extraordinary physical and mental strength and endurance that the accident cost him.

Tennis rehab for brain injury

Tennis was the other thing that James decided he wanted to try. He raised the idea more and more frequently from about 2012. He hoped to use me and our friend Jill as tennis partners. I was not keen to try tennis. I had not been on a court since about age 11, when I had found the game boring and its scoring incomprehensible. From approximately age 35 I had been prone to tendonitis and a variety of chronic and acute muscle injury. I had it under some control but if I feared that the cycle would start up again if I tried tennis. Jill loves tennis but already had a tennis group close to home, 40km away. I thought that James would have problems playing anyhow, due to his extreme slowness, his left-sided weakness, his lack of stamina and his impaired coordination. It was easy to ignore his efforts to organize a game.

James persisted. He bought two rackets. In 2013, in summer, he had booked a tennis court in advance, but the weather was too hot for anyone to play. In September 2014, we finally went to hit a few balls at the Overport Tennis Club courts, in Frankston. It was not actually a game, since neither of us knew or cared about the rules. It was like a slow motion study in total incompetence. We could barely hit a ball. We could barely pursue one. My inertia was such that running felt like forcing my body as if it were submerged up to my waist. On this particular occasion, James seemed to have slightly more stamina than I, which is to say, not much at all. Our ages were, respectively, James 55 years old and me, 62 years old. After less than 20 minutes on the court, I pulled a groin muscle. (I have noticed that this detail always elicits a laugh.)

My groin muscle settled after a few days and James and I decided to have another hit. After ten minutes, I nearly pitched over, as my right knee somehow wrenched out of position, with surprising pain. I was able to push it back into position and the pain abated after a minute or two. I continued to play, when the same thing happened, but the pain was more severe. After a pause, I continued to hit balls back, but only if James could send them right to me. That gave me very few balls to hit and we gave up after ten minutes. This was the beginning of my own battle with unstable knees, which I have consigned to an endnote, which will only interest fellow sufferers.[81] I did continue to have hits with James but my participation was unreliable.

My knee problems, however, led to us making a very valuable new contact in the Overport Tennis Club coach, Cameron Milne. Cameron turned out to be one of those rare gifted people we have met since James's accident, who were able to see how to help specific areas of his problem. It was a great relief to find him just as James's physical fitness and outside activity seemed to be dwindling in the absence of any formal therapeutic program.

Neuropsychological and physical effects of tennis

Cameron had superb teaching skills and an interest in and knowledge of brain injury that went beyond the psychological and took in the physical. Some of his knowledge and enthusiasm came from having a brother with Downs Syndrome who was active in sports. Cameron also coached another student with brain injury - Sam Howe - who featured in a *Catalyst* program on ABC Australia television. Sam played competition tennis before his injury and Cameron hoped he would again. Cameron worked with another tennis coach as well, Andrew Scott, which helped with continuity over vacations and busy periods.

We filmed some of the sessions, which began on 15 September 2014. The films are on my YouTube channel (queeniealexander2000). Here is a link to the seventh in December 2014: https://www.youtube.com/watch?v=OpBW2kt3Sk8

Tennis is a neurologically very global sport as well as a social sport. Not only did it seem to have a positive effect on James's overall function, but he really enjoyed learning the skills and interacting with his coaches. Cameron dropped his fees so that James could afford to have two lessons a week; one with him and one with Andrew. Cameron focused on movement and building up weak areas and developing symmetrical strength. Andy focused on 'rallying' – maintaining a longer and longer series of shots before someone missed a ball return. Both coaches were keen to find other people who would benefit from these activities and wanted to make this as affordable as possible and accessible to several students. The Overport Tennis Club in Frankston was very supportive of their project.

The theory of muscle strength and coordination being reflected in brain development is an obvious basis for recommending tennis, but tennis also has direct social and cognitive benefits.

The sport has a number of advantages to help with its challenges. Unlike gym sessions, lessons can be closely monitored by a professional sports coach who interacts one for one with the student. There are immediate rewards (serving to target and receiving) which are controllable by the coach, who directs the ball skilfully. There are simple rules. Ball action and mirroring by the teacher can provide external cues for any parkinsonian-like inertia or broken rhythm. Skill acquisition can easily be broken down into steps. There is a challenge of speed and distance to cover which can be varied according to number of players, or, if only two, distance from net.

The benefits are various. There is the possibility of wider social interaction with other players and in club activities. For carers and families there is the possibility of wider social interaction with other carers and families as well as playing with the person they care for. The game challenges attention - to follow ball, estimate trajectory etc. It challenges energy and possibly retrains in its deployment. It enjoyably builds stamina. Tennis challenges concentration: An easy measuring tool is to count how many balls can be returned before concentration lapses. Tennis encourages self-awareness, through the student tailoring their own responses to meet the challenges, and, if filmed, seeing themself on film. Posture should improve as a by-product of technique, coordination, anticipation etc, feedback from coach and self-awareness. Alertness increases in response to enjoyable challenges and friendly interaction. A film record can provide the possibility of measuring and comparing reaction speeds over time. The student's muscle development and coordination are also measurable by sports physiologists. The student's cognitive status is also measurable by a neuropsychologist.

In James we noticed improvements in self-confidence, posture, self-awareness, concentration, attention, ability to finish things, executive function, and joyfulness.

We could not ascribe all the improvements in James to tennis alone. Other factors would also have been helping to improve James's brain and fitness. One of these was obviously thyroxine. James had been receiving thyroxine for hypothyroidism since July 2012, although it was not until he was treated by his endocrinologist, whom he first consulted on 13 June 2013, that his thyroxine regime and Vitamin D were optimised in accordance with the progress of his Hashimoto's disease. Using a two year's strong improvement rule of thumb for brain injury recovery, the effective thyroid treatment would have contributed to some of his improvement during tennis training. Change of diet to decrease simple carbohydrates had also led to weight loss and fewer sleepy periods, although decreasing glucose tolerance, a precursor to diabetes and a frequent traveller with Hashimoto's disease, remained a lurking danger.

What I believe I can say about the tennis training with confidence is that it provided stimulus in many areas which might otherwise have remained relatively dormant even with the administration of thyroxine. It gave James an opportunity to maximise his response to thyroxine, albeit at the tail-end of the first two years of this treatment. Potential improvement must also have been diminished by any permanent injury incurred during the period after his accident when his hypothyroidism remained untreated.

There was some increase in James's physical stamina, but it remained fairly limited. Although both our staminas improved after we had greatly reduced simple carbohydrates in our diet, my stamina improved well beyond James's. Even though I played infrequently, at the end of an hour I was keen to go on, whereas James usually wanted to stop before the hour was over.

Hitting tennis balls with me was not nearly so easy for James as doing it with a tennis coach, because the coaches aimed the balls so well back to the student. I served quite well from the start, but James tended to lob the balls too high and, if I actually hit one, my returns tended to go widely around him. One had to enjoy the challenge. It would have been hard to get worse, so we expected to improve. And so we did.

I played several more tennis 'games' (we never actually tried scoring) with James, however the unpredictability of my knees removed much of my enthusiasm. I found that I would go for weeks without pain, as long as I avoided a particular small pivot motion on the right foot. Every few weeks some situation would occur, however, where I would do exactly this motion and pull my right knee part-way out of its socket. The consequences varied from no pain after I forced the knee back into position, to a few days of pain and despair at the setback.

Another factor that tended to keep me off the court was the time factor. After several years James's legal affairs, involving many medical consultations and much form filling, as well as his formal rehabilitation, had come to an end. James's overall improvement meant that he did not need me to accompany him to tennis, singing or art therapy. I could now spend more time on writing, research, and art, as well as attending to the maintenance of two houses and two sets of financial responsibilities. James continued to attend tennis training sessions twice weekly at Overport Tennis club with Coach Cameron Milne, who had also taken over the filming of the sessions.

Peripheral Neuritis

One should never forget peripheral neuritis because by its very nature, it suppresses normal pain warnings of injuries to feet. What many people, including doctors, do not realise, is that it can be the first indicator of a deadly disease which has, however, a simple and cheap treatment. Unfortunately, James was a victim of this widespread medical ignorance.

The reader may remember that James wore his calf down to the muscle on an exercise machine at VicRehab in 2011 because his defective neural feedback failed to warn him of when only the skin was breached. Whilst James had been able to protect his feet before the traumatic brain injury, despite fluctuating peripheral neuritis, after the car accident, his lessened self-monitoring meant that he was at higher risk of foot injury than before.

But it was so easy to forget about someone else's feet.

Perhaps early in 2011, James and I had gone for a long walk along a beach, wearing gumboots. After about a kilometre, James said, "I suppose one cannot avoid getting blisters and foot pain when wearing gumboots."

"On the contrary," I replied, "Gumboots are really comfortable."

I asked him to sit down and he took one gumboot off, revealing bleeding blisters on his feet and ankles. He had not been wearing socks. I can only assume that the accident had erased the ordinary knowledge that one should wear socks with gumboots from his brain. He told me that he had not worn socks because, for some reason, he thought that one did not wear socks with gumboots.

After a few days it became obvious that the blisters were badly infected. I was unable to accompany him, so I sent James to the doctor with a note explaining the situation, knowing that something in writing signed R.N. would make the doctor more careful about examining James's feet and monitoring successful treatment. A course of antibiotics fixed the problem that time.

In December 2014 we were caught out again by peripheral neuritis. After his sixth tennis lesson, James mentioned that he had a dull pain in his heel. We went and bought new tennis shoes that day and he mentioned it again. We surmised that he had bruised his heel slightly. I did not think to look. It was not strong immediate pain like the one he had experienced with his gumboots, a hour into wearing them, and perhaps that is what made me discard the thought that it could have something to do with his peripheral neuritis.

Two or three days later we were in a small park with the two dogs, Harvey the miniature Pomeranian and Suzi, the miniature Tenterfield Terrier and our human friend, Jill. Harvey was off-leash. We were all sitting on the grass.

A woman entered the park with a young Labrador on a leash. Harvey, at an eighth the size of the Labrador, chased the dog out of the park and up a street. Because I was recovering from another knee incident, I could not run. I expected James to pursue Harvey, but James said his foot hurt too much for him to run. That left Jill, who could still run, fortunately.

After Jill returned with Harvey, we looked at James's foot at a bad angle in failing light. Jill thought it was probably infected, but I could not see it properly. James said it wasn't *that* bad and was against a long wait at the GP clinic that day. The next day I insisted because there was the beginning of a red streak up his ankle, indicative of severe infection. The doctor hardly looked at the foot and did not take in the issue of peripheral neuritis (which should have alerted him to a possible diabetic complication among other possibilities). He simply prescribed antibiotics, which I hoped would suffice. By the next evening James was in agony, but still did not want to go to the doctor again. The infection had worsened, with a wide red streak approaching the middle of the calf. It subsequently needed some basic surgery under local anaesthetic, different antibiotics, daily dressing and drainage and skilled bandaging for about four weeks.

The clinic GP with surgical duties organised to have James designated a 'complex patient' and to be allocated a specific practitioner. We were told that, as a complex patient, James had the right to regular access to a podiatrist. No-one was particularly interested in working out why James had peripheral neuritis or whether it was getting worse. Assumptions were probably made that this was a result of James's traumatic brain injury from May 2010, although no GP we met there seemed ever to familiarise themselves with the details of that injury.

But we would eventually discover that there was more to peripheral neuritis than risk of injury to feet and hands, or loss of balance and that the cause was ongoing. Although a test had been done that revealed the likely cause and there was a simple cure for the cause, the many GPs, several neurologists and other doctors whom James had seen for legal and personal health purposes, completely failed to take note of the results of the test and thus to treat James. Because of this gross systemic medical incompetence and negligence, James incurred ongoing brain and neurosystem related muscular-skeletal damage.

Property next door to James's sold and moonscaped:

James's father owned a house in Queensland next to James's house. It stood on stilts about 20 feet high, which had been badly re-stumped so that it had swayed ever since, making it uninhabitable. If you want to see what it was like, from the inside, there is a you-tube video of it here:

https://www.youtube.com/watch?v=a1V2l-eUbOY.

For years it had been completely concealed from the road by feral bamboo, planted by a bygone tenant. Beginning with an order to remove the bamboo, Brisbane City Council kept the pressure up. Sadly, the rating system meant that the property was costing too much to leave undeveloped in this already overdeveloped part of the world. With the help of another architect, one of his sons-in-law, James's father attempted to design a couple of apartments that could be built with minimum damage to the forest, which reached up the slope and dotted the site with old Chinese elms and various figs.

The task was very difficult because the land sloped so steeply and because any low building would receive very little light, due to the looming multi-story apartments across the road that blocked the sun, along with a built-up block next door. Potential buyers recognised these disadvantages. The rates must have been huge.

Sometime in September James's father broke the sad news that there was a buyer. On the 11th of December James's tenant told us the land had been moonscaped from the street right down to the gorge, with the exception of one tree near the road. That meant that established rainforest twenty minutes from Brisbane CBD had been destroyed. It was another depressing testimony to a zero-sum system of land-use planning that forced people to ruin their environment by clearing land, subdividing, and otherwise intensifying use in order to accommodate constantly rising prices caused by Australia and Brisbane's artificially engineered population-pressure.

We thought of all the animals now without homes: the spangled drongos, the kookaburras, the owls, the ringtail and brushtail possums and the brush turkeys. We pictured the forest floor funghi and the fig and black bean trees among the many others it had sustained, now steaming, dying mulch.

I remembered my first visit in 2006, when the council had first issued an order to remove the feral bamboo and how, for three days, Ian, a son-in-law and "Number One Son" James, had swung machetes in ferocious heat and humidity all day long, coming in for baths at the end of the day. I overheard Ian saying of James, "That boy sure can work. Once he gets going he just doesn't stop." Gradually the old house on stilts was stripped of its concealing bamboo forest revealing James's bedroom windows, so he had to curtain them for the first time. The geckos stopped chasing bugs on the exposed windows and went round to the other side of the house.

James spends two weeks without me in Brisbane Christmas 2014

Over Christmas 2014, James booked to fly up to Brisbane. I was due to join him two weeks later. It was his first time there without me in nearly four years. He stayed with his new tenant and could call on his father, who lived within a long walk. I anticipated James would run into some mild problems but survive with some help until I arrived two weeks later. My chief concern was that he would fail to prepare his own meals and to clean up after himself and so his tenant might feel obliged to feed and clean for him. Or that the two might not get on without me as a buffer. If that had become a reality, I would have had to have flown up there earlier than I had intended.

The dawn to dusk construction next door that we had dreaded, was temporarily suspended but another unpleasant situation arose. Just before James left, his tenant emailed us to say that the old multi-trunked black-bean tree had calved onto the back of the house, damaging the balcony and the kitchen during a terrible hailstorm. I contacted James's insurers who took a couple of weeks to get the fallen trunk and branches sawn up and removed, due to the competing needs of the many other houses around Brisbane damaged in that storm.

A day or two before I arrived to join him, James said he was 'just managing'. He had difficulty working out how to use a new mobile phone, so felt very isolated, although he did manage to figure it out just before I arrived. When I got there, it was very hot and James was walking extremely slowly, with drooping shoulders and a pale, blank face. I worried that his Hashimoto's (hypothyroid) disease was accelerating. After a week, however, he began again to stand up straighter, and to smile and walk more briskly. My conclusion was that he was fundamentally exhausted, as only a severely brain-injured person can be. Maintaining relative independence – sharing cooking with his tenant, finding clothing to wear, putting things away, very basic organisation without my help – simply used up all his available energy and left him none for fun. This is the essence of diffuse axonal brain injury; connections are very precious; you function much better with a helper to organise the basics in your life. It is lonely to have to work so hard on primary chores, hard to stay oriented, hard to get out of bed.

Brain injury and dealing with heat

There was a spate of very hot, humid days. We would wait until evening to walk to the Brisbane City Beach – an artificial lagoon off Brisbane River and swim. Now that there was nothing between us and the next house but a yawning canyon – both physical and socioeconomic - I could watch the neighbour's children in their beautifully landscaped swimming pool. I mentioned how I found this amusing to James and he said it made him suffer all the more because he could not help being jealous of them for having a swimming pool. Jealousy was very uncharacteristic of him and knowing this made him feel even worse.

James felt and looked so poorly that we went to his old doctor in Brisbane. She ran some blood tests to rule out low thyroxine, high sugar or an infection and concluded that he simply wasn't coping with successive days of heat and humidity. This seemed like a reasonable explanation. Brain-injured people typically lose some of their body temperature-regulating system.

I covered James with wet towels and pointed two electric fans on him until he felt somewhat better and kept repeating this treatment for as long as it was hot. We also used our friend Greg's swimming pool twice, standing in the water up to our necks for hours, chatting enthusiastically with Greg and his wife. It was amazing how, within minutes of immersion, one's fevered brain began to function again. We also began to drive instead of walking to the Brisbane City Beach artificial lagoon. It seemed that the seasonal heat wave was actually the main cause of James's deteriorated state. We would learn later that there was another explanation.

What carers mean by 'stress'

I also found the heat hard to cope with and experienced overload similar to James's.

Formal rehabilitation, neuropsychological tests, visits to specialists and form-filling had all ended with an amount of money that had to be wisely invested. It took me several months to research then organize this. We consulted five banks and two 'independent' financial counsellors. Finally I obtained a phone appointment for 'financial information' with Centrelink (the Australian Government social security service). This service cost me nothing but confirmed my impression that the lowest fees and best returns were in worker superannuation funds. Despite having heard of these, I was still amazed at the extent of hidden costs in the bank financial investment services. As well as the up-front fees that the financial managers charged, the funds they then invested a client's money in also charged a percentage for 'managing' that money. A client could finish up with as little as two per cent return - in the belief that they were invested at 6 or 8 per cent - as agents in every investment took their cut – like an old fashioned hierarchy of bureaucrats that a person had to bribe before they would act. I was still trying to retrain James to be conscious of budgeting and there was a constant tendency to overshoot his modest real income, but he was managing overall and the future looked reasonably secure.

Unfortunately, after two weeks in Melbourne without James, attending mostly to my own needs, I had to confront huge problems of partial insurance refusal and the need to repair the severe storm damages involving the split black bean tree to the back of James's house on his limited income. The insurers paid for the removal of the fallen trunk, but their advisers regrouped on the question of repairing the balcony, built as a kind of two-legged lean-to against the house, now hanging knocked-kneed from a post that formed a junction to several important supporting beams for the main building.

The house sits on stilts on a steep incline, so repairs meant perching on long ladders or paying for scaffolding and builders. It was built around 1891 and contemporary Brisbane building codes tended to question its very existence.

The insurers eventually decided that they would pay part of the damages, but their quoting builder then decided he did not want to do the job. We delayed leaving Brisbane whilst we tried to find other builders. Four people promised to come. A building maintenance manager came and expressed sympathy but did not think he would be able to convince his builder to have a look. Only one actual builder came. He stayed for a moment, then fled back up the stairs throwing over his shoulder that he would send his quote. We never heard from him again.

The cost of repairs seemed an elastic and increasingly scary proposition, only preferable to the possibility of the house falling down, which seemed to grow in light of the reluctance of builders to attend the matter. I was anxious about being delayed in Queensland away from my elderly parents and the dogs. I had booked electricians early the next month to rewire the house in Melbourne. I had daily duties meeting the demands of the website and of working on various books and articles. I found this breadth of responsibility and financial juggling mind-boggling: That is, there was just too much for my brain to deal with reliably. I would wake at night worrying about the consequences of failure to give all matters close attention, knowing that it was beyond me. This kind of responsibility is probably what many carers mean by 'stress'. It is intellectually and emotionally stretching. It makes you want to run away, but run away you cannot, because so much depends on you.

In desperation I emailed Rob W., an architect in Sydney, who had house-sat James's place for six months in 2010-2011, in case he knew of any builders who might deign to take on the job. Rob emailed back that he was coming up to Brisbane the next month and would have a look at the problem for us at no charge. In the meantime he asked me to email the insurer's engineer's report and photos of all the stumps under the house.

In the remote case of a builder consenting to work there after we left, James and I hired a skip and began carrying loads of rotting wood and broken furniture and other rubbish from the back yard in anticipation of it becoming a builder's site. As we were doing so, a large jovial man who we assumed was the owner called out to us from the side of the magnificent house and pool across the canyon. He said he had a friend who wanted to buy the pile of old bricks at the side of our house. He offered a dollar a brick.

But when we asked him, James's father said he was very fond of the bricks and would not sell. James wrote a note to the owner of the house across the canyon explaining this and dropped it in his letter box.

Before breakfast a few days later a younger slimmer man – Damien - knocked on the door and asked about the bricks. It turned out that he was the head of the team of carpenters building a superior-looking deck over the superb L-shaped swimming pool at the magnificent house across the canyon. He wanted to construct a wood-fired pizza oven with old bricks and the bricks at the side of the house would be ideal. The real owner of the house, who it turned out we had not met, had not passed James's note on to him.

We naturally fell upon this young carpenter and begged him to have a look at our semi-collapsed balcony. He had a look and said that the task would probably not take more than a day and his team had all the equipment at hand. He charged by the hour and would use two or three men. But he really wanted the bricks. He offered to exchange four hours of labour for the pile of bricks.

We explained again about the problem of ownership of the bricks, saying we would rather pay in dollars. Damien looked wistful at this. He said that he was not sure when he would be able to fit us in… maybe some time in the next week or two.

Two weeks went by and the time came for us to leave for Melbourne, where we had a deadline with James's tennis coach who had been booked by a commercial tv channel that wanted to do a special on brain injury and tennis.

I left messages for the third time on Damien's phone answering machine. He rang back when we were on the road to Melbourne via Sydney and we exchanged further details, including email addresses. As soon as I got back to Melbourne, I sent a detailed email documenting our requirements and formalising our desire to engage his carpentry services.

In Sydney we visited James's uncle and stayed at Rob W.'s house overnight. In the morning Rob uploaded the photos I had taken of James's house. When I mentioned that the house was insured for $880,000 to rebuild in case of an accident, he told me it would only take about $175,000 to rebuild and was totally overinsured. I had not known any better. In fact the insurer's agent had told me we needed to upgrade the insurance. The annual payments had been so steep that I had negotiated a $5000 excess. This excess was so excessive that the insurers had needed to check in case it was a mistake when James made his claim for the fallen tree-trunk damage.

Back in Melbourne, a week later, James reminded me of some unopened mail from the insurers. I had supposed it was simply hard copy of email correspondence we had already received. I had a careful look now and gasped. The insurers had sent James a check partially reimbursing his premium because they had unilaterally cancelled all his house insurance. Their grounds were that the house was wildly overinsured, that it would not cost more than $200,000 to rebuild. Futhermore, they implied it was uninsurable due to lack of maintenance and or non-adherence to current building regulations. I tried to get them to tell me what they would require in terms of repairs and maintenance to make it insurable, but they refused to be pinned down.

I emailed Rob about this, asking if, when he went up to Brisbane, he could consider what might be necessary to make the place insurable. He came back with an affordable estimate for restumping the house and reconstruction of the front veranda out of contact with the soil to deter termites. I figured out that I could loan half to James and he could close his slush-fund to pay the rest. In this way we should be able to preserve the capital in his superannuation that pays his small pension. The $200 weekly rent he gets would pay me back within a few years, but James would have to go without that $200 weekly, half of which goes in rates and maintenance anyhow. You never knew, we might make some money out of this book, but really, we needed a second tenant, which would mean that James and I would not be able to visit the house for a few years. And the amount estimated did not cover a coat of paint for the house or rebuilding the bathroom. We still had to get actual quotes.

But the problem of the balcony hanging off the house was still not fixed. After investigation, I found that my email to Damien the carpenter had bounced. I tried another spelling and this one didn't come back. The next evening James's tenant emailed me to say that the carpenters had just repaired the balcony. He commented further that they had done something strange: they had carried all the bricks from the back of the house to the front and had taken most of them away. The whole job had taken about four hours.

One year later, repairs under the house had not yet proceeded. As I prepare this book for print, back in Brisbane, we have three sets of plans which vary from repairing some stumps to total replacement. It is not that simple.

UPDATE AUGUST 2015

Early in 2015 I came to the conclusion that James was pretty near as good as he was going to get, as far as recovery from his brain injury went. I remember giving up on the super-longshot that he would somehow recover most of his abilities and that I would regain the companion who could read and proof-read my work, who could add up many figures in his head, who knew most scientific definitions, measures, examples and dates, and could even carry my fat little person up a hill if my strength faltered.

How wrong I was. If James could not be restored to his previous very high levels, there was still plenty of potential there, dwindling for lack of proper diagnosis and treatment.

Pernicious Anaemia and Peripheral neuritis

On Thursday 14 May 2015 James's endocrinologist informed us that James was suffering from pernicious anaemia (a chronic inability to efficiently absorb from oral intake of vitamin B12) which meant that he would need to have Vitamin B12 injections regularly for the rest of his life in order to avoid dementia, paralysis and death.

One of the signs of pernicious anaemia was peripheral neuritis. I had not known this and I knew next to nothing about pernicious anaemia.[82] I did not realise it was the result of a deficiency in Vitamin B12 (cobalamin). When I began to understand what it was and the damage a deficiency could do, my blood boiled for about three days and nights as I had vivid flashes of what might have been for James and what had been stripped away by the empty sophistication of insurance-tailored medicine.[83]

James had suffered from peripheral neuritis of mysterious origin for years. The first time I realised that James had peripheral neuritis was probably around 2006 when I visited him in Brisbane for the first time. I remember him mentioning to me how he preferred to wear shoes rather than sandals because it was easier to walk. Although he had enjoyed walking around in bare feet as a child, he no longer did. He explained that he had a mysterious problem where the sensation in his feet waxed and waned.

Although peripheral neuropathy or neuritis is a well-known risk of diabetes, after he received a warning about his blood results indicating a pre-diabetic state, James changed his diet and rode a bicycle to work daily and his blood results returned to normal. I did not think that his pre-diabetic tendencies could have been sufficient to have caused peripheral neuropathy. As a psychiatric nurse who had started work in brain injury wards in the 1970s, I knew about peripheral neuritis as a feature in Wernickes encephalopathy and Korsakov's psychosis, diagnoses associated with alcohol toxicity and B Vitamin depletion, notably Vitamin B1 - thiamine. I also knew that heavy metal poisoning could produce the problem. And I had learned in a union course that many substances used in factories, such as degreasers like trichlorethelene, could cause peripheral neuritis as well as the brain injury involved in Wernickes and Korsakov's. Since James hardly ever drank alcohol and never took any other mind-altering substances, I supposed that he might been exposed to some other toxin, perhaps in one of his factory jobs, in some unspecifiable and distant event.[84] I hoped that he would recover from the problem, which did not seem very troublesome.

He had a referral to a neurologist and was trying to find time to organise to see the neurologist when he was hit by the car. I read the referral soon after the accident and the doctor had suggested that James's 'glove-anaesthesia' might be neurotic in origin. How misleading that letter was.[85]

In 2011, months after the accident, we attended a Melbourne neurologist to try to find out what the cause of the peripheral neuritis was. We wanted it to be diagnosed and treated if possible. The neurologist tested James quite extensively and took blood samples but failed to make a diagnosis. In retrospect, this was incredible, because one of the blood samples showed that James's B12 level was 220, a low level that protocol said required treatment in the presence of neurological signs.[86] How could a neurologist have missed this?

It was not until the endocrine specialist that James saw for his hypothyroidism (Hashimoto's Disease) that James was tested for Pernicious Anaemia. One of the indications of Pernicious Anaemia is low Vitamin B12 in the blood and neurological problems. James' B12 was low. The specialist then tested him for intrinsic factor antibodies. He tested positive for this and was diagnosed with Pernicious Anaemia. We were told that James would have to have Vitamin B12 injections for the rest of his life at a rate to be determined by his response to them – usually one monthly. We were referred by the specialist back to the GP Clinic for this treatment. The GP we saw wanted to give James four weekly injections then stop for one month, test and start monthly injections. However I had read on the subject and found that it was recommended that injections be given every second day where there was neurological damage. I said this to the GP, who said that this did not seem to apply in James's case because the damage was old. This was an illogical assumption because, if James had pernicious anaemia, then damage had to be ongoing. I argued that James's state was still fluctuating and that we should treat him intensively to see what could be retrieved and repaired. I unfolded the information accompanying the NEOB12 injections in the packet James had been instructed to provide for the nurse to inject and showed the doctor that it recommended injecting every second day for two weeks. The doctor then accepted this.

It then occurred to me to ask whether James had ever had a previous test for B12 levels. The GP had a look and, with some disquiet on his face, said that James had first been tested in 2011 for B12 by his neurologist. The tests had shown low B12 at the time and the GP clinic had received copies.

So, the neurologist and the GP clinic had had the means to diagnose and treat James for this crippling disease that demyelinates spine and brain, and ... they had done **nothing**.

After reading more about pernicious anaemia and B12 deficiency, especially how the risk increases after the age of 60, I asked whether my mother – who had very suddenly demented in 2012 – had ever been tested.

She had, in 2012, and her results had been just over 200pg/mL, lower than James's. Looking back, I wonder whether clostridium difficils (a gut bacteria overgrowth) which developed in response to broad spectrum antibiotics for a chest infection preceding her memory loss, destroyed the intrinsic factor that permits absorption of Vitamin B12. Like James, Mum was eventually diagnosed with hypothyroidism, after suffering symptoms for years. This was treated very conservatively. She had also, in the past, been diagnosed as diabetic, although this diagnosis had been overturned later. (High blood sugars often accompany thyroid disorders).

After she spiked temperatures and lost her memory overnight and permanently, I had taken her to the GP clinic asking that she be hospitalised as her condition had worsened and had been told that it was 'dementia, and you have to accept it'. Apparently it did not matter that she had high temperatures, since she was old. In other words, because she was old, she was not entitled to a rational diagnostic procedure. Another factor I would eventually understand, was that the GP Clinic employed foreign doctors on contract and these doctors lacked their own 'provider numbers'. (Most overseas trained doctors are subject to Medicare provider number restrictions.) This meant that they could only work under the supervision of Australian doctors with provider numbers. They could not refer people to a private hospital or make difficult diagnostic decisions. They were obliged to toe in-house lines on treatment decisions. No wonder so many of these doctors seemed depressed and resentful.

And that was not all. It turned out that my father also had B12 deficiency. He had developed severe peripheral neuritis, with footdrop, and walked on 'tippytoes'. His balance had deteriorated, requiring him to use a stick. His doctor (same clinic) had told him that this must be due to spinal stenosis arising from broken vertebrae when Dad was in his early 30s. Suddenly, after an athletic life with no back pain!

For many years I worked with doctors as a nurse in hospitals. I also met doctors from many specialties when I helped James to negotiate a legal settlement for his injuries. I did not think I was under any great illusion about the quality of medical treatment. Medicine seemed to be a group-think occupation where most practitioners were terrified of breaking rank. However, after experiencing the total failure of any timely diagnosis of pernicious anaemia over years of obvious symptoms and actual indicative test results in three people, I have come to the conclusion that western medicine is profoundly flawed. I can now understand why people with serious illnesses run to 'cranks'. In many cases, they might just as well.

B12 treatment for Pernicious Anaemia and electrical brain stimulation

Diagnosed on 14 May 2015, treatment began a couple of weeks later. At time of writing James has been receiving hydroxycobalamin injections for nearly four months, initially every second day, now weekly. I could detect an improvement in his ability to take in verbal information and to keep up with me when walking within the first two months. I was also very impressed by a pencil self-portrait in a beanie that he did at Artwell in July 2015, which I have included in this book.[87] I think this may have inspired me to rejoin the McLelland Artists' Guild in Frankston, and we both started going to their portraiture self-help group. James seemed to have no problems with three hours of concentration on the model, with short breaks. He has applied to attend a portraiture class, partly to explore new techniques.

In the third month of his treatment with B12, Monash Alfred Psychiatry research centre, where I made contact a few years ago in the hope that he might get access to new treatments on an experimental basis, and thus to specialist advice, contacted James and engaged him in a clinical trial. The trial involved 'Combining transcranial Direct Current Stimulation (tDCS) with Cognitive Training for Cognitive Impairment following Head Injury. Although the trial involved a control group, James was told at the end of the trial, that he had been part of the 'active' or 'real' group. Here is a short report from the research centre.

James Sinnamon's self-portrait 8 July 2015

"As part of the trial we conducted cognitive assessments at the beginning and completion of the treatment course. At your initial assessment you performed largely at expected levels for your age, with performances within the average range for speed of information processing, attention, working memory and reasoning and problem solving. You had some difficulties with learning both verbal and visual information, performing below expected levels for your age. Following the treatment course your performances on tests of speed of information processing, attention, and reasoning and problem solving all improved slightly, remaining within the average range. You showed a greater improvement in your verbal and visual learning, both of which were in the average range at your final assessment."(Email for James from Researcher at Monash Alfred Psychiatry research centre, mid August.)

He had to attend the hospital five days a week for a month and I was impressed that he was able to do this and still manage to do a little at home as well. I felt that his memory improved slightly and, particularly, that he was more able to plan and follow through, where a little while before, fatigue would tend to interrupt any projects. I had hoped for such an improvement with the B12 injections, but suspected that the clinical trials also fed into it. About a week after the clinical trials, James felt weak and had to go to sleep in the afternoon a couple of days running. I played a game of tennis with him and he was too tired and slowed to keep up a volley for long, actually missing easy shots.

We kept an appointment with his endocrinologist, who told James that he had now clinically crossed the line from pre-diabetic to diabetic. James was to monitor his blood sugars now several times a day. The doctor also increased James's thyroxin dose. James's blood glucose baseline is too high, even though it does not change much after meals. He has increased his exercise, going for walks daily or more frequently to see the effect on his blood glucose.

I continue to give him hydroxycobalamin injections (B12) and am considering accessing the harder to get methylcobalamin (methyl B12) injectable because it is further up the chain in vitamin

B12 processing and has been linked to the reversal of some long term neurological damage.

Three other things I have noticed are that James is able to relate recent events in far more detail than prior to the hydroxycobalamin injections and that he now serves me tea in matching tea cup and saucer. We have several giant cups and saucers in horse, dog and flower themes. Before the B12 injections, the level of detailed observation required to match themes in cup and saucer seemed to require so much effort that James constantly overlooked it. I was used to getting horse cups on dog saucers and pastel coloured tulip cups perched on garish fields of red poppies. Suddenly I began to get coordinated cups and saucers. At the same time James's study is much tidier, the couch is mostly clear, and the dogs and I can join him there to watch the RT and Iranian PressTV news on his computer.

Late in 2014, we had given up on the immediate project of James trying again to install Linux and download a working copy of *candobetter.net* onto a new PC with a 2010 Microsoft operating system, due to inbuilt new obstacles. Once James might have worked out how to overcome these problems, but now it seemed too difficult. Abandoning the computer to collect dust, we recontacted VG Computers and asked them to quote us for migration of candobetter.net to an ordinary PC with a CENTOS operating system and a second boot to Linux Mint.

On 28 August 2015, about three and a half months since B12 treatment had commenced, James succeeded in removing Microsoft from the hard disk on the computer he was told would not accept Linux. He had initially attempted to install Linux via a cd and failed. Perhaps then his impoverished problem-solving abilities had relied on remembered routines from before USB memory drives and had prevented him from a flexible contemporary review of the situation. In the meantime, his brain function had begun to improve. Now, about eight months later he went back to the problem and realised that he should use USB memory drives instead of CDs. He attempted to install Linux mint on a new partition he had made, in order to preserve Windows, but Linux mint refused the user-defined partitions, so he just allowed it to install it with the default method, in which the whole disk was used, unpartitioned. Windows was destroyed but he had what he wanted, a functioning Linux system on a new powerful computer.

The other day he downloaded Drupal and Mysql onto his new computer, with the intention of revisiting the creation of a testbed for *candobetter.*

In the meantime the work done by VG computers has turned out to be useless and inaccessible, despite a price of around $2000. We had asked for candobetter.net to be downloaded on the computer in a manner whereby we could update it. We since found that there was no graphical interface provided for the computer and have not been able to install one.

A couple of years ago James bought a scientific calculator, but found he was only able to use its most basic functions. As he explains to me, he could not work out how to store and retrieve values from its memory or find the arbitrary numeric root of another number, for example, the cube root of 27, or even roots that include fractions. Early in September 2015 he discovered that he could now use the trigonometric and logarithmic functions and process fractions.

Chiropractic update

Since receiving treatment for pernicious anaemia, James has responded to chiropractic treatment from George Cipurovski, in Mt Eliza Chiropractic, for two persistent injuries which we assumed were associated with his left sided weakness (which harks back to his traumatic brain injury.) Mentioned previously in this book was an injury to his shoulder which we linked to activities undertaken in a gym and which another chiropractor, Will, had tried to fix. James had a second problem which I may not have then described separately, because it became more noticeable as time went on. This was pain in his left knee if he lowered himself into a squat, and meant that he had to get down on all fours rather than squat down to pick something up from the floor, or do weeding, for instance. It also made it hard for him to rise. Sometimes his leg nearly gave way unexpectedly. He began to worry that the condition might continue to worsen until he would have difficulty in merely walking.

Although the first chiropractor, Will, had made some difference to James's posture and gait, the improvements did not last. Will cannot be held responsible for this because, since James's Hashimoto's disease and pernicious anaemia were still undiagnosed and untreated at that time, there was a point where the healing from his traumatic injury would meet the unsuspected but inevitable deterioration in his neuro-musculo-skeletal system, caused by Hashimoto's disease and Vitamin B12 deficiency (caused by pernicious anaemia). Certainly the two were intersecting or on a collision course at the very time that Will was attempting to correct James's posture and left-sided weakness problems.

Before the diagnosis and treatment of pernicious anaemia, but after the treatment of hypothyroid (in Hashimoto's) both the left shoulder problem and the left knee problem became more noticeable and interfered with James's enjoyment of tennis. James started going to a physiotherapist at a sports clinic. After multiple $80 visits and trying to perform exercises meant to feed back to his brain on the assumption that the shoulder problem was the result of the left-sided weakness, James ceased the exercises because of the pain. The knee problem also persisted.

When James was diagnosed with pernicious anaemia and treated with Vitamin B12, his improvement was visible within days, most notably in the increased speed of his walking and, over a few weeks, greater endurance. But he was still complaining of his shoulder problem and his knee, although with philosophical resignation.

If the knee and shoulder were caused by old neuronal damage and they had not been helped by thyroid treatment, could they be attributable to some degree of demyelination or its beginnings from pernicious anaemia? Will, with his love of wild spaces, had left the Mt Eliza Clinic and Victoria for a better life in Tasmania, our greenest and least populated state, home of the Tasmanian devil. I made a visit to the same chiropractic clinic for a problem of my own and had an appointment with George Cipurovski. He asked me how James was doing and, when I told him, indicated that he would be interested in trying his hand on the problems now that James was receiving treatment for two relevant diseases which had not been diagnosed when Will had been dealing with him.

When I brought James for an appointment, I was amazed that George vastly increased James's arm mobility almost instantly, working on his pelvis. As James was rising from the couch, he mentioned his knee problem, saying he wished that could also be fixed by chiropractic. George next worked on James's knee. After about two sessions, that problem totally resolved. James's shoulder mobility is about 85-90% fixed. Where he had been unable to raise his arm above his shoulder without quite severe pain, he could now raise it almost vertically. There is still some discomfort at the highest point, but nothing like the previous pain. James also walks better, holding his head straighter, but he still has a tendency to stoop, especially when fatigued and the left-sided weakness is visible in a left-sided shoulder-droop when he walks in a fatigued state. My own experience, as related in a endnote about my knee, is that thyroid illness often also manifests in joint and muscle problems and responds to treatment.

Another complicating factor which is still being investigated is the nature of his diabetes and the best treatment for it. It looks like he has autoimmunity to his pancreas, which means that he is probably developing late onset adult diabetes, rather than a diet-related problem.

Mortality and inconsequence

The changes to my life and to James have made me very aware of my mortality and of my inconsequence vis a vis the universe and time. James experienced this sense of mortality at a fundamental level – when he nearly died and when he became aware that his mind was changing. Mine has been more a philosophical and perhaps age-related realisation. It is profoundly humbling. My comfort is life itself – living nature - as the only force that combats thermodynamic dissipation. Lately I have come to understand better James's concerns about and the increasing distance between objects in the universe. That will make some readers smile – or snort – it sounds so arcane, I know. I wrote this looking out on the magnificent Morton Bay fig-tree and its community of mature Australian palms in James's Queensland backyard, that stand so bravely, linking times past and times future, against the general destruction wrought to this beautiful natural environment by successive soulless Queensland governments. I wonder whether the self-centered and economically religious leaders we now have will one day realise that they too are mortal and regret that they wasted their lives destroying life and bullying nature in hopes of appeasing the gods of the 'market'.

At the same time, my artistic intellect temporarily dimmed my horror at the moonscaping next door with a fascination for its revelation of James's house like a damaged dollhouse in profile backed onto a slice of jungle on a cliff. Instead of rebuilding the balcony, we are trying to plan financially to landscape with new soil and replant James's back yard which has for years been used for the storage of relic building materials.

While the heat that sapped our strength and we struggled sweatily to move an accumulation of old furniture and wood up the hill to a skip at the top, I could not help noticing the many species of lizard inside and outside the house who were enjoying precisely this weather. It became a game to count the number of tails disappearing as I entered a room, to note the sizes and markings of various couples of lizards in love on the lawn, and to smile at the excited progress of a gecko stalking insects too small for me to see, or with its head blurred by the movement of a large flapping moth in its jaws. I realised that there was this whole substrata populated by smartly patterned lizards awaiting my investigation on our next visit to Brisbane.

Back in Melbourne I took a load of laundry out to the washing-line and was suddenly struck again by the activity of the birds in the back yard that my mother carefully cultivated to be wild and bushy before she became an invalid. Harvey, the miniature orange Pomeranian we rehomed three years ago blinked in a sunbeam at my feet. With high fences all around it is our island.

The mystery of James's traffic accident

From the time James landed in hospital and various people heard about it, we all wondered, of course, how the accident had come about.

The question we framed without saying so was more mystical: What quirk of destiny had produced such an apparently meaningless tragedy?

For months I was under a misapprehension that it had occurred on the opposite side of the freeway from the usual route I knew James took. My confusion was the result of people pointing from James's hospital bed to indicate where they thought the bicycle track was, and this seemed to me to be far from the track I had walked along with him on a couple of occasions, and which I regularly walked to visit him in hospital. A few weeks after James became an outpatient we went to see a solicitor I had engaged and the solicitor suggested that we walk the route of the accident to see if any memories would be stimulated. No memories were stimulated at the time, but many questions were. The police investigator had interviewed James when James was still an inpatient, wildly and somewhat hilariously confabulating in his pyjamas. She emerged no more enlightened. She seemed to feel at that stage that no-one had been at fault – except maybe a wire fence which had been overgrown by a vine creating a blind corner – and whoever was responsible for maintaining it. Perhaps one or the other or both driver and bike-rider had been too close to the middle of the road, which anyway had no median strip. But she did trail a slight question as she left, "What was James doing on the road instead of on the bicycle path?" The obvious answer, that the bicycle path must have been blocked, seemed for some reason to find less favour than the idea that James had made an illogical or dangerous choice to ride on the road instead of on the bike path. The reader should keep in mind that there was nothing illegal about riding a bicycle on that road.

Even though James didn't really grasp what the interview was all about at the time, the message slowly sifted through his synapses over the next day and he became really upset at this idea, which he felt was entirely uncharacteristic of his cycling behaviour.

There were of course other reasons to find out where, how and why he had been injured, and these were legal and practical – for James needed expensive treatment and an income and the only way to get this was to pursue any available party for damages.

Unfortunately for readers who were hoping to educate themselves on the law and financial settlements in similar accidents, the outcome of these matters is outside possible discussion in this book.

It is high time there were books by lawyers for the public telling people what their rights are and what they can expect in terms of financial compensation, blame and evidence. Financial compensation should be a matter of public record, although perhaps names could be disguised. There are, of course, numerous websites touting various legal firms and barristers for accident claims, but there seems to be nothing for the general public that would give a good overview of recent cases and financial outcomes in different states of Australia. One reason for this is probably confidentiality clauses in actual settlements.

The concept and reality of a traumatic brain injury epidemic, mentioned earlier in this book, in connection with increased car use, speed, commuting distances, traffic and population, highlights the public need for a more transparent system.

Grief

Editing this book has been complicated by the fact that I am in it as well as James. I usually dodge self-revelation, preferring to record things outside me rather than to confess. Will I appear a rather cold and arrogant person, dealing with James apparently at arm's length, as if he were an academic subject of research?

Did I never cry? What were my emotions?

About three years after James's injury, I was driving alone when I suddenly remembered getting out of a taxi on the night of May 19th 2010 and walking, dragging my little suitcase on wheels behind me, to the hospital where James had been admitted the day before to ICU. The front doors of the hospital were closed, so I went round to Emergency, where I expected to find a waiting-room full of patients and an irritable ward clerk attending a queue at the desk. Instead, the rows of chairs were sparsely occupied and there was no queue at the ward clerk's window. It was as if I were the only person in the room. The woman at the window looked up at me and smiled kindly as if she had completely understood why I was there. I said, "Can you tell me how to get to the ICU?" She said, "Certainly. You go to the front entrance and you press a button. The doors will open, then you take the lift." Inexplicably, every time I think of this I cry. That is just about the only time I ever cry. It makes me cry to write about it. When I stop thinking about it, I won't feel sad again. It is like something I can open and shut like the cover of a different book from this one.

Once, under questioning from a very kindly compensation psychiatrist, who asked me to describe James's treatment in hospital, it was as if I turned over the first page of that book of my emotions. I began by telling him that of course, I remembered James's treatment in detail. Then I cast my mind back to the beginning, where I met the kind ward clerk, and then to standing at the doors of ICU. I was suddenly flooded by a rapid series of impressions like speeded up film frames and by an immense and crushing sadness that threatened to wipe out all my present environment and remove me from actual time and space. I sensed my eyes filling, but we had come late for this crucial appointment and I needed to give as much information as I could, so I pulled myself back from that very strange alternative universe of profound emotion, and related the facts.

An unfinished story

Ending a biography of someone still alive is difficult. The story continues. I tell myself that I will write a sequel if this book arouses any interest, and I probably will. I know from experience how important books like this are to people who are dealing with a brain-injured person. I have read nearly every one that has been recently published. The book was written for James so that he would have a record of what had happened. It was an attempt to capture lost time for him. As I mentioned early in the book, he wanted people to know what had happened to him and that he still existed and is living as well as he can. We also really hope that this detailed record of our experiences will help someone else who may be struggling through similar situations. Please get in touch with me and James with any feedback or questions at astridnova@gmail.com and countershockpress@gmail.com. And check out https://candobetter.net and a possible new site modernheresy.net, something we are planning in 2017.

END

APPENDICES

APPENDIX ONE: Films of James Sinnamon before the accident

You can see a couple of videos of James shortly before the accident, in 2009, to do with his campaigning. They are "James Sinnamon's privatisation speech at Dave Zwolenski's campaign opening," https://www.youtube.com/watch?v=yC4qGs0epUg and three subtitled videos of James interviewing the Queensland State treasurer Andrew Fraser at a community forum. These are subtitled due to the very noisy fan placed, perhaps on purpose, in the near vicinity. The first one is here: https://www.youtube.com/watch?v=EaMYQRGM5Yk . If you are interested in seeing James just having fun, I have also uploaded a short poorly edited compilation of fragments from films of James and me from 2006 to 2009, in Queensland, at: https://www.youtube.com/watch?v=ISWSTSsu0Zs&feature=yout u.be

The most part takes place on the Obi Obi Creek in Queensland on James's sister and brother-in-law's property. It shows James and me enjoying the creek and you can see how well James looked at the age of about 48. I was 55. The films with the baby possum are a record of how much James liked to hold and help little creatures. The last clip, about a burned saucepan is up the hill from the creek, at a time when James was particularly confident and energetic. I suspect that his thyroid was running hot during this period. In fact I took hours and hours of film in company of James, but much of it was of our surroundings and James only featured sporadically in these films, sometimes lengthily doing little. There is a film over an hour long of James and me having tea in a downpour on a ledge over the Obi Obi creek, filmed from a static tripod, which I have obviously not included. When I look at it I am reminded of how ill I felt at the time. As time goes on I feel that the gap between the James of 2006-2009 in the films and the James of 2016 is lessening, in part because he has lost weight and gained muscle again, but also because he has kept his values, his sense of humour, his openness of character and his kindness and sweetness, which were his most attractive qualities. He may not be able to do what he could technically, but he is still able to assess things logically and to provide moral support, as long as he is not rushed. I also find, when looking at such old films, that I am able to see James more three-dimensionally; to remember what he can no longer enable about his mind, but which I sense is still there, in his smile, in his eyes, and which forms the basis of his trust in me.

APPENDIX TWO: Correspondence with RBWH

2 July 2010

Dear RBWH Feedback,

Would you please help expedite the forwarding of requested MRI scan on CD to Dr R- of the Mater Private Rehab Hospital, Annerley Campus, 41 Annerley Rd, South Brisbane, 2747, and to myself, Sheila Newman, 60 C- St., Red Hill, QLD, 4059.

We had a meeting following my first contact with you (see my email below this one) and it was agreed to have an MRI scan done on James Sinnamon.

It took two weeks following that MRI scan for me to be able to get information about the results, but that information was restricted to a conversation with a resident - Chris (very pleasant and very helpful and very kind but his information was limited) and a very short report.

On Tuesday 29 June, James Sinnamon was discharged to Mater Private Rehab as an inpatient. I had expected scans to be provided to me to take with us, but this was not done. I therefore asked B-L, the Unit Manager for 8AS to have copies on CD sent to the registrar treating James at the Mater Private Hospital and to myself. I was told that there might be a charge for this, of around $50, which I said I was willing to pay. B-L also said to ring on Friday, in case she needed reminding. I understood at the time that sending a CD to Dr R- of Mater Private was in hand.

Today is Friday. B-L is on leave. Apparently there is a message about the CD in the diary. A person I spoke to about this said that I could phone again on Monday.

This is not good enough.

That CD has much more information on it than the very short general report that both Dr Rois and myself have received. It probably contains information about damage to the brain stem, for instance, involving James's balance. It would also have more information than the report contained about the locations of axonal damage in the left side of the brain, affecting social behaviour, for instance, and the corpus callosum transmission of information from left to right. And more.

When James was examined at the Mater Private I had the impression that the found more damage to his balance than they had been led to expect. I was surprised to see that he displayed Babinski reflex, therefore my desire for detailed brainstem info from MRI.

I am asking you in this email to have this matter expedited. James has already been at the Mater Private for nearly a week. The staff there deserve the best information available in order to treat him. The delay is wasting valuable time.

Apart from that, please be aware that, despite my complaints, I am grateful for the good work of the ward staff and the residents and the Social Worker, Veronica. Veronica has been outstanding. The nurses and the assistant nurses were very professional and nearly all of them were very very kind despite the safety and observation limitations I perceived due to their training and lack of authority.

Please also accept my thanks for your professional attention to the earlier matter. (I am cc'ing Veronica M., Social Worker 8AS for her information since she was present when I raised the matter with B-L on Tuesday.)

Sincerely,

Sheila Newman, NOK for James Sinnamon

>>> Sheila Newman 27/06/2010 9:46 am >>>
    Sir/Madam,

James Sinnamon is in Bed 21 8ASouth. He was admitted with a head injury to RBWH on 18th May, subsequent to being knocked off his bicycle by a car. He is employed by your hospital and is on Workers Compensation.

Firstly, I want you to realise that I do understand that head injuries cause confusion and various forms of agitation, indeed many different behaviours, that every head injury recovery is different etc. However I have read widely and asked questions of doctors and nurses with knowledge of head injuries in Melbourne in the last two days, who agreed with me and encouraged me to voice my strong concerns.

I am concerned about 8ASouth response to James Sinnamon's slow progress and ongoing confusion, the fact that he has not had an MRI for this, only a repeat CT scan, and the fact that the neuroleptics he is receiving from time to time may be exacerbating the situation in that they could be a cause of what the doctors describe in a loose manner as 'agitation' but in fact involves a series of repetitive actions which I would call akathisia, and which my colleagues would also call akathisia, which from decades of psych nursing I know well as side effects of the drugs he has been on and which he continues to get from time to time: haloperidol, risperidone and olanzapine. Despite older advertising around the 'atypical' antipsychotics they do - each of them - cause akathisia. Not always in everyone but sometimes in some people. It does not matter how small the dose in a sensitive person and brain injury makes a person much more sensitive to these drugs.

Today I rang the ward at 3.30 am and asked about James's behaviour that night. I was trying to verify what exactly was justifying the ongoing claims that James is 'agitated'. Apparently he had been restless earlier that night. I asked for a description of the restless behaviour.

Here is what the nurse assistant said: "He gets up and out of bed, asks for a drink, goes to the toilet, then goes to bed, then gets up again to the chair, then gets up and goes to the toilet or back to bed - all night long until he goes to sleep. Also, if we don't let him get out of bed, he repeatedly sits up, gets on all fours, kneels, then flops on one side and lies down again, only to sit up, get on all fours, kneel, flop down on one side, lie down ..." (almost endlessly.) That is, he tries to increase the area and activities in his circuit to make them more meaningful, but if confined, he continues to cycle in a circumscribed fashion like an animal in a cage because he cannot stop this.

This is exactly what I have observed and what other visitors have observed and what we have recorded. I have not seen any literature describing anything like it in acute brain injury under the heading of 'agitation', except where antipsychotics were involved. I have, however, seen this kind of behaviour many times in back-wards and aged care and Acute Management Units and it disappeared when the neuroleptics were removed, reduced, or sometimes when benztropine was administered. (Yes, I know there is controversy about whether benztropine works in akathisia, but the literature I have read on this issue says that the definitions and stats are not reliable enough to say whether it works or not in such cases).

In the end, these neuroleptics don't seem to be working, anyway. Why don't we try stopping them?

NEUROLEPTICS ASSOCIATED WITH SLOWER RECOVERY FROM CONFUSION

Let me say at the outset that I initially had no objection to James receiving small doses of anti-psychotics. I only objected when I saw how he appeared to be responding to them. My first query about them was to the night duty charge on Wednesday 26thMay.

Importantly, some research has also associated these medications with slower recovery from confusion. Here is one very clear example (among many) which showed that "the neuroleptics affected cognitive recovery with almost 7 more days required to clear PTA in the neuroleptic treated group." i.e. confusion lasted 7 days longer on average than in patient groups where neuroleptics were used :

Brain Inj. 2006 Aug;20(9):905-11.

The impact of acute care medications on rehabilitation outcome after traumatic brain injury. Mysiw WJ, Bogner JA, Corrigan JD, Fugate LP, Clinchot DM, Kadyan V., Department of Physical

Medicine and Rehabilitation, The Ohio State University, Columbus, OH 43210, USA. mysiw.1@osu.edu

Abstract

OBJECTIVES: To examine the impact of medications with known central nervous system (CNS) mechanisms of action, given during the acute care stages after traumatic brain injury (TBI), on the extent of cognitive and motor recovery during inpatient rehabilitation. DESIGN: Retrospective extraction of data utilizing an inception cohort of moderate and severe TBI survivors. METHODS: The records of 182 consecutive moderate and severe TBI survivors admitted to a single, large, Midwestern level I trauma centre and subsequently transferred for acute inpatient rehabilitation were abstracted for the presence of 11 categories of medication, three measures of injury severity (worst 24 hour Glasgow Coma Scale, worst pupillary response, intra-cranial hypertension), three measures of outcome (Function Independence Measure (FIM) Motor and Cognitive scores at both rehabilitation admission and discharge and duration of post-traumatic amnesia (PTA)). MAIN OUTCOME AND RESULTS: The narcotics, benzodiazepines and neuroleptics were the most common categories of CNS active medications (92%, 67% and 43%, respectively). The three categories of medications appeared to have no significant outcome on the FIM outcome variables. The neuroleptics affected cognitive recovery with almost 7 more days required to clear PTA in the neuroleptic treated group. The presence of benzodiazepines did tend to obscure the impact of neuroleptics on PTA duration but the negative impact of neuroleptics on PTA duration remained significant. CONCLUSIONS: The results suggest that the use of neuroleptics during the acute care stage of recovery has a negative impact on recovery of cognitive function at discharge from inpatient rehabilitation. Due to the paucity of subjects with hemiplegia in this cohort, conclusions could not be drawn as to the impact of acute care medications on motor recovery.

P.M.ID: 17062422 [PubMed - indexed for MEDLINE]

James's restlessness can be stopped by walking him or by wheeling him in a wheelchair. Because he doesn't have much strength yet, a wheelchair is necessary. The hospital would not provide one, so I hired on from the Chemist in the hospital court on 1st floor. It is due back there on Wednesday.

Obviously this repetitive behaviour waxes and wanes, depending on the amount of neuroleptics and their breakdown in his system at any time, and possibly in response to environmental factors.

I asked the nurse or nurse assistant to record what she had said at 3.30am. She said that she could ask the nurse to do so and that she would. At 0800 I called O-, who is in charge today, and asked her if the observations had been recorded. They had not been recorded. What had been recorded was a 'settled night' - so the period of restlessness had not even been mentioned.

I have spoken to one registrar and one resident. I have asked to speak to the Consultant, Mr Campbell, but got the impression that that simply doesn't happen. I have spoken to Betty-Anne, the Nurse Unit Manager, who said that a registrar would call back (several days ago). No-one called. I talked with the new resident in 8ASouth, who repeated what the registrar had told him, telling me that I was not seeing what I and others are seeing, albeit in a nice way. He also said that a single room was being considered. Today I asked to speak to the consultant again but he was not on duty because it was a Saturday. I asked for an email address to write my concerns to him, but I was told there was no email address. It was suggested to me that I wait for a family meeting. I said we have one scheduled on Thursday, but I consider the matter too urgent to wait for that long.

I asked O- to tell the doctors how concerned I am and to insist on an MRI. She said she would carry my message but could not guarantee that anything would happen. She read from some notes that an MRI was 'planned' but there was no date. I asked that the MRI be scheduled for today. I don't care if it is Saturday. I consider the matter urgent. I consider it negligent that no MRI has been done already when James is not getting better and no-one knows why. An MRI and a few days trial without the neuroleptics seems the least that the ward should do.

CT SCAN AND PROGRESS REPORT:

See at the end of this letter the most recent CT scan report, which I have transcribed.

This is the third CT scan that James Sinnamon has had, for the purpose of comparing progress. What it says is that two small bleeds have been resolving over time, which is good news, and that there are no new bleeds. The scan could find nothing remarkable at all. But it cannot show what is happening at a microscopic level in the brainstem or anywhere else.

What James's new resident (i.e. Intern doing 3 month rotation on the ward, who started on Thursday) told me (after he spoke to the registrar) was that they still cannot rule out axonal damage in the region of the brain stem due to a possible contusion (twisting) of it in the accident, although they don't actually expect to find axonal damage. Axonal damage is bad news because it is permanent damage. They would expect to be able to see such axonal damage with an MRI scan, but he said that they don't see the point in doing one in the short term since it wouldn't change their treatment of James's symptoms and it would be hard to get him to be still.

Discussions I have had with doctors who have trained in head injury at the hospital where I work have said that they would expect an MRI to have been done well before now. "If you don't do one, how can you know there is nothing else wrong?" "How can you say that you wouldn't change your treatment after an MRI if you haven't done one?" are two remarks that doctors where I work have made.

The Nurse Unit Manager also indicated to me one evening verbally that there was involvement of the Reticular Activating System (RAS) near the brain stem in the accident and this part of the brain controls consciousness, therefore the slow resolution of James's delirium could be explained by this. Until the delirium resolves, James will only have tiny windows of clarity in which to lay down new memories. When the delirium improves or resolves, his ability to remember should improve.

But the CT scan tells us nothing in particular about this; without an MRI it is all conjecture. Perhaps an MRI would not tell us much but without one, how can we know? An excuse has been made that James is too restless for an MRI but I am reliably informed that he could be sedated for one and that this is done frequently, just as was done with the CT scan.

Whilst there is hope that he would improve back to near normal, that is not a given and the chances deteriorate statistically with the length that recovery from confusion takes. As the CT Scan request reason shows, we don't know what is causing the confusion. As well as the confusion, there is the ongoing trauma. I know that James is scared and isolated, in pain and crippled, and that he has insight into his memory loss. I consider that his human rights have been neglected.

James, who until the accident, was a tall, strong, fit 50 year old man who engaged in rock climbing, walking, cycling and various kinds of manual work, now has some left sided motor weakness, most noticeable in his inability to walk unassisted, to stand alone, and his tendency to drag his left leg.

His continuing confusion is preventing him from using the gym and having physiotherapy there. He is growing weaker every day.

RESTLESSNESS, MEDICATIONS, WHEELCHAIR:

Various visitors and other observers have noticed how James restlessness has recurring features, such as the tendency to bury his head in his pillow, then flop down on one side, turn over on his back, groan and sit up, then get on all fours, then kneel, then sometimes to try to stand up, then, from the kneeling position, flop down on one side and repeat this sequence continually, at different rates. Sometimes he does the circuit rapidly; other times it is so slow that you are unlikely to pick the pattern. Most recently he has also been making involuntary vocalisations, notably "Alright, alright", as he sits up. When he goes through this routine rapidly and noisily, the staff describe it as 'agitation' and he gets neuroleptic tranquillisers.

Recently the problem got worse and James avoided eye contact or direct interaction because it set off these involuntary vocalisations and some new ones including echolalia which were part of the evolving sequence of the repetitive cycle. He told me this when he was more relaxed as we wheeled him around and around the corridors of the hospital a few nights ago.

Perhaps because they do not spend long periods watching him, the doctors have not spotted this repetitive phenomenon and think that I am making it up when I report it. A ?registrar came to see James when he was not 'agitated', then claimed that the phenomenon described did not occur. Perhaps I have also confused a doctor by describing the behaviour as 'akathisia' when it doesn't fit a narrow definition of this term, such as dancing on one spot.

I have told them that I am concerned that the antipsychotic medications- called 'neuroleptics' - that James is on cause this behaviour as a side-effect. I cannot get anywhere with this because the repetitive nature of the behaviour is not acknowledged by the doctors. This really bothers me because, if I am right, the behaviour will be made worse by the medications. It is indeed possible that I am wrong and that this pattern is part of his injury behaviour, but in that case why aren't the doctors investigating with MRIs?

Olanzapine causes a very high rate of diabetes and morbid obesity. James has become very sedated and food focused since he has received doses of this medication and his orientation has dropped even further, to the extent that he has got his own identity confused at times and has become emotionally flat. This is an overall deterioration. He is also on Respiradone, which has similar side effects, although not such a high rate. James might spend several more weeks in RBWH and if he remains on these meds, especially olanzapine, I don't like to imagine the consequences.

TRANSCRIPTION OF CT REPORT
CT Head JAMES, JAMES PATRICK*Final Report"
Reason   for   Exam
1901879 8BS
CT Read
HISTORY:
PBA. Slow recovery. Ongoing agitation. Progress of SAH/TBI.
TECHNIQUE:
Non contrast CT brain.
FINDINGS:
Comparison is made with the previous CT dated 20/5/10

There is interval resolution of the subarachnoid blood that was previously seen in the left lateral ventricle and at the foramen of Monro. There is also a decrease in size of the hypodense lesion at the foramen of Monro, which now measures 5mm in width (previously 9mm). This would be in keeping with a resolving haematoma.

No acute intracraial haemorrhage detected. No extra-axial surface collection. The cerebral ventricles, cortical sulci and basilar cisterns are normal in size. No midline shift. Grey/white matter differentiation preserved.

Cervicomedullary junction unremarkable.

Interval resolution of ethmoidal air cell opacification. The paranasal sinuses are now clear. No skull abnormality.

COMMENT:

Interval resolution of subarachnoid haemorrhage. No acute intracranial pathology detected. No ventriculomegaly.

Dr M C-, Registrar

Result type: CT Head
Result date: June 02, 2010 15:07
Result status: Auth (Verified)
Cosigned by: A, V. on June 02, 2010 17:55
Verified by C, M on June 02, 2010 17:20
Encounter info: 1901879-1, RBWH, Inpatient, 18/05/2010
Printed by: S, C
Printed on : 03/06/2010 12:12

Signature Line
Electronically Signed [Some identifiers removed here.]

Sincerely,

Sheila Newman
NOK for James Sinnamon
Tel: 03 97835047, 97834556
07 33690819

RBWH-Feedback RBWH-Feedback wrote:

Good Morning,

Thank you for your email. It has been forwarded to the appropriate area.

If the matter is urgent, please contact the hospital directly via Phone: +61 7 3636 8111.

Kind Regards,

The Royal Brisbane and Women's Hospital
Metro North Health Service District
Queensland Health
Australia

Ph: +61 7 3636 8111
Fax: +61 7 3636 4240

5 June 2010: Email to family re family meeting: News about James Sinnamon including recent head scan - update

Please convey the news of the family meeting on Thursday and the contact number to Anne James and Bill for whom I do not have email addresses handy.

Contents:

1. Family meeting

2. CT Scan Report

3. Restlessness, medications and wheelchair

4. Outside opinion.

(Just adding today 5-6-10) that I talked with doctors and nurses familiar with head injuries at the hospital in Victoria where I work last night and they all felt that James should have had an MRI by now. I therefore made a strong request to that effect to the nurse in charge at RBWH this morning, and also let the hospital feedback system know of my concern. I also asked them to start making detailed notes about James's 'restlessness'.)

1. FAMILY MEETING:

Veronica, James's social worker, has scheduled a Family Meeting for Thursday at 11am. Her phone number is ......

2. CT SCAN AND PROGRESS REPORT:

See at the end of this letter the most recent CT scan report, which I have   transcribed.

This is the third CT scan that James has had, for the purpose of comparing progress. What it says is that two small bleeds have been resolving over time, which is good news, and that there are no new bleeds. The scan could find nothing remarkable at all. But it cannot show what is happening at a microscopic level in the brainstem.

What James's new resident (i.e. Intern doing 3 month rotation on the ward, who started yesterday) told me (after he spoke to the Registrar) was that they still cannot rule out axonal damage in the region of the brain stem due to a possible contusion (twisting) of it in the accident, although they don't actually expect to find axonal damage. Axonal damage is bad news because it is permanent damage. They would expect to be able to see such axonal damage with an MRI scan, but they don't see the point in doing one in the short term since it wouldn't change their treatment of James's symptoms and it would be hard to get him to be still.

The Nurse Unit Manager also pointed out to me that there was involvement of the Reticular Activating System (RAS) near the brain stem in the accident and this part of the brain controls consciousness, therefore the slow resolution of James's delirium could be explained by this. Until the delirium resolves, James will only have tiny windows of clarity in which to lay down new memories. When the delirium improves or resolves, his ability to remember should improve.

Whilst there is hope that he would improve back to near normal, that is not a given and the chances deteriorate statistically with the length that recovery from confusion takes. That still doesn't mean that he won't recover well.

James has some left sided motor weakness, most noticeable in his inability to walk unassisted and his tendency to drag his left leg. There is hope that this may resolve soon, with physiotherapy.

3. RESTLESSNESS, MEDICATIONS, WHEELCHAIR:

Some of you have noticed how James restlessness has recurring features, such as the tendency to bury his head in his pillow, then flop down on one side, turn over on his back, groan and sit up, then get on all fours, then kneel, then sometimes to try to stand up, then, from the kneeling position, flop down on one side and repeat this sequence continually, at different rates. Sometimes he does the circuit rapidly; other times it is so slow that you are unlikely to pick the pattern. Most recently he has also been making involuntary vocalisations, notably "Alright, alright", as he sits up. When he goes through this routine rapidly and noisily, the staff describe it as 'agitation' and he gets tranquillisers.

Perhaps because they do not spend long periods watching him, the doctors have not spotted this repetitive phenomenon and think that I am making it up. Perhaps I have also confused them by describing the behaviour as 'akathisia' when it doesn't classically fit a narrow definition of this term.

I have told them that I am concerned that the antipsychotic medications- called 'neuroleptics' - that James is on cause this behaviour as a side-effect. I cannot get anywhere with this because the repetitive nature of the behaviour is not acknowledged by the doctors. This really bothers me because, if I am right, the behaviour will be made worse by the medications. I hope I am wrong because I am not getting anywhere on this matter with the doctors, although they have tried to respond to my concerns by coming and looking at him for short periods of time when he has not been so restless. It is indeed possible that I am wrong and that this pattern is part of his injury behaviour, but in that case I don't see why the doctors have not recorded the repetitive cycling through movements. The doctors say that he is only on small doses of respiradone and olanzapine but dosage with brain injured patients are much more potent than with ordinary people. Olanzapine causes a very high rate of diabetes and morbid obesity. James has become very sedated and food focused since he has received doses of this medication. He is also on Respiradone, which has similar side effects, although not such a high rate. James might spend several more weeks in RBWH and if he remains on these meds, especially olanzapine, I don't like to imagine the consequences. Importantly, some research has also associated these medications with slower recovery from confusion. Here is one example which showed that "the neuroleptics affected cognitive recovery with almost 7 more days required to clear PTA in the neuroleptic treated group." i.e. confusion lasted 7 days longer on average than in patient groups where neuroleptics were used :

The impact of acute care medications on rehabilitation outcome after traumatic brain injury. Department of Physical Medicine and Rehabilitation, The Ohio State University, Columbus, OH 43210, USA. Mysiw WJ, Bogner JA, Corrigan JD, Fugate LP, Clinchot DM, Kadyan V., Brain Inj. 2006 Aug;20(9):905-1. mysiw.1@osu.edu

Abstract

OBJECTIVES: To examine the impact of medications with known central nervous system (CNS) mechanisms of action, given during the acute care stages after traumatic brain injury (TBI), on the extent of cognitive and motor recovery during inpatient rehabilitation. DESIGN: Retrospective extraction of data utilizing an inception cohort of moderate and severe TBI survivors. METHODS: The records of 182 consecutive moderate and severe TBI survivors admitted to a single, large, Midwestern level I trauma centre and subsequently transferred for acute inpatient rehabilitation were abstracted for the presence of 11 categories of medication, three measures of injury severity (worst 24 hour Glasgow Coma Scale, worst pupillary response, intra-cranial hypertension), three measures of outcome (Function Independence Measure (FIM) Motor and Cognitive scores at both rehabilitation admission and discharge and duration of post-traumatic amnesia (PTA)). MAIN OUTCOME AND RESULTS: The narcotics, benzodiazepines and neuroleptics were the most common categories of CNS active medications (92%, 67% and 43%, respectively). The three categories of medications appeared to have no significant outcome on the FIM outcome variables. The neuroleptics affected cognitive recovery with almost 7 more days required to clear PTA in the neuroleptic treated group. The presence of benzodiazepines did tend to obscure the impact of neuroleptics on PTA duration but the negative impact of neuroleptics on PTA duration remained significant. CONCLUSIONS: The results suggest that the use of neuroleptics during the acute care stage of recovery has a negative impact on recovery of cognitive function at discharge from inpatient rehabilitation. Due to the paucity of subjects with hemiplegia in this cohort, conclusions could not be drawn as to the impact of acute care medications on motor recovery.

P.M.ID: 17062422 [PubMed - indexed for MEDLINE]

James's restlessness can be stopped by walking him or by wheeling him in a wheelchair. Because he doesn't have much strength yet, a wheelchair is necessary. The hospital would not provide one, so I hired on from the Chemist in the hospital court on 1st floor. It is due back there on Wednesday. Please push him round in it. Bill and I did this for three hours one night and it had a really good effect. We took James down to the food court. I did the same thing the next night by myself. Could someone please return it for me because I won't get into Brisbane until after the Chemist closes.

4. OUTSIDE OPINION:

I think that we should get a second opinion on James from someone outside the hospital for this reason and I will try to find someone who is really independent.

I am not canning the hospital. I think they are doing their best and are really good at some things, but you cannot just leave a patient in the hands of a public hospital and expect them to get everything they need. We need to represent him if this restlessness continues. There are alternatives to medications, such as having extra nurses (called 'specials') to guide his movements, single room, mattresses on floor, and walking aids and wheelchairs. It is also important for him to be able to get away from noise and bright lights and people. The hospital will talk about staff safety and they have some right on their side there, but some of us feel that James is not truly aggressive and can be managed with careful supervision by constantly present staff or relatives and friends. The best probable outcome here would be a big reduction in the perceived need for medications.

Below is the head scan report.

CT Head *Final Report"
   Reason for Exam
   1901879 8BS

CT Read

HISTORY:
PBA. Slow recovery. Ongoing agitation. Progress of SAH/TBI.

TECHNIQUE:
Non contrast CT brain.

FINDINGS:
Comparison is made with the previous CT dated 20/5/10

There is interval resolution of the subarachnoid blood that was previously seen in the left lateral ventricle and at the foramen of Monro. There is also a decrease in size of the hypodense lesion at the foramen of Monro, which now measures 5mm in width (previously 9mm). This would be in keeping with a resolving haematoma.

No acute intracraial haemorrhage detected. No extra-axial surface collection. The cerebral ventricles, cortical sulci and basilar cisterns are normal in size. No midline shift. Grey/white matter differentiation preserved.

Cervicomedullary junction unremarkable.

Interval resolution of ethmoidal air cell opacification. The paranasal sinuses are now clear. No skull abnormality.

COMMENT:

Interval resolution of subarachnoid haemorrhage. No acute intracranial pathology detected. No ventriculomegaly.

[This part altered for editorial purposes] Result type: CT Head, Result date: June 02, 2010 15:07,

Encounter info: 1901879-1, RBWH, Inpatient, 18/05/2010, *Order by C-, DR S. -- PPSC on June 01, 2010 14:56

APPENDIX THREE: First Neuropsychological Report Post Hospital Therapy

"[…]
Name: Mr James Sinnamon
[…]
Thank you for referring Mr James Sinnamon for a comprehensive neuropsychological assessment and provision of a full report outlining his current cognitive status. Thank you for providing me with the following relevant documents and reports which were reviewed prior to completing the assessment and report:

WorkCover referral form

Discharge report from the Victorian Rehabilitation Centre (VRC) dated 4 April 2011

Driving assessment report by Ms Sue S. (Occupational Therapist) dated 24 May 2011

Medical practitioners Certificate of Capacity dated 01 August 2011

Report from Ms Sarah J. (Exercise Physiologist) dated 10 June 2011

Mr Sinnamon was interviewed and underwent a psychometric assessment in Brisbane on 7 September 2011. The following background history was taken from the aforementioned documents or was provided by Mr Sinnamon in interview. His friend/carer (Ms Sheila Newman) also kindly provided some collateral information in the form of questionnaires enquiring about possible functional and executive problems in his everyday functioning.

1.0 RELEVANT BACKGROUND
1.1 Circumstances of Injury and Treatment

Mr Sinnamon is a 52-year-old *General Clerk* at the Royal Brisbane and Women's Hospital (RBWH) who sustained a severe traumatic brain injury when he was hit by a car while riding his pushbike to work on 18 May 2010. He had no recall of the accident with his last memory being riding his bike to work after being delayed by a punctured tire. He had patchy recall of his friend Ms Newman visiting him in hospital but he had no recall of any of the other visitors who came to see him.

Hospital records were not available however the discharge summary from the Victorian Rehabilitation Centre stated that Mr Sinnamon was rendered unconscious for two minutes and he registered a Glasgow Coma Score (GCS) of 11/15 when the ambulance arrived on the scene. Apparently he was transported to the RBWH where a CT brain scan indicated facial fractures and traumatic subarachnoid and intraventricular haemorrhages. A subsequent MRI brain scan on the 12 June 2010 also highlighted diffuse axonal injury. He was reported to emerge from post-traumatic amnesia (PTA) approximately six weeks after his accident.

Mr Sinnamon commenced his rehabilitation at the RBWH before he transferred to the Mater Private Hospital on day 42 post-injury. He participated in multidisciplinary rehabilitation at the Mater hospital for three and a half weeks before he moved to Melbourne where Ms Newman took over the role of his carer.

Mr Sinnamon continued his rehabilitation in Melbourne as an inpatient at The Victorian Rehabilitation Centre (VRC) where he participated in regular physiotherapy and an exercise program. He also attended to speech therapy and underwent a neuropsychological assessment which revealed impairments in his processing speed, attention to detail, self monitoring, working memory and verbal new learning. He also saw a social worker to address issues around self esteem and assertiveness. Following his discharge from VRC on 3 March 2011 he continued his therapy as an outpatient.

Mr Sinnamon reported that he underwent psychotherapy with a psychologist to address symptoms of depression and trauma associated with the accident.

1.2 Return to Work and Driving

Mr Sinnamon has not returned to any form of paid employment. He reported that mental/physical fatigue was his primary barrier for returning. He indicated that he wanted to retrain/upgrade his computer skills so he could return to working with computers.

After failing two on-road driving assessments he passed his test in an automatic car in May 2011, allowing him to return to driving an automatic car only.

1.3 Reported Ongoing Symptoms/Functional Changes Post Accident:

At the time of the present assessment (nearly 16 months post-injury), Mr Sinnamon reported that he was experiencing the following ongoing accident-related symptoms and functional changes:

Physical:

High level balance difficulties
Less acute sense of smell
Less acute eyesight
More frequent urination

Reduced libido
Heightened fatigue

Cognitive:
Slower speed of thinking
Reduced concentration, more easily distracted
Short-term memory problems
o Forgets names/faces
o Repeats himself
o Forgets what he is told
Needs extra clarification of unfamiliar information
Harder to find the right words when trying to express himself
Slower speed of reading and needs to reread text
Greater difficulties completing tasks
o Puts off harder tasks
o Gets distracted or fatigues
Agitated by any changes in his routine
Finds planning and organising himself more mentally taxing
May become confused in unfamiliar and busy situations
Finds it harder to orient himself in unfamiliar places

Emotional/Behavioural:

Less physically and mentally active
Depressed mood from time to time and tends to worry more
Has difficulty relaxing due to worrisome thinking
More irritable and mood swings due to frustration
More reliant on Sheila

2.0 RELEVANT PERSONAL INFORMATION
2.1 Family Background and Current Living Circumstances:

Mr Sinnamon was born in Rome and he has three younger siblings and two paternal half-siblings. He was not aware of anything out of the ordinary about his birth or develop.m.ent. His parents separated when he was seven years old which he described as a disruptive time in his life. Both of his parents (mother 71 years and father 75 years) are still living in Brisbane. He reported that his relationship with his mother was strained however he maintained a reasonably close relationship with his father.

Mr Sinnamon reported that he was single and he has been living with his friend/carer (Ms Sheila Newman) in Frankston Victoria since the time of his accident. He has never been married and he does not have any children.

2.2 Education and Employment:

Mr Sinnamon completed his secondary education at Holland Park High, Brisbane Grammar, and Marden Senior College (South Australia). He described himself as an average to above average student who was not academically driven because of his disruptive family life. He listed Science and Art as his strengths and Physical Education as his weakest subject. There were no reported learning difficulties however he noted that he repeated Year 11 which he attributed to his unstable home life.

Ms James's work history included:

1978-1985 Steel worker (loading trains and cranes)
Shunter, Locomotive Driver and Station Assistant for Queensland Rail
1985-1988 Taxation Office (1985 and 1988)
1988-1994 Programmer for TNT (retrenched)
1996-1998 ISP Manager at Warwick TAFE College
1999-2001 Studies in Applied Mathematics and Information Technology
1999-2001 Tutoring at University of Southern Queensland
1999-2001 Programming at the University of Southern Queensland
2002-2004 Research Programmer at University of Southern Queensland
2003-2010 Web Content Administration

2004-2005 Various casual positions (e.g. factory work and rubbish collection)

2005-Ongoing Patient Support Officer (General Clerk) at RBWH (2005 – ongoing)

In his role as *Patient Support Officer* Mr Sinnamon primary duties include cleaning offices and providing patient support services as required.

2.3 Relevant Medical and Psychiatric History:

Previous loss of consciousness in a bike accident at age 12 years
o Brief loss of consciousness but no persisting neurological problems
Peripheral neuropathy in the lower limbs
Mild hypothyroidism

There were no other reported prior significant medical, neurological or psychiatric events in Mr Sinnamon's background. He is not taking any prescribed medications.

Mr Sinnamon has not consumed any alcohol since his accident and described himself as a very light drinker prior to this. He is a non-smoker and has no significant history of illicit drug use.

3.0 FORMAL ASSESSMENT
3.1 Clinical Presentation and Mental Status:
3.1.1 Clinical Presentation

Mr Sinnamon presented as a tall balding average-framed 52-year-old right-handed man dressed in casual attire and wearing glasses. Besides a degree of restlessness, there was no other unusual psychomotor activity. He remained socially appropriate and cooperative. His mood and affect during the interview and assessment seemed anxious, particularly when he felt he was not performing as well has he should have been. He exhibited minimal facial expression and only gave intermittent eye contact. He started to fatigue towards the end of the assessment.

3.1.2 Mental Status

Mr Sinnamon was fully oriented to person place and time and he provided a reasonable autobiographical history. His speech was of normal tone and rate. His basic receptive/expressive language processes in conversation seemed intact, and he had no trouble understanding written or auditory instructions/directions. He closed his eyes to help focus his auditory attention and a degree of visual inattention was apparent. He confabulated when he was unable to freely recall story details and this later led to confusion between real and confabulated details in the recognition trial. He rushed unnecessarily at times leading to some careless errors. His thought processes were coherent and there were no indications of perceptual disturbances.

3.2 Subjective Assessments of Affective Functioning, & Everyday Executive Behaviour:
3.2.1 Depression, Anxiety and Stress Scale (DASS-42)

The DASS is a widely used psychological self-report instrument for assessing psychological distress in the domains of depression, anxiety and stress. The range of scores is *Normal, Mild, Moderate, Severe* and *Extremely Severe*. On the DASS Mr Sinnamon's self reported levels of depression (D = 10), anxiety (A = 8), and stress (S = 17) all fell in the *Mild* clinical range.

3.2.2 Beck Depression Inventory (BDI)

The BDI is a more specific quick self-report measure of depression severity. The scale consists of 21 items relating to symptoms and attitudes characteristic of depression where each item can be rated from 0-3 in degrees of intensity. The range of scores is *Normal, Mild* to *Moderate, Moderate* to *Severe* and *Severe*. On this more specific measure of depressive symptomatology Mr Sinnamon's responses again fell in the *Mild* clinical range (BDI = 18).

3.2.3 *Dysexecutive Questionnaire (Self and Informant Version)*

The Dysexecutive Questionnaire (DEX) is a 20-item questionnaire that enquires about common problems (e.g. changes in emotional regulation, personality, motivation, dysinhibition, planning, etc) encountered in dysexecutive syndrome, often observed when there has been disruption to the frontal regions of the brain. Mr Sinnamon's scores in the self report (Score = 35/80) and informant (Friend/Carer's rating = 33/80) versions of the DEX questionnaire were consistent with a moderate frequency of problems in his executive behavior.

Some of the more frequent problem areas (occurring "fairly often" or "very often") noted included:

Trouble understanding what people mean unless simple and straight forward
Acts without thinking, doing the first thing that comes to mind
Has difficulty thinking ahead or planning for the future
Has trouble realizing the extent of his problems and is unrealistic about the future
Does or says embarrassing things when in the company of others
Finds it hard to stop repeating saying or doing something once he's started
Finds it hard to keep his mind on something and is easily distracted
Has trouble making decisions, or deciding what he wants to do
Is unaware, or unconcerned about, how others feel about his behaviour

3.2.4 Frontal Systems Behaviour Scale (Informant Version)

The FrSBe is a 46-item rating scale that assesses for the presence of everyday frontal behavioural problems in the areas of:

1. Apathy – possible problems with initiation, psychomotor slowing, spontaneity, drive, persistence, loss of energy/interest, and blunted affective expression.

2. Disinhibition – potential problems with inhibitory control such as impulsivity, hyperactivity, socially inappropriate behaviour and difficulty modulating emotions.

3. Executive Dysfunction – potential problems with sustained attention, working memory, sequencing, organization, planning, future orientation, mental flexibility, self-monitoring and modifying errors with feedback.

There are self- and family-rating versions for rating both before and after the injury/illness to allow for premorbid personality. The before and after informant ratings by Mr

	Pre-injury	**Post-injury**
	T score	**T score**

Sinnamon's friend/carer highlighted marginal increases on all three subscales, with the Executive subscale now falling just within the clinical impairment range (see in Table below). **Partner**

Apathy	41	59
Disinhibition	43	62
Executive Dysfunction	47	***77**

3.3 Psychometric Assessment Results:
3.3.1 Tests Administered

o Formal Tests of Effort
o Wechsler Test of Adult Reading
o Speed and Capacity of Language Processing
o Wechsler Adult Intelligence Scale – Fourth Edition (WAIS-IV)
o Wechsler Memory Scale – IV (WMS-IV)
  □ □ Logical Memory
  □ □ Visual Reproduction
  □ □ Spatial Addition
o Wechsler Memory Scale – III (Mental Control only)
o California Verbal Learning Test - II
o Rey Complex Figure
o Delis Kaplan Executive Function System (DKEFS – selected subtests)
  □ □ Trail Making
  □ □ Verbal Fluency
  □ □ Color-Word
  □ □ Tower |
  □ □ Proverbs

3.3.2 Format of Test Results

Table 1: Descriptive Terms for Standard Scores and Percentile Rankings DESCRIPTIVE TERM	STANDARD SCORES	PERCENTILE
Very Superior	130 and above	>98th
Superior	120-129	91st – 97th
Above Average	110-119	75th –90th
Average	90-109	25th – 74th
Low Average or Mild Deficit	80-89	9th – 24th
Borderline or Moderate Deficit	70-79	2nd - 8th
Extremely Low or Severe Deficit	60 and below	<2nd

The assessment results are mainly presented in 95% confidence intervals and as percentiles. The 95% confidence interval represents the chances that the participant's true performance level falls within the stated interval. The percentile value indicates the participant's percentage ranking relative to other adults their age with a similar educational background and no history of a brain injury or cognitive deficits. The descriptive terms for the different standard scores, percentile rankings of intellectual abilities are listed in Table 1.

As indicated in Table 1, the average performance range is conventionally considered to fall between the 25th and 74th percentiles. The classifications of low, moderate and severe deficits are only relevant for individuals who have previously functioned in the *Average* range. That is, people who have performed in the *Above Average* or better intellectual range premorbidly may experience significant relative declines in their abilities even though their performance still falls in an unimpaired range. Likewise, for individuals with *Low Average* or below premorbid intellectual abilities, a score falling in the deficit range may either reflect no change or a further acquired impairment in intellectual functioning.

3.3.3 Reliability of Cognitive Test Performance

An important component of psychometric assessment is ensuring that an individual is applying their maximum effort to tasks; otherwise poor performance may be wrongly attributed to cognitive deficits that may not exist. Similarly, it is important to stress that evidence of suboptimal effort does not necessarily equate to a total absence of cognitive impairment as embellishment of genuine deficits is common and often originates from a desperate "cry for help or acknowledgement of an injury". Likewise, evidence of suboptimal effort may not reflect a deliberate attempt to feign impairment for external gain (such as financial compensation) as other factors (such as organically-based apathy, poor motivation due to low affective state, non-engagement in testing as a form of protest, and preserving deep-seated emotional needs) may also be responsible for under performance.

Mr Sinnamon appeared to apply his full mental effort to tasks. He achieved an acceptable score in a formal measure specifically designed to assess the authenticity of a memory complaint. Mr Sinnamon's overall performance profile can therefore be assumed to represent a reasonable reflection of his intellectual functioning on the day.

3.3.4 Premorbid Abilities

Without a baseline measure, it is often difficult to accurately estimate a person's prior level of intellectual functioning. The most reliable estimation is based on a collection of indicators, including educational and occupational history, pronunciation skills, and performance in tasks assessing other relatively stable abilities.

1. Although formal reports were not available, Mr Sinnamon's reported educational and occupational history is consistent with someone who is likely to have functioned in the [accidentally omitted here was the word "*Above*"] *Average* ability range.

2. Providing there is no prior history of reading difficulties, language disorder (developmental or acquired) or alexia; the Wechsler Test of Adult Reading (WTAR) is commonly used to estimate an individual's level of pre-morbid intellectual functioning, particularly verbal abilities. In addition to being less vulnerable to the effects of most head injuries, clinically significant levels of depression have been found to have no significant impact on test performance.

The WTAR requires the examinee to read 50 irregularly spelt words that become increasingly complex. Mr Sinnamon's overall pronunciation score (Total = 47/50) also suggested that his premorbid abilities are most likely to have fallen in the [accidentally omitted here was the word "*Above*"] *Average* range of intellectual functioning (estimated Full-scale IQ = 114, 95% Prediction Interval = 93-135, 82nd percentile).

3. A person's crystallised abilities refer to the knowledge that they have acquired over their lifetime and it typically remains unchanged following most brain injury, although they may be reduced by selective language deficits and in the instance of extremely severe head injuries. Mr Sinnamon's scores on tests of vocabulary and general knowledge (both = 99th percentile) placed him in the *Very Superior* performance range.

Thus, there is sufficient evidence to suggest that Mr Sinnamon functioned in the *Above Average* ability range (74th to 90th percentile) in terms of overall general intellectual abilities. The usual variation of strengths and weaknesses would be expected in his profile. His performance in the current cognitive assessment is presented in Appendix I and will be described further below.

3.4 Psychometric Test Results:
3.4.1 Current Overall Intellectual Functioning

The Wechsler Adult Intelligence Scale – Fourth Edition (WAIS-IV) was used to assess Mr Sinnamon's current overall intellectual functions. The Full-Scale IQ score is the overall summary score which estimates an individual's general level of intellectual functioning. The four purer cognitive index scores of the WAIS-IV include: Verbal Comprehension (VC) - acquired general knowledge, understanding and expression of word meanings, the formation of abstract concepts, and flexible thinking; Perceptual Reasoning (PR) – perceptual organisation and reasoning skills; Working Memory (WM) - the ability to take in, hold and manipulate oral verbal information in one's head; and Processing Speed (PS) - the capacity to rapidly process visual information using graphomotor coordination skills.

In the present assessment Mr Sinnamon's overall Full-Scale IQ summary score fell in the *Above Average* performance range (FIQ = 118, 114-122, 88th percentile), and was consistent with the reading-based estimate of his premorbid intellectual functional level however there was considerable variability amongst his index scores. For instance, while his *Verbal Comprehension* (VCI = 98th percentile) and *Perceptual Reasoning* ability (PRI = 92nd percentile) were impressive, both his *Working Memory* (WMI = 63rd percentile) and visuomotor *Processing Speed* (PSI = 42nd percentile) were significantly and unusually weaker than these scores.

3.4.2 Summary of Current Functioning in Specific Cognitive Domains

Further analysis of Mr Sinnamon's neuropsychological profile (see Appendix I) revealed that while most of his cognitive skills were impressive and well preserved, there was evidence of likely residual relative declines (considered to be scores falling at the 50th percentile and below) and deficits (bold) in his:

1. *Processing speed and efficiency in visual and auditory attention tasks*

(especially in complex, novel and sustained attention tasks) e.g.
- Grapho-motor speed = 37th percentile, *W
- Single target scanning = 50th percentile, 0 omissions
- Dual Symbol Search = 50th percentile, 1 error
- Numerical sequence tracing = 37th percentile, 0 errors
- Color naming = 50th percentile, 0 errors
- Word reading = 37th percentile, 1 error
- Copying complex visuospatial design = **30/36, 2nd to 5th percentile**
- Tower mean first move and time per move ratios = 50th percentile

2. *Immediate auditory attention and working memory, e.g*

- CVLT List A = 6 words, 31st percentile; CVLT List B = 5 words, 31st percentile
- Auditory attention to numbers = 50th percentile (5 forwards and backwards)
- Letter-Number reordering = 37th percentile

3. *Immediate and delayed recall of auditory and visual information*, e.g.

- Immediate Story recall = 50th percentile (with marked confabulation)
- Delayed free and cued retrieval of CVLT word list = 50th percentile
- Delayed free recall of multiple visual designs = 50th percentile

4. *Novel sequential problem solving, self monitoring, response regulation, and mental flexibility, e.g.*

- Tower building = 50th percentile, Move Accuracy ratio = 25th percentile
- Word recall = 6 repeats, 31st percentile; 7 intrusions = 31st percentile
- Verbal fluency = 3 repeats, 50th percentile
- Category word switching = 50th percentile, 1 error
- Number-letter switching = 63rd percentile, **2 errors**

It is important to remember that Mr Sinnamon was assessed in a relatively controlled environment that contained few distractions/stressors with the aim of measuring his optimal performance. In particular, the formal testing conditions may underestimate difficulties in attention, concentration and executive processes/behaviour which tend to be more pronounced in less structured everyday environments or in novel/challenging situations in which previous routines or strategies cannot be applied. His performance difficulties will also be more prominent when he is stressed or overly tired, and brain injuries commonly result in both reductions in stress tolerance and excessive fatigue.

4.0 RESPONSE TO SPECIFIC QUESTIONS
1. History as related by Mr James Sinnamon.

See Sections 1.0 and 2.0
2. Relevant past medical history.

See Section 2.3
3. Current medication

See Section 2.3

4. Examination findings, including tests or investigations required if the diagnosis is still uncertain

See Section 3.0 and Appendix I

5. Diagnosis of all work-related conditions, please include if any of these are an aggravation of a pre-existing condition.

Mr Sinnamon's clinical presentation and reported emotional/behavioural difficulties are consistent with him having developed an *Adjustment Disorder with Mixed Anxiety and Depression* as a result of his persisting cognitive functional difficulties however he current symptom levels appear to be relatively mild.

The neuropsychological assessment results indicated that Mr Sinnamon is someone of high intellect who appears to have made a relatively good cognitive recovery from the severe brain injury he sustained when he was hit by a car on the 18 May 2010. There was however some objective evidence of likely residual relative declines and deficits in his speed of processing, visual and auditory attention and working memory, auditory and visual memory, and some executive functions (e.g. novel sequential problem solving, self monitoring, response regulation and mental flexibility). He and his friend/carer also acknowledged the presence of a moderate frequency of problems in his day to day executive behaviour (see Section 3.2.3 and 3.2.4).

With respect to DSMIV Axis 1 diagnosis, I am of the following opinion:

Axis I: *309.28 Adjustment Disorder With Mixed Anxiety & Depression (mild)*

799.9 * Organic brain Injury (diagnosis deferred, recovery not stable and stationary)

6. Relationship of the current work related diagnosis to the stated mechanism of injury. Whether the work-related condition(s) is an aggravation of any pre-existing condition(s).

There was no evidence of any significant pre-existing psychiatric disorder or head injury. *Neuropsychological Assessment: Mr James Sinnamon,*

7. Any treatment, rehabilitation, and return to work recommendations for the work-related injury.

Mr Sinnamon may benefit from further adjustment to injury counseling which should also include a review of common problems that associated with head injuries, and strategies to help him compensate for his ongoing difficulties. His progress in therapy may be accelerated with antidepressant medication however I will leave this to the discretion of his treating medical professional.

It has been nearly sixteen months since Mr Sinnamon's mild head injury and so while most physiological brain recovery has already taken place he may achieve further physiological recovery over the next 6 to 12 months. In addition to ongoing natural organic recovery, Mr Sinnamon is likely to achieve further functional gains through successful treatment of his noted adjustment difficulties, through further adoption of cognitive and behavioural compensation strategies (see Appendix II), and through him maintaining a stable, healthy and fulfilling lifestyle.

It is recommended that a reassessment of Mr Sinnamon's intellectual functioning is performed in 6 to 12 months time.

Semantically Related Word List	48 words, 66 percentile	Average
Total	50th percentile	Average
Use of Semantic Structure	84th percentile	Average
Use of Serial Structure	35% 69 percentile	Average
%Primacy	33% 16 percentile	Average
%Middle	31% 69 percentile	
%Recency	69th percentile	
Learning Slope	69th percentile	
Across trial Consistency		
Immediate free recall	10 word, 69 percentile	Average
Cued Immediate recall	12 word, 69 percentile	Average
Delayed Free recall	10 word, 50 percentile	Average
Cued delayed recall	11 word, 50 percentile	Average
Recognition recall List A	16 words, 84 percentile	Average
	4 words, 31 percentile	Average
False Positives		
Visuospatial Reproduction	63rd percentile	Average
Immediate	50th percentile	Average
Delayed	Total = 6/7	
Recognition		
Visuospatial Design Immediate	21/36, 66 percentile	Average
Delayed	22/36, 73 percentile	Average
Recognition	23/24, 92 percentile	Superior

ACQUIRED KNOWLEDGE, LANGUAGE AND VERBAL CONCEPTUALISATION/REASONING

Word Reading	Total = 47/50, 82 percentile	Above Average
Lexical Word Knowledge	75th percentile	Above Average
Word Knowledge	99 percentile *S	Very Superior
Semantic Knowledge	99 percentile *S	Very Superior
Concept Formation	63rd percentile	Average
Reasoning – Proverbs Overall	63rd percentile	Average
Common	63rd percentile	Average
Uncommon	63rd percentile	Average
Accuracy	50th percentile	Average
Abstractness	75th percentile	Above Average
Proverbs – Multiple choice	100% correct	

HIGHER ORDER FRONTAL EXECUTIVE FUNCTIONS

Planning and Problem Solving	Copying = 30/36, 2 –5 p'tile	Borderline/
Copying of Complex	Good structure, but careless	Moderate Deficit
Visuospatial Task	with poor attention to detail	
Tower Building	50th percentile	Average
Rule Break Errors	1 error, 65 percentile	Average
Mean First Move Time	5.4 secs, 50 percentile	Average
Time per Move Ratio	3.6 secs, 50 percentile	Average
Move Accuracy Ratio	25th percentile	Average (lower limit)

8. Is the work-related condition now stable and stationary? In your opinion, when will there be a capacity to upgrade to full hours and/or work duties?

Mr Sinnamon's work-related psychiatric condition and brain injury is not stable and stationary, but should stabilize with appropriate treatment over the next 6 months.

Mr Sinnamon has not returned to any form of paid employment, stating that he considers physical and mental fatigue as his primary return to work barrier. I am of the opinion that Mr Sinnamon should be capable of making a graded return to suitable duties at his former workplace (or commence vocational retraining if he so desires) under the supervision of his general practitioner or an occupational therapist. With appropriate treatment and workplace accommodations for his heightened susceptibility to fatigue and his residual thinking difficulties, he should be able to increase his work hours and return to full hours and work duties over the next three to six months.

Neuropsychological Assessment Results OVERALL INTELLECTUAL ABILITIES

Intellectual Abilities	IQ Score, 95% Confidence. & Percentile Rank	
Overall Full Scale Abilities (MQ)	*118, 114-122, 88 percentile*	*Above Average*
Verbal Comprehension (VC)	*130, 123-134, 98th percentile*	*Very Superior*
Perceptual Reasoning (PR)	*121, 114-126, 92 percentile*	*Superior*
Working Memory (WM)	*105, 98-111, 63 percentile*	*Average*
Processing Speed (PS)	*97, 89-106, 42 percentile*	*Average*

PROCESSING SPEED, SEQUENCING AND VISUAL ATTENTION

Basic Motor Tracing Speed	*75 percentile, 0 errors*	*Above Average*
Grapho-motor Speed	*37 percentile, 0 errors *W*	*Average*
Single Target Scanning	*50 percentile, 0 omissions*	*Average*
Dual Symbol Line Search	*50 percentile, 1 error*	*Average*
Numerical Sequence Tracing	*37th percentile, 0 errors*	*Average*
Alphabetical Sequence Tracing	*75th percentile, 0 errors*	*Above Average*
Counting Forward 1-20	*4 seconds, 0 errors*	*Above Average*
Counting Backward 20-1	*6 seconds, 0 errors*	
Alphabet	*5 seconds, 1 error*	
Days of Week Forward	*2 seconds, 0 errors*	
Days of Week Backward	*3 seconds, 0 errors*	
Months Forward	*3 seconds, 0 errors*	
Months Backward	*7 seconds, 0 errors*	
Counting by 6s + plus days	*10 seconds, 0 errors*	
Overall Mental Control Sequencing	*84 percentile, 1 error*	
Color Naming	*50th percentile, 0 errors*	*Average*
Word Reading	*37 percentile, 1 error*	*Average*
Letter Word Fluency	*98 percentile, 3 errors*	*Very Superior*
Category Word Fluency	*75 percentile, 2 errors*	*Above Average*
Sentence Comprehension	*84th percentile, 0 errors*	*Above Average*

AUDITORY ATTENTION AND WORKING MEMORY

Attention (Words) CVLT 16 Words - List A	*6 words, 31 percentile*	*Average*
CVLT 16 Words - List B	*5 words, 31 percentile*	*Average*
Auditory Attention (Numbers)	*50 percentile (SF, SB) *W*	*Average*
Letter-Number Reordering	*37th percentile*	*Average*
Mental Arithmetic	*75th percentile*	*Above Average*
Spatial Addition	*63rd percentile*	*Average*

PERCEPTION AND VISUOSPATIAL ASSEMBLY / CONSTRUCTION

Perceptual Reasoning	*95th percentile*	*Superior*
Perceptual Analysis and Synthesis	*91st percentile*	*Superior*
Visuospatial Assembly	*75th percentile*	*Above Average*

NEW LEARNING AND MEMORY

Story Recall *Logical Memory I*	*50th percentile*	*Average*
Logical Memory II	*63rd percentile*	*Average*
Recognition	*Total = 25/30*	

Self-Monitoring Word Recall Verbal Fluency	6 repeats, 31 percentile 3 repeats, 50 percentile	Average Average

Response
Regulation/Selective
Attention
   Word Recall
   Verbal Fluency
   Color-Word
Inhibition
	7 intrusions, 31 percentile
	1 set loss error, 63 percentile
	84 percentile, 0 errors
		Average
		Average
		Above Average

Mental Flexibility/Alternating Attention Color- Word Switching	84th percentile, 0 errors	Above Average
Category Word Switching	50 percentile, 1 error	Average
Number-Letter Switching	63 percentile, **2 errors**	Average

Neuropsychological Assessment: Mr James Sinnamon:

Strategies to Accommodate Common Problems Following Brain Injury

1. The intake and processing of information is often less efficient following brain injuries. For instance, it may take longer to understand sensory information and make sense of what is going on in a situation, or only fragments of the information presented may be processed because of a slowed mental speed.

To combat such problems:
Allow extra time to complete tasks.
Encourage information to be written down or recorded to provide extra time for improved comprehension/understanding.
Slow down and repeat information.
Encourage organisation of information.
Concentrate on accuracy first and speed later.
Encourage single task completion rather than focusing on multiple tasks.
Limit the amount of new pieces of information to be learned at one time.

2. Attention and concentration problems may be demonstrated by difficulties in staying alert, trouble maintaining focus on task (especially in the presence of competing distractions), and a tendency to lose train of thought.

Some strategies to deal with attention problems include:
Be sure that you have the person's attention before giving them an instruction.

Where possible, minimise competing noises (e.g. close doors and windows), visual distractions (e.g. remove colorful items from his field of view and reduce glare), and interruptions.

Verbal and visual cuing can be used to sustain focus on tasks (such as teaching them to talk themselves through tasks or set timers/alarms to remind themselves to refocus).

Direct them back to the topic if their conversation becomes unfocused.

Improved visual attention can be achieved through teaching them to scan left to right and top to bottom while using their finger to anchor their eyes.

When they have finished a task encourage them to check over their work to correct errors and make sure they have done what was required.

3. Memory problems may reflect disturbances at one or more of the three phases of memory (i.e. the initial intake phase, the storage phase, or the retrieval phase).

Some ways to overcome memory problems include:
Create as much structure and routine in the workplace as possible.

Use external strategies (i.e. calendars, lists, planner, and timers).

Repeat and encourage mental rehearsing of new information.

Provide visual as well as verbal instruction for new tasks.

Check to clarify understanding and retention of information.

Encourage important details to be recorded in a consistent location and organised fashion (e.g. in a diary, on a memo tape, or in a pocket notebook).

Present important, lengthy, or complex information in an organised format that enables them to adopt a stepwise approach and helps to break large tasks into manageable chunks – e.g. such as using a checklist.

Help develop associations/mnemonic devices to improve carryover of information learned (e.g. associate old information with new information, group or link information to prompt recall etc).

Place items in a common location (i.e. keys in a bowl by the front door, bills in a tray etc).

Neuropsychological Assessment: Mr James Sinnamon, DOB:

4. *Executive functions:* Planning, identifying priorities, organising information, sequencing steps in goal-directed behaviour, switching attention/responses, reasoning and problem solving, and monitoring one's own behaviour/emotions are referred to as the supervisory executive functions of the frontal regions of the brain. To some extent executive processes control and manage performance in all tasks (including memory processes) and are particularly relevant in novel or challenging tasks in which previous routines or strategies cannot be applied; in tasks requiring planning, organisation and decision making; in tasks that involve error correction; and in situations that involve overcoming a strong habitual response.

Some basic strategies to help compensate for executive functional difficulties in the workplace include:

Simplify – break overly complex activities into simple step by step tasks that can be checked off.

Routinise - develop standard operating procedures for all job tasks with clear step by step instructions on how to carry out each specific task.

Structure – organise and structure your time (ie., day, week and month) and work area (such as uncluttering your desk so it only contains material you are working on).

5. *Disinhibition* is a syndrome marked by difficulty properly directing and controlling energy and emotions. Some signs of dysinhibition include impulsively saying or doing things without considering the consequences, reacting to small things with too much emotion, restlessness, low frustration tolerance, and emotional flooding during which the individual feels overwhelmed

and is temporarily unable to think clearly or act purposefully. Potential problems with disinhibition may be better managed by:

Cueing them to slow down, plan and think (use the "stop, think, act, and evaluate" sequence).
Building in automatic pauses between tasks.
Encouraging them to seek feedback before acting.
Tactfully let them know when their behaviour is inappropriate and how they are coming across ("when you say I do this, it makes me feel…").
Encourage them to take breaks to relax, calm down, and reattempt communication when ready.

6. Organically-based *neurofatigue* is also a common aftermath of brain injury. It is not due to excessive activity or abnormal sleep patterns and can emerge suddenly without warning, especially after engaging in cognitively demanding tasks. Some signs and symptoms of neurofatigue include a lack of energy to engage in activities, low arousal (such as difficulties waking up or staying awake during the day), and decreased alertness, awareness or response to surroundings.

To combat potential higher susceptibility to physical/mental fatigue, stress and headaches (all of which will exacerbate any thinking difficulties) they should be encouraged to take regular short breaks. Assure them that taking "time out" is needed and okay. Where possible they should also alternate between easy and more challenging tasks and physical and mental tasks."

APPENDIX FOUR: MRI and Neurology Reports June-July 2012

In an interpretation of an MRI done late June 2012, a radiologist wrote:

"MRI BRAIN
Clinical History:
Previous severe brain injury. Memory disturbance. ?Progressive change.

Findings:

Slight cerebral atrophy involving frontal and temporal lobes. No space occupying lesion.

Minor non haemorrhagic foci of T2 hyperintensity posterior right frontoparietal corona radiate and subcortical opercular right frontal consistent with mild leukoaraiosis.

SWI demonstrates multiple tiny punctuate paramagnetic foci scattered through the brain, occult on other sequences and due to old microhaemorrhagic damage from the brain injury.

The microhaemorrhagic change is unaltered compared with the previous Royal Brisbane MR of 12/6/2010 however it is shown to much better advantage on the current study as the previous SW1 series was somewhat compromised by movement artifact.

Distribution:

1. Multiple tiny subcortical and coronal radiate posterior left frontal and parietal.

2. Subcortical and coronal radiate posterior right parietal.

3. Multiple foci within the corpus callosum body particularly on the right side and also within the surpra adjacent cingulated gyrus.

4. Multiple foci around the left lateral and ventral margin of the body of the left lateral ventricle.

5. Multiple bilateral tiny basal ganglia foci.

6. Linear and focal punctuate lesion in the subcortical white matter of the right frontal orbital gyrus due to a longitudinal shear.

7. Multiple tiny foci in right and left temporal pole cortex and subcortical white matter.

8. Larger focus in the mesial left temporal lobe involving hippocampal head and adjacent uncus with the axial FLAIR and T2 studies showing patchy hyperintensity of this area suggesting significant local damage/post traumatic atrophy.

9. Similar relatively large focus mesial right temporal lobe in uncus. Both these mesial temporal lobe lesions have diameters around 10 x 5mm plus additional microhaemorrhagic change in the area however the hippocampal damage/post traumatic atropy is more obvious on the left than right side on axial views.

10. I note that the original CT of 18/5/2010 shows bilateral forniceal haemorrhages particularly in the right fornix at the level of the foramen of Monro. The current study shows residual minor microhaemorrhagic staining of the left fornix and probably also

right at the foramen of Monro and of course fornix damage causes significant short term memory pathology.

Conclusion:
Very extensive paramagnetic microhaemorrhagic foci related to previous severe brain trauma. With respect to memory issues there is extensive old haemorrhagic damage of both mesial temporal lobes and left hippocampus head and left probably right fornices.

The extent of microhaemorrhagic change is unaltered since the previous Royal Brisbane MR however is much more obvious on the current study (this is just a reflection of movement artifact on the initial examination).

The patient will be recalled for additional coronal studies to further evaluate mesial temporal lobes." [End Report.]

And, after an additional clarificatory scan the next day:

"There is a slight degree of atrophy of the head of the left hippocampus compared with the right with associated slight prominence of the adjacent choroidal fissure and temporal horn. Signal intensity is normal at FLAIR. Right hippocampus normal, slight left parahippocampal gyrus atrophy manifest by a little asymmetric prominence of the adjacent sulcus however the changes are relatively subtle given the severity of the haemorrhagic sequelae shown at SW1.

Right and left fornix morphologically normal in spite of previous haemorrhage (refer CT at time of injury) and residual haemorrhagic staining at contemporary SW1 study.

Conclusion:
The previous study of 31/5/12 showed very extensive trauma related microhaemorrhagic sequelae as described in detail, with particular involvement of mesial temporal lobes and fornices. The current study shows a rather minor degree of left mesial temporal atrophy only and with no associated signal change on FLAIR."

On 27 June 2012, neurologist, Dr P. wrote:

"I reviewed Mr Sinnamon at Monash Neurology Clinic today. He attended the clinic with his supportive friend/carer.

Since he was last reviewed he was completed his one unit that he was studying for his master's course in information technology. His carer noted that he was probably suboptimal but with help and regular teaching and input, he has just managed to pass the unit. Generally, normal students apparently can do four units in about a three month period while he has taken four months just to pass one unit studying full time. This demonstrates that his capacity to learn new things is low but it is understandable in the context of significant previous brain injury.

He has also had his MRI scan of the brain which shows extensive changes related to severe brain trauma. With respect to memory issues there is extensive old haemorrhagic damage of both mesial temporal lobes and left hippocampus as well as left and probably right fornices. The extent of the haemmoraghic changes is generally unchanged compared to the previous scan from the Royal Brisbane Hospital. I have explained the findings to the patient and the carer.

Neurological examination is unchanged.

Mr Sinnamon has an information technology background before he had the accident. But would encourage him to continue with memory stimulating task as well as physical exercises to improve his physical ability as well as his mental capacity. I will leave ongoing follow up with rehabilitation team and I have not organised any further neurology follow up at this stage. They also noted that he had elevated TSH level consisting for a period of time and I would recommend that he has an endocrinology assessment to consider appropriate treatment. I would also appreciate your follow up as required."[TSH stands for Thyroid Stimulating Hormone, which tends to be elevated when the thyroid is not putting out enough thyroxine for normal metabolism.]

APPENDIX FIVE: Second Neuropsychologist's report July 2012

Dr S- wrote:

"Presentation

James presented at the current review as a warm and friendly man. He was somewhat socially awkward, and made minimal eye contact. He was also observed to lack some awareness of social behaviour and boundaries, being at times blunt and loud when coming into reception. He was otherwise very polite and gentle in his manner. He was evidently passionate about his computing and his website and enjoyed talking about this in detail. In doing so, he did tend to be over-inclusive and tangential in answering questions however, he was able to pick up on this himself and apologise whilst reigning himself in. He showed some insight into his disorganisation and distractibility as well but to some extent sees this as a longstanding part of himself that is not necessarily problematic. There did not seem to be any anxiety present during assessment or on any sessions with James. His affect was mildly blunted but he could be reactive and spontaneous at times. He described himself as happy. Overall, he was an articulate and obviously intelligent man with a wide range of interests that he found it hard to prioritise and manage.

Neuropsychological review results and observations

On review, verbal and nonverbal intellectual testing was not repeated given that this was not showing any signs of deficit on prior testing – in fact it was at a very high level. The only subtest carried out in the nonverbal domain a measure of attention to visual detail as this had been raised as a source of concern by Sheila. However, this remained strong on testing.

Attention & working memory

Attention and working memory skills were reviewed and found to be broadly in keeping with past testing, with the exception of one improvement.

Compared to 2011 James's digit span (reciting 5 digits forwards reliably; six unreliably; 5 digits backwards reliably), remained similar, with him showing an adequate capacity for taking in and mentally manipulating material in a brief auditory store. He also continued to perform in the high average range on a mental forward and backward sequencing measure. However, he showed considerable improvement on a task requiring him to mentally reorder numbers and letters (superior range; previously average range).

Memory and learning

Memory and learning skills also showed variable results, with James continuing to perform in the average range of ability in the verbal domain, but showing marked improvement in his recall of a complex spatial design. This may represent a practice effect as alternate forms were used for verbal memory but not for visual. However, even so – that in itself is an indicator of retention of visual material over time.

Executive functioning

A questionnaire tapping into higher level/executive skills continued to show that this domain caused the most significant issues at a functional level for James. Organisation and planning, as well as time management were of predominant concern, with distractibility and self-monitoring remaining somewhat problematic but at times manageable. He was also considered to be slow to process information, and therefore he tended to miss details. On top of this it was reported that he would often forget instructions unless he wrote them down, and he would frequently lose things. In addition, group conversations were observed to be hard for him, whereby he would miss the point. However, one on one he could manage. In positive terms, James was seen as flexible with areas of insight, and some capacity for problem-solving. He was also able to logically think through ideas and come up with well developed discussions for his political blog.

On testing, James was able to show good organisation and planning in his drawing copy of a complex design which was contrary to past descriptions, and he could plan ahead effectively on a verbal task. Mental flexibility and set shifting were also found to be improved to premorbid levels on a range of measures (superior range). In addition, verbal fluency had improved significantly (superior range). Qualitatively, James continued to show some impulsivity, mild self-monitoring weakness and distractibility. He could also be tangential and lose his train of thought.

On testing, James was able to show good organisation and planning in his drawing copy of a complex design which was contrary to past descriptions, and he could plan ahead effectively on a verbal task. Mental flexibility and set shifting were also found to be improved to premorbid levels on a range of measures (superior range) Qualitatively, James continued to show some impulsivity, mild self-monitoring weakness and distractibility. He could also be tangential and lose his train of thought.

Conclusions and Recommendations

In summary, review assessment has demonstrated that James has continued to make gains since his head injury in 2010. It is evident that he had a high premorbid intellect (superior to very superior range), despite a history of disruptions to his capacity to reach his potential at school and in the work place. Moreover, this intellectual capacity has remained largely intact. In addition, it is now apparent that many of his other cognitive skills are returning to that high level, with test results conveying significantly improved organisation and planning, mental flexibility, and verbal fluency, as well as some stronger (albeit variable) working memory performances, and visual recall. Verbal memory remains below expectation but within a sound average range.

James' high premorbid capacity has been an important buffer to the traumatic effects of his brain injury, such that the deficits he has sustained have meant that skills that were high average to superior —although suffering a major drop – have not gone below average level function. This will however, be experienced by him as a deficit and he will continue to require psychological as well as cognitive adjustment to his changed brain.

The residual concerns pertain most prominently to the aspects of executive functioning that are more difficult to test, such as managing competing demands, time management, tangential thinking, and distractibility -and impulse control was also having a continued functional impact. These difficulties were observable during sessions and have been consistently reported over time as problematic. James also remains slow to process large amounts of verbal information and his capacity for taking information into a brief auditory store is unreliable. Therefore, he is vulnerable to misunderstandings or missing key details unless the information is broken down and repeated. Importantly, with the right approaches he is capable of learning but will need to structure things to enable him to meet his personal expectations which are often overly demanding given his deficits. In effect, the focus subsequent to this assessment has been to work with James and Sheila in changing his approach to goal setting, study, time management and organisation in particular, with tailored strategies. These strategies are outlined below. It is hoped that James can practice these techniques and in so doing adapt to his changed brain through the development of new habits and approaches. It came across through the sessions that one of the biggest obstacles for James as well as his most important buffer is his retained high intellect, as it means that he finds it difficult to adjust to the concept that he needs to process information and carry things out in a new way. However, it also means he retains the potential to improve and learn – with effort.

She provided Queensland Workcover with the following summary:

-What are the cognitive and behavioural deficits that James currently exhibits?

As detailed in the report, James continues to display a range of consistently identified executive deficits including reduced organisation and planning, distractibility, tangentiality and mild problems with impulse control.

James also shows significantly impaired verbal memory and learning as well as working memory/immediate auditory attention span compared to premorbid expectations.

- Are these deficits consistent with the nature of the injury sustained on 18/05/2010?

These deficits are consistent with the nature of his head injury and have been reliably demonstrated over time.

- What are the implications arising from these deficits?

The implications are that James finds it very hard to attend to, process and therefore store new information, compared with his premorbid self. His skills are at an average level, but this is not in line with his intellect, and therefore his personal need and expectations. Overall, the process is likely to feel very effortful and stressful for him, and he is easily distracted from his path. He needs to learn to cope with this major shift, and adapt to new approaches and strategies.

James also becomes easily overwhelmed by large amounts of information, or competing demands, as well as novel tasks. He shows a reduced ability for imposing structure and his impaired strategic organisational skills impact on his ability to learn new strategies effectively. Adding to this, he often goes off on tangents, so tends not to stay on task. He completes tasks less efficiently - if at all.

James' conceptual reasoning is also not as strong as it used to be, so he may appear relatively rigid or concrete at times, which can cause confusion and misconstrual of information. Communication needs to be appropriately tailored to minimise problems in this regard. In addition, he needs ongoing development of his impulse control/self-regulation so that he does not get as easily distracted or remain vulnerable to misunderstanding.

James needs to learn to adapt to his changed brain, which is made possible by his retained high intellect, but at the same time hindered by this as he expects to be able to achieve more than he can. He also needs intellectual stimulation, therefore takes on too many demands than he is yet capable of managing or organising, and he will respond on impulse to something that piques his interest or sense of social justice.

In effect, all of James' deficits interact and impact on the process of learning strategies, and developing new techniques. Having said that, he has important intellectual potential which can be tapped with the right, albeit, necessarily consistent and involved support over a period of time before this can gradually be reduced to allow greater independence.

- What are the rehabilitation needs for James to help to increase his level of function?

A range of strategies are outlined below that have been commenced with James. He needs external support to encourage consistent practice of these techniques to move them towards becoming habitual. He also needs help structuring and ordering his environment. (see also comments above).

- What is the prognosis regarding a return to the workforce?

James would be capable of repetitive low skilled work at present, but would probably find it hard to maintain his concentration, due to the lack of intellectual stimulation and distractibility, whilst fatigue continues to be a problem. He could attempt a graded return to work of a higher level (albeit in a structured environment)-with support. Support via CRS or case management through ARBIAS or CBDATS at Royal Talbot would be of benefit if he is eligible. Ultimately, if he is given the opportunity to develop his skills and adapt to his changed brain, he will be capable of a full time work load in an appropriate domain - at a cognitive level. His fatigue levels however, may continue to be a problem due to the effort required.

- What type of work would be suitable for James based upon his cognitive and behavioural abilities as they relate to James work-related injury?

James may be best to attempt a graded return to his well-learned skill in computing, as this remains a strong skill, is familiar to him, and can be routine. This may potentially involve him undertaking a one-to-one tutoring role where he could be selective on type of student, level of demand, and times allocated (to minimise fatigue and maximise attention). Research into the options available in this domain of work would be needed. Given his desire and interest in social issues, he may also want to look into a graded return to work in the health care domain, such as being a patient advocate. He will need a flexible and accommodating work environment.

- Is the work-related condition now stable and stationary?

At this point in time two years post injury, his injury is likely to be fairly stable, but not stationary, in the sense that he has the capacity to develop increased functional ability through the right cognitive retraining and practice. He has the intellect to carry this out, but needs to streamline his activities, and environment in order to enable this to happen. He will also need guidance and support in ensuring he carries out the strategies consistently, with a gradual stepping away, as he develops them into useful habits that no longer require external support. More intensive cognitive support would be ideal as he is still adjusting to his changed processing skills, and gradually gaining better insight into his limitations in amongst his retained abilities.

Rehabilitation for James

The primary goal for rehabilitation continues to be around executive function – including organisation, planning and time management. James has been attempting to keep a daily diary of his plans, timing estimates and actual times to try to develop insight.

Training has been commenced around developing checklists and daily plans as well as task related plans (see attached).

James needs to learn to set realistic goals and try to manage his time in a way that allows him to see them through. A time management chart has been implemented to develop insight and enhance this skill (see attached).

James needs to take more effective and organised notes, so that he does not miss important details and so that he can then readily refer back and utilise the material. A structure for doing so has been provided to him (attached).

James needs to make regular use of a diary. It needs to be filled out in an orderly manner and checked at routine intervals to be of use.

James needs to have a memory/organisation corner at home where all important items are kept including a white board for messages, time tables and lists.

External help is needed to work with him towards developing order and structure in James' environment so that he can start with a clean slate towards creating routines and organisation.

To minimise the impact of his distractibility and avoid him feeling overwhelmed, James needs to tackle large, complex projects by breaking them into parts and setting himself very small manageable goals. That way his progress and achievement is more tangible and he will be better able to plan ahead.

Goal setting needs to be more appropriate and targeted, with anticipation of potential obstacles and progress monitored (see attached).

James needs to be encouraged to develop visual flow charts and diagrams when planning his approach or learning something new. This suits his way of thinking and information processing /tech knowledge.

Motivation can be enhanced by improving insight and recognition of the utility of the strategies suggested- with support and monitoring charts. That is, encouraging James to set realistic and meaningful goals, and seeing those achieved through guidance in effective planning and organisation, and charting his progress. As his confidence and motivation in using these techniques increases, the support can be gradually removed.

Eventually – repetitive practice of these techniques should help them become habitual. He has the cognitive ability to store these new strategies –with focus and repetition.

This will require a frequent and constant program for a significant period of time. However, James has a great deal of potential given his intellectual ability so this is deemed extremely worthwhile and important.

It was reported that James has gained a great deal of mental stimulation through music and participation in a choral program. This was considered to have helped his mood, and sense of achievement as well as stimulating his thinking and learning. Therefore, it is suggested that this should continue.

All of these techniques at this stage require someone working closely with him to ensure he puts them into place correctly and consistently, and learns to monitor and therefore see the outcomes. At present, he seems to be using somewhat of a trial and error approach, and will apply a technique only briefly or in an incomplete way."

INDEX

ENDNOTES

[1] Cathy Crimmins, Where is the Mango Princess, Vintage Books, Random House, New York, 2001. Tragically, this brilliant woman and wonderful writer died aged 54 in 2009 of complications from surgery.

[2] Yes, I made this word up. Felt that English needed it.

[3] Despite some pejorative uses, the term 'spastic' is a technical term, which, as I describe in the text, means that muscles are involuntarily tensed as opposed to relaxed.

[4] The term 'grand mal' comes from the French, 'grand', meaning 'big' and 'mal' meaning, 'illness', 'pain', 'indisposition' etc. It may have become current in English medicine as a euphemism for a condition that carried a stigma. It is considered a bit old fashioned to use it these days, when the official term 'tonic-clonic seizure' tends to be substituted. 'Tonic' here refers to the involuntary stretching of the muscles and 'clonic' to their relaxing. When the muscles are stretched the patient cannot draw breath. The tonic stage is often preceded by 'the cry', when the sudden universal tensing of all the muscles drives breath out of the body at the same time as the patient loses consciousness. The cry is followed by 'the fall', as the patient, losing consciousness, falls to the ground. The 'clonic stage' follows and then the person can breathe. If there is no further fit, the patient usually goes into a sleep and often feels pretty good when they wake up. Sometimes the fit does not end there and there may be 'automatic behaviour' instead, which is a prolongation of the original seizure, taking a different form, with impairment of consciousness rather than total unconsciousness. The most dangerous thing that can happen apart from injury as the patient falls, is repeated tonic-clonic seizures that prevent the person from breathing over a long enough period to cause brain injury. Repeated tonic-clonic seizures are called 'status epilepticus' and are a medical emergency. They are usually treated with a benzodiazepine, such as clonazepam or diazepam. When a person is in status epilepsy they cannot take medications by mouth because their teeth are clenched during the tonic phase and they are not conscious and have no voluntary control of swallowing during the clonic stage, even though the teeth unclench at this point.

[5] Lyrysa Smith, *A Normal Life: A Sister's Odyssey through Brain Injury*, February 2014 https://www.createspace.com/4227185.

6   This is certainly like the process that Lyrysa Smith describes in *A Normal Life: A Sister's Odyssey through Brain Injury*, Op. Cit.

7   The Glasgow Coma Scale was developed in 1974 at Glasgow University by professors of neurosurgery, Graham Teasdale and Bryan J. Jennett. You can find copies of this standard test in many places on the internet.

8   The source of most of this information is N.E.V. Marosszeky, L. Ryan, E.A. Shores, *The PTA Protocol, Guidelines for using the Westmead Post-Traumatic Amnesia (PTA) Scale*, published by the Department of Rehabilitation Medicine, Westmead Hospital and the Department of Psychology, Macquarie University, http://www.psy.mq.edu.au/pta/page8.html

9   For instance from the Department of Psychology, Macquarie University, NSW, where a form for scoring was also available in December 2010.

10 David Talbot, *Brothers, the hidden history of the Kennedy years*, Free Press, 2008

11 Oliver Stone, *JFK*, 1991.

12 Kurt Vonnegut, *Slaughterhouse-Five*, Dell Publishing, 1972.

13   "The motor signs that can be seen in akathisia consist of an increased, abnormal frequency of movements. The movements are often complex and stereotyped, that is, repeated in the same pattern over and over." Stewart A. Factor, Anthony E. Lang, William J. Weiner, *Drug Induced Movement Disorders*, Blackwell Publishing, Massachusetts, 2005, Chapter 10, p 271.

"Motor activity may be inappropriate, purposeless or semi-purposeful and non-goal directed, but it is not abnormal in the way it clearly is in the other extrapyramidal syndromes associated with antidopaminergic drug exposure. By this, what is essentially meant is that movement does not appear disrupted by *involuntary* mechanisms. [Here the author would be referring to obviously involuntary tics and spasms of muscle groups, such as produce cramps and torticollises.]

The distinction between what is voluntary and what is involuntary in motor behaviours is, one has to admit, sometimes more gut reaction than keen observation. It certainly does not spring from the application of specific rules. For example, the fact that sufferers from akathisia can, at will, temporarily suppress or abandon their aimless overactivity, helps us little, for the same is true of those with tardive dyskinesia, a syndrome comprising features universally – and, indeed, by definition – considered *in*voluntary. By the same token, no-one suggests that the motor activity of those with depression or generalized anxiety is involuntary, yet in some such cases one would have to tie the patient to a chair to prevent hi or her from moving."

"The restless movements most commonly associated with the subjective experience of akathisia are not dyskinetic or stereotypical, but resemble normal patterns of restless movement. Most typical are lower-limb movements, such as rocking from foot to foot and walking on the spot when standing, and shuffling and tramping of the legs or swinging one leg on the other when sitting. Pacing rapidly up and down is a characteristic of severe akathisia, and in the worst cases, patients are unable to feel comfortable in any position, such as sitting, lying or standing, for more than a few minutes. In addition, trunk rocking and fidgety movements of the upper limbs may be seen. The severity of the movements in akathisia can vary according to the situation and the patient's degree of arousal. For example, the movements may be less obvious during an interview or while concentrating on some mental task, and more evident when standing engaged in conversation on neutral topics (Barnes, 1992). Thus, limiting the assessment of akathisia to a brief, formal examination runs the risk of underestimating the presence and severity of the condition." Stewart A. Factor, Anthony E. Lang, William J. Weiner, *Drug Induced Movement Disorders,* Blackwell Publishing, Massachusetts, 2005, Chapter 3, "Spontaneous movement disorders in psychiatric patients." This chapter also gives some idea of the difficulty in separately defining stereotypical behaviour, compulsive behaviour, complex tics and akathisia. The choice of term often finally relies relatively arbitrarily on the setting and the patient's diagnosis.

See also heading, "Acute and Chronic Akathisia," in an article by Maurice Gervin and Thomas R.E. Barnes, "Assessment of drug-related movement disorders in schizophrenia," in *Advances in psychiatric treatment,* http://apt.rcpsych.org/content/6/5/332.full for a quick run-down, or the chapter Akathisia in D.G. Cunningham Owens, *A Guide to Extrapyramidal Side-effects of Antipsychotic Drugs,* Cambridge University Press, UK, 1999, pp 130-162 and also Chapter 11, "An overview of some standardized recording instruments," which shows how primitive the science of describing and diagnosing these problems is.

14   "[The] syndrome can be so distressing that it has been associated with aggressive behavior and violence or suicide attempts." Stewart A. Factor, Anthony E. Lang, William J. Weiner, *Drug Induced Movement Disorders,* Chapter 7, Blackwell Publishing, Massachusetts, 2005, p143.

15   For information about this life-threatening syndrome which occurs in conjunction with antipsychotic medications, also known as major tranquilizers, see, for instance: Nissar Shaikh, Ghanem Al-Sulaiti,1 Abdel Nasser,1 and Muhammad Ataur Rahman, Neuroleptic malignant syndrome and closed head injury: A case report and review, Asian J Neurosurg, v.6(2); Jul-Dec 2011, P.M.C3277062http://www.ncbi.nlm.nih.gov/p.m.c/articles/P.M.C3277062/

"Abstract

Neuroleptic malignant syndrome (NMS) is a rare, but potentially lethal neurological emergency. Fifty percent of traumatic brain injury (TBI) patients will have emotional disorders and post-traumatic agitations. Haloperidol is a neuroleptic antipsychotic medication commonly used in the traumatic brain injury patients due to its advantage of no effect on respiration and conscious level. But it is one of the common medications causing NMS. ..."

"Risk factors

The main risk factors for the develop.m.ent of NMS are[12] (i) high doses of neuroleptics (ii) rapid increase in dose of neuroleptics (iii) parentral neuroleptics (iv) use of highly potent neuroleptics (v) preexisting central nervous system disorders (vi) dehydration (vii) young patients (viii) history of NMS (xi) concomitant use of predisposing medications (lithium) (x) acute medical and surgical illness (infection, trauma and surgery) (xi) anti-Parkinsonism medication withdrawal. Recently, a case is reported where high protein enteral feeding causing decreased levodopa (anti-Parkinsonism medication) concentration leading to NMS or neuroleptic malignant like syndrome."

"Pathphysiology

NMS is a hypodopaminergic state of the brain. Dopamine has a major role in autonomic cardiovascular stability, hypothalamic temperature regulation, maintaining the conscious level, and normal muscle tone. NMS results of either altered dopaminergic transmission or blockade of dopaminergic pathway or changes in pre/post synaptic dopamine signal activity. Neuroleptic causes blocked of the dopamine receptors, adding to it the traumatic brain (TBI) injury patients, due to diffuse axonal injury had a decreased dopamine neurotransmission. This combine effect leads to hypodopaminergic state and causing signs and symptoms of NMS.[11]"

"Literature about NMS associated with head injury patients is limited to the case reports only. Total nine cases of NMS are reported with use of haloperidol ... in the traumatic brain injury patients. ... Younger patients are at higher risk for the develop.m.ent of NMS, it is twice more common in male, the reported incidence of NMS with use of haloperidol is ranging from 0.02 to 12.2%."

A comprehensive chapter on Neuroleptic Malignant Syndrome can be found in Stewart A. Factor, Anthony E. Lang, William J. Weiner, *Drug Induced Movement Disorders*, Blackwell Publishing, Massachusetts, 2005, Chapter 8, pp174-212. The book may be downloaded for a daily fee from http://www.scribd.com/doc/34790337/Drug-Induced-Movement-Disorders-2nd-Ed#download

16   Since the rise of HIV infection risks, hospitals have instituted yellow plastic bins for safely storing sharp instruments that have been in contact with body fluids. There was a bin like this on the wall next to James's bed and locker.

He simply mistook it for a safe receptacle and stuffed his glasses in the one-way aperture in the top. Although everyone knew the glasses were in there, no-one was willing to do anything to retrieve them. Although there must have been a protocol for opening the bins in cases of emergency, hospital routines and bureaucracy meant that doctors, unit managers, nurses and nurse assistants felt it was not their job. It took James months to find his old prescription and to get new glasses, until which time his view of the world was rather like Mr Magoo's, although he was able to squint to see things close up and to read still.

17   Drupal is a kind of software architecture for websites. It is open source and freeware.

18   Linux is an alternative computer environment to Microsoft Windows. Whereas you pay for Windows, Linux is often available as 'free ware'. Whereas it is illegal to alter or resell Microsoft Windows without a commercial licence, Linux is open source and anyone can do what they like with it. Linux and Microsoft Windows are just two of many different 'operating systems'.

19   A bit more technical stuff about website management and different operating systems: Websites may exist inside their owners' computers but that is not usually how you and I access them. They are generally uploaded to and then 'hosted' on relatively powerful remote specialised computers, called virtual servers or 'v-servers'. These may be non-profit sites owned by governments or universities, but many are commercial and make their money from operating many v-server computers which they rent out to people who want to publish websites, which the host service then maintains for a fee.

A lot of web-hosting companies offer services whereby anyone can simply enter information into pre-formatted internet pages – like wikipages - or upload Microsoft-formatted html pages. That was not the case with candobetter.org. It wasn't one of those one-size-fits all that can be cheaply and easily maintained by a commercial service. James did not use the Microsoft operating system. He used 'opensource' programs to design a unique format only accessible to technicians familiar with the Unix operating system – more popularly identified with Linux. (Unix is the original open source, free operating system from which Linux is derived. Linux is the name under which a lot of open source operating systems are developed. Most, but not all, are free.) Drupal was his open source content management system. It was written in PHP. PHP is a general-purpose "server-side" scripting language. (Some thanks are due to various Wikipedia definitions, which I paraphrased here.) Such sites are relatively difficult to hack compared to commercial Microsoft.

In James's case, he maintained http://candobetter.org himself for several reasons. One was to control access to the identity and contact details of any users and contributors to protect them from political persecution. Another was because it is much harder to hack Linux operating systems and Drupal than popular commercial ones like Microsoft. Yet another reason was to reduce management costs from thousands of dollars to hundreds of dollars annually for such a large, complex and unique site. We both spent our own time and money to keep the site non-commercial because sponsorship and ads inevitably cause compromise. This is why, for us, time was so much more important than money, and the primary importance of our paid employment elsewhere, was to pay for our time spent writing, editing, managing and developing candobetter.org., for which we also had plans to extend to an international cooperative publishing house and institute of higher learning. The only commercial part of the site was the virtual server provider, known as the 'host', to whom we paid bills. Sometimes the other contributors kicked in too. The hosting was provided by a firm in Melbourne, located in a university, who rented out the 'V-servers' for the purpose of providing space to site builders and developers with computer programming and software design knowledge. The Melbourne firm would provide professional management, but only for a very high price – like several thousand annually - since management of a non-standard non-microsoft site required special skills and was quite complex and time consuming.

Prices were only just beginning to come down as more and more people entered the business of web-hosting in different countries. For instance, you could host websites in Australia, the United States and Canada using low-paid staff in Russia, former Yugoslavia, or India. Big Websites where people have gone to a lot of trouble to store information and provide a reliable interactive environment often have 'mirror'-sites in several different locations in order to be able to restore sites that have been hacked or suffered other mishap. Despite the fact that Linux operating systems are much safer from hackers than Microsoft, the political nature of our site meant that Candobetter had suffered a number of hacking attempts and we had recently rented a prospective mirror site located in another country to minimize potential damage and secure our data.

[20] His failure to recognise his half-brother at first could have been because James remembered D. as a young boy, whereas D. was now a mature adult.

[21] Cunningham Owens, D.G., 1999. A Guide to the Extrapyramidal Side-effects of Antipsychotic Drugs, Cambridge, Chapter 11.

[22] Olanzapine comes in doses of 2.5mg up to 20mg, with the recommended dose not to exceed 20mg in one day, however that dose is frequently exceeded, typically at 30mg and above in acute psychiatry, with many consequences, of which diabetes, obesity and associated heart and vascular problems are well represented.

[23] Mysiw WJ, Bogner JA, Corrigan JD, Fugate LP, Clinchot DM, Kadyan V., 2006, August. The impact of acute care medications on rehabilitation outcome after traumatic brain injury. Brain Inj. 2006 Aug;20(9):905-11. The authors were located at Department of Physical Medicine and Rehabilitation, The Ohio State University, Columbus, OH 43210, USA. mysiw.1@osu.edu. Here is the entire abstract:

"OBJECTIVES: To examine the impact of medications with known central nervous system (CNS) mechanisms of action, given during the acute care stages after traumatic brain injury (TBI), on the extent of cognitive and motor recovery during inpatient rehabilitation. DESIGN: Retrospective extraction of data utilizing an inception cohort of moderate and severe TBI survivors. METHODS: The records of 182 consecutive moderate and severe TBI survivors admitted to a single, large, Midwestern level I trauma centre and subsequently transferred for acute inpatient rehabilitation were abstracted for the presence of 11 categories of medication, three measures of injury severity (worst 24 hour Glasgow Coma Scale, worst pupillary response, intra-cranial hypertension), three measures of outcome (Function Independence Measure (FIM) Motor and Cognitive scores at both rehabilitation admission and discharge and duration of post-traumatic amnesia (PTA)). MAIN OUTCOME AND RESULTS: The narcotics, benzodiazepines and neuroleptics were the most common categories of CNS active medications (92%, 67% and 43%, respectively). The three categories of medications appeared to have no significant outcome on the FIM outcome variables. The neuroleptics affected cognitive recovery with almost 7 more days required to clear PTA in the neuroleptic treated group. The presence of benzodiazepines did tend to obscure the impact of neuroleptics on PTA duration but the negative impact of neuroleptics on PTA duration remained significant. CONCLUSIONS: [My emphasis] The results suggest that the use of neuroleptics during the acute care stage of recovery has a negative impact on recovery of cognitive function at discharge from inpatient rehabilitation. Due to the paucity of subjects with hemiplegia in this cohort, conclusions could not be drawn as to the impact of acute care medications on motor recovery."

24 Transciption of second CT report:

CT Head JAMES, JAMES PATRICK [...]
*Final Report"

Reason for Exam
1901879 8BS

CT Read

HISTORY:
PBA. Slow recovery. Ongoing agitation. Progress of SAH/TBI.

TECHNIQUE:
Non contrast CT brain.

FINDINGS:
Comparison is made with the previous CT dated 20/5/10

There is interval resolution of the subarachnoid blood that was previously seen in the left lateral ventricle and at the foramen of Monro. There is also a decrease in size of the hypodense lesion at the foramen of Monro, which now measures 5mm in width (previously 9mm). This would be in keeping with a resolving haematoma.

No acute intracraial haemorrhage detected. No extra-axial surface collection. The cerebral ventricles, cortical sulci and basilar cisterns are normal in size. No midline shift. Grey/white matter differentiation preserved.

Cervicomedullary junction unremarkable.

Interval resolution of ethmoidal air cell opacification. The paranasal sinuses are now clear. No skull abnormality.

COMMENT:

Interval resolution of subarachnoid haemorrhage. No acute intracranial pathology detected. No ventriculomegaly.

Transcribed : 02/06/10 16:27
Signed: 02/06/10 17:55

*Order by [Consultant] on June 01, 2010 14:56

25   Valium is a brand name. Diazepam is the generic name. The drug comes in adult doses of 2mg, 5mg, 10mg.

26   Cathy Crimmins, Where is the Mango Princess? A journey back from brain injury, Vintage Books, 2001.

27   The term 'nocte' just means 'at night'.

28 I am inserting this note in March 2016 as I conclude my last draft, because I cannot remember whether I included the very important difference in attitudes to drugs between neuropsychology and psychiatry. I apologise if I mention this elsewhere in this book. Psychiatry, during my experience of it from the 1970s to 2016, has increasingly relied on psychotropic medications and it has been argued that the *Diagnostic and Statistical Manual of Mental Disorders* (DSM) has recrafted many diagnoses to reflect Big Pharma's promotion of drugs designed for specific diagnoses. Neuropsychology and neurology, however, tend to use neurological, environmental and behavioural explanations and theories to help learning, and to modify attitudes, behaviour, mood and affect. This approach also tends to involve the patient much more. In my view psychiatry would benefit greatly by taking a similar approach, exercising much more restraint in the use of medications.

29   For instance, Microsoft itself uses Drupal platforms for security on its websites, but makes available a lot of automatized software to users.

30   Although he recognises the routine now, in August 2012 scissors in his pencil case were confiscated at the airport prior to a flight to Brisbane. The same

thing happened on the way back, because he had forgotten the problem and packed a new pair of scissors in the same pencil case.

31   In August 2012 he read 38 pages of Phillip Roth's, *The Plot against America*, 2004 on the same flight route. On the bus to the city, he described the plot so far. Coming home on the train (one hour's ride) I noticed that he had got up to page 68.

32   Tibrogargan is the name of a gorilla-head shaped mountain in the Glasshouse Mountains in Queensland. This visually fascinating range was named by Captain Cook, who spotted them from the coast. Tibrogargan is formed from a volcanic plug of reolite about 27 million years old.

33   Kazak was Winston Nite Rumsfoord's dog in Kurt Vonnegut's *Sirens of Titan*.

34   Grammos Mountain in Greece was a major communist stronghold and location of the provisional democratic government during the Greek Civil War. The United States which backed the Greek Government of the time used its new weapon, Napalm B, on the Greek fighters located there.

35   Perhaps it is similar to the searches people make to explain more ordinary suffering, wondering why they are being victimised. The disturbed consciousness of acute brain injury however must add special isolation and loneliness.

36   Cathy Crimmins in her book, *Where is the Mango Princess*, Op Cit., also gave importance to the subject of her daughter's dog and her brain-injured husband's changing response to it. I hasten to emphasise that James always tried to be kind to our dogs even when he was still very unwell.

37   RJ45 cables are Ethernet cables that bundle wires together to connect computers or telephones. They have those little transparent clips you can insert in the backs of computers and into routers. They come in various shades but James generally used blue ones, about half a centimeter thick. Because he liked to have computer access everywhere in his house in Brisbane, these wires snaked along the floor, walls and ceilings from his study at the back of the house, through the kitchen, the living room and the hallway, to his bedroom at the front. Because they were so visible, particularly in an era where people tended to replace cables with wireless, we identified them with James's thinking. As well as preferring cable for security reasons, he also tried to avoid incremental exposure to different kinds of radiation. Acknowledging that wireless transmission was not known to carry dangers he was nonetheless aware that any wave had to act on body cells to some degree. Since he also used a mobile phone, his idea was to limit exposure as much as possible.

38 Rereading this in March 2016, when we are visiting James's Brisbane residence with view to restumping, I can say that James has only one diary now, but calendars in several rooms. He still has multiple notebooks in Frankston, mostly defined by room and use. So he has one for the tv room, to document controversial news items, and another that should live in his bag, which he uses for notetaking when away from the house. All the notebooks at the moment are identical red-covered ones, which does not help when one goes missing. However he has only brought his diary and one notebook to Brisbane and I sense that he is becoming more reliant on his memory. We have forgotten our main calendar and we have had to replace the cheap wall-clocks which have gone missing in Brisbane. My father, now aged 92, is more on the ball than anyone in the household because he uses Google diaries – although if he loses access to his computer this can cause chaos. He downloaded Microsoft Windows 10 in a free download and when he restarted his Toshiba laptop (a generous gift from me) it would not restart. He had to pay a professional to recover his data, but found that his the windows email handling system he had relied on for his google mail and diary was somehow inaccessible, due to it failure to recognize his new operating system or something. The expense, stress and missed appointments for my father and my mother that resulted from this have severely tested my father's faith in Microsoft and merely confirmed my lack of faith in it. However, when he was finally able to restore the google calendar or diary system, we all heaved a sigh of relief, because, when we absolutely have to remember a date, we tell my father to put it in his google calendar. In those circumstances we defeat our attempts to remain unsurveilled. In truth, however, since we both now use gmail, which monitors headings and content of emails, we are kidding ourselves about security. We need more secure emails. However, since James uses a mobile telephone, he is also trackable. Since I use automatic tellers and pay most things with my keycard, I am also trackable. Increasingly I consider google calendar with desire, but have so far not succumbed.

39   In 2006, in reponse to Liberal Government minister Ruddock threatening to codify a federal law on the subject, every state in Australia began to liberalise its draconian defamation laws. Ruddock is widely hated for his response to asylum seekers – probably undeservedly – but his great positive contribution to Australian freedom of expression has unfairly gone almost unheralded. One reason might be that the old laws suited the Australian newsmedia magnates. For most writers and researchers, the old laws had been so punitive and unreasonable that most public debate was, until 2006, effectively censored. Biographies and autobiographies were absurdly lacking in detail, and only journalists who could rely on wealthy media owners to protect them from law-suits by politicians dared report anything that might affect a person wealthy enough to sue. Several prime ministers were said to make a good living from legal suits. The mainstream newsmedia in Australia, then as now, was the exclusive province of wealthy individuals who almost always had an interest in preserving the status quo and who might use their ability to publish to bring politicians into line with their commercial interests. Until 2006 only parliamentarians might speak freely, under 'parliamentary privilege' but what they said might not be repeated outside parliament. Writers should remain aware of the rules of defamation for each state, but once aware, it is easily possible to report almost anything fairly and not run the risk of losing your home.

40   "Basically, the main reason I've stood was I wanted to ensure that people had something better to vote for than what was on offer from the major parties or from what I regard as the flawed political party known as the Greens which has some worthwhile policies. Perhaps if I had got together with Dave, before the elections, we could have worked it out and there would be one independent candidate instead of two. You know I'm not particularly standing for my own sake. But my own concerns, besides honesty are effectively the environment. I'm a person whose deeply worried about the state of the environment and I have a deep abiding fear we may be facing global warming and all sorts of terrible things. I think we've got a chance now to probably turn back from that but I don't think we stand much chance, unless we get rid of the political leaders that are now leading us to the abyss, so that is what I'm standing for. It's obvious that we are not a democracy in the sense of government of the people by the people for the people. The reason for that is if you just think of how many decisions are made by politicians that are opposed by the majority and ask how that is democracy. The obvious example is privatization. They have sold, since the Beattie Government came to power – SGIO - that was the State Government Insurance office, against an explicit promise by Beattie not to fully privatize it. Then, later on, they privatized TAB, and they privatized Dalrymple Bay coal-loader. Probably a lot of other things that they privatized even I haven't heard about because another comes up every couple of days, like the Gladstone power station flogged off in 1994. They flogged off a couple of the airports last year. And the list goes on. In spite of the fact that every public opinion poll shows that the majority of public opinion is emphatically opposed. So there's something wrong. And one thing I've tried to do is to force the politicians that probably have secret plans to go and flog off queensland rail and our water grid and probably our electricity generators, in the coming years. I've forced them to actually defend their election policies during an election campaign. I've actually challenged Laurence springborg and anna bligh and the treasurer, whom we are both standing against, to publicly defend – either commit themselves to not privatizing or publicly defend their arguments for privatization. And I have got no response. They will not debate the issue. They just want to wait until they're safely in power and go ahead and do it, regardless of what we think. I'm also deeply worried about the insane policy of digging up as much coal as we can in Queensland. We dig up a record amount of coal. We're contributing far more than our fair share to the coming global ecological holocaust, and yet our Premier anna bligh wants to dig up, to triple our coal exports, by 2030, which is just totally utterly mad. We have to at least stabilize and aim to reduce it, not increase it, whilst we are watching the polar ice-caps melt. I do run a website. It's really important that with this election campaign. The name of the website is *candobetter.org*. It's an easy one to remember. It's the one that I started up a couple of years ago. A lot more people are participating in it from around Australia and around the world, writing articles and people, if they log on, you can contribute your ideas, you can get accounts, you can even start up discussions about the election."

41   I explain this further in a 2015 speech here:
http://candobetter.net/node/3876 and
https://www.youtube.com/watch?v=xCodkZub2-I&x-yt-ts=1421914688&x-yt-cl=84503534.

42   Thanks to Greg Wood for summarizing this argument for me.

43   I have changed the name because there has been a change of ownership.

44   Rsync was an open source free software program for copying and updating websites, whereby one site adds its new files to another secondary site. Generally an rsync site is not as fully interactive as the main site, but is used for some supplementary function, such as providing an additional download source.

45   Servers don't invariably host websites; they sometimes host other internet applications including email and name-servers.

46   'Open source' software and even hardware refers to software and hardware which the user may re-engineer. Whole operating systems like Drupal Linux and Ubuntu come under this heading as well as small programs, such as 'e-cleaner'.

47   'Freeware', as the word implies, doesn't cost you anything.

48   Installation programs for Linux test what hardware the computer is comprised of and usually can automatically install the necessary software drivers.

49   It is of course possible to put an upgrade off for so long that it becomes impossible to integrate it into the latest version or to any version that is current. Scripts to automatically upgrade from version 6 are available, but earlier versions have to be manually upgraded by the user.

50   Normally globalization and its depressant effects on Australian wages and local industry would have been a source of political consternation for us, however there really is some chance of geeks participating in global competition. The thing that prevents a so-called 'level playing field' in this is huge variations in costs of land for housing, as rent or mortgages, in different countries. Australia's costs are amongst the highest in the world, making all but corporate enterprises uncompetitive. We also had a lot to say about energy costs in criticism of so-called economic efficiency and globalization, but you can read these at http://candobetter.net if you are interested.

51   He had mistaken the B3 for a home server.

52 Sheila Newman (Ed.) *The Final Energy Crisis*, Pluto Press, UK, 2008.

53 Because I was in a hurry I did not get James to check my final version of the chapter, "Nuclear Fission Power Options," and I managed to mix up the definition of atomic number, to my lasting embarrassment. When he looked at the published book, he picked this mistake up immediately, although no-one else ever did.

54 I was taught that hypothyroidism is completely reversible but recent reading has led me to believe that this may not always be so. (The same rules apply as for any brain insult.) Reversal may take somewhere around two years and it may still be limited. Reasons for varying outcomes and expectations may be poor testing of subjects, too superficial familiarity with patients and so underestimation of dysfunction and personality changes, and lack of close long-term follow up. On the brighter side, some reports seem to indicate that where long-term follow up occurs, patients who seemed not to be improving, did improve. Some research reports are:

Davis JD, Tremont G., "Neuropsychiatric aspects of hypothyroidism and treatment reversibility," Minerva Endocrinology, 2007, Mar;32(1):49-65, http://www.ncbi.nlm.nih.gov/pubmed/17353866

55 There are many references for this statement about hormones and brain injury. Here is just one: http://www.lapublishing.com/blog/2009/traumatic-brain-injury-pituitary-hormones/

56 Source: *Genetics Home Reference,* http://ghr.nlm.nih.gov/condition/hashimoto-thyroiditis

57 RAM stands for 'Random Access Memory', which is a term for a computer's working memory, i.e. what it keeps immediately available for tasks at hand, rather than large volumes of data less accessibly stored in files on the hard disk. So, for instance, if you are working on a file in Word, all the aids are there on the screen. If you want to call up another word file, you have to look for it on your hard disk and bring it up onto the screen, where it becomes available to the RAM.

58 Oliver Sacks, *Musicophilia*, Vintage Books, New York. I had read this and every single other one of Oliver Sacks's books before James was ever injured. They were wonderful books, and great resources for appreciating life and in psychiatric nursing, notably for mental status observations and descriptions. Now that knowledge was helpful for me in dealing with James.

59 That loss of control is much worse in the modern units attached to general hospitals, because there is no 'outside'. You are just stuck in a complex of walled rooms, with windows mostly looking over carparks. In the old mental hospitals there were extensive grounds with many trees and even some wild spaces between the buildings. The wards were nearly all on ground level and had verandahs where patients who could not walk, could still sit and see the passing parade of ambulant patients and staff with business outside their wards.

60
    I do not have a copy of Durham's book to check my recollection because it is out of print. I read a copy on an interlibrary loan in 2010. I later heard that Durham was doing a doctorate on brain injury and traveling round doing talks on the subject. There are resources associated with her work on the internet: http://www.bia.net.au/docs/keys_abi_bird_cage.pdf; http://hstalks.com/main/browse_talk_info.php?talk_id=2118&series_id=610&c =252

61
    Ghost Boy, Simon & Schuster Ltd, July 2012, is the autobiography of a man of very superior intelligence who went into a vegetal state when he was a young boy. Incredible is really the term to describe it. Martin emerged with help as a young adult and was able to tell what it was like and how he was treated during his apparently comatose state. He went on to become a computer engineer, specializing in making computers accessible to the brain-injured. His website is http://www.martinpistorius.com/. His autobiography is powerfully and beautifully written. One feels he would have been a great writer, even without his extraordinary story; his gift for clear-thinking and emotional expression is so unusual.

62
    A $ sign pre-pended to a string in PHP causes unpredictable problems.

63
    Peiyi Tang and James Sinnamon. Interaction and coordination for distributed grid computing. In *Proceedings of the 2004 IEEE International Conference on Services Computing (SCC'04)*, pages 383--390, Shanghai, China, September 2004. (http://www.ualr.edu/pxtang/papers/SCC04.ps http://www.ualr.edu/pxtang/papers/SCC04.pdf)

James Sinnamon and Peiyi Tang. Multipi: A java implementation of multiparty interaction for distributed computing. In Post-Conference Proceedings of the 2004 International Conference on Parallel and Distributed Processing Techniques and Applications (PDPTA'04), Las Vegas, USA, June 2004.

64
    On Wednesday October 12, 2011, James emailed his old professor, Peiyi Tang:

"I think you may remember me from when I implemented your algorithm for distributed coordinated processes in 2000 as your Honours Year student.

I hope things are going well for you at the University of Arkansas (http://ualr.edu/computerscience/about/faculty-and-staff/peiyi-tang/).

Since 2002 when I moved to the Australian National University my life has taken all sorts of turns, only a few of which you would be aware.

In 2004 after I had worked for two years very hard implementing a distributed java virtual machine (http://djvm.anu.edu.au) I was not given the PhD scholarship I was expecting and my contract was not renewed.

This did not help my professional self-esteem. It would have helped me if I was told that a Canadian student, Jennifer Baldwin (http://webhome.csc.uvic.ca/~jbaldwin/) used the DJVM for her Master's project. She commenced using it in April 2004 only weeks after I finished working at the ANU.

In 2008 I think, I learned from Jennifer that she experienced no problems with the DJVM.That was surely testament to the DJVM having been properly implemented and thoroughly tested by me, but no-one from the ANU bothered to inform me nor inquire about my well-being. I only learned some years after (2008?) that the DJVM had been used and had not been consigned to the shelf to gather dust.

I ended up performing low-paid menial work until May last year (although, to be fair, my hospital cleaning job at the Royal Brisbane and Women's Hospital kept me fit and the people I worked with were very nice.)

In my spare time I set up, administered and wrote for a (Drupal/PHP/Mysql) content-managed web site. (http://candobetter.net). It expresses poliitcal and environmental views that I think you will find quite novel. (It needs better structure, which I am working at providing now.)

I have also stood as an Independent candidate for Lord Mayor of Brisbane in 2008 and in the Queensland state elections in 2009. I didn't get many votes in either, but I still think that my stance has put some important facts in the public record, which those I stood against will find hard to walk away from in future.

On 18 May, I was almost killed in a road accident on my way to work. (I have no recall of the accident itself, but the available photographic evidence and the contradictory statements made by the driver and his passenger seem to confirm my own belief that I was not a cyclist given to taking stupid risks, Hopefully, I will obtain some settlement that will go some of the way towards compensating me for the damages.)

I have recovered pretty well although I am suffering diffuse axonal brain damage. That means I have lower mental and physical stamina, less ability to coordinate and some difficulties with memory. They tell me my intellect seems as good as it was but my memory problems and lower mental stamina cannot help.

Still, I am hopeful I can at this late stage apply my skills towards getting some income as a skilled worker and achieving something that others will notice.

I think I may be able to get multipi to work robustly and implemented in a form that will allow your algorithm to support more powerful distributed applications on the Internet (unless someone has already dne that but I haven't seen any evidence of it).

BTW I found multipi at

http://www.pubzone.org/pages/publications/showPublication.do?publicationId=538195 and

http://dblp.uni-trier.de/db/conf/pdpta/pdpta2004-3.html#JamesT04.

It's said to be a 'discussion paper', but I see no evidence of anyone having discussed it or having done anything further with it.

Hope I will be able to hear from you soon.

Thanks again for all your encouragement and support."

On Thursday, October 13, 2011, Peiyi Tang wrote back:

"Dear James,

Of course, I remember you. I was trying to find [word left out, could have been "you"] when I finished the paper about MULTIPI and submitted to one of the conference here. Actually, I wrote two papers as follows:

Peiyi Tang and James Sinnamon. Interaction and coordination for distributed grid computing. In Proceedings of the 2004 IEEE International Conference on Services Computing (SCC'04), pages 383--390, Shanghai, China, September 2004 (Acceptance Rate: 24.3%). (ps , pdf)

James Sinnamon and Peiyi Tang. Multipi: A java implementation of multiparty interaction for distributed computing. In Post-Conference Proceedings of the 2004 International Conference on Parallel and Distributed Processing Techniques and Applications (PDPTA'04), Las Vegas, USA, June 2004.

You can find both these papers in http://www.ualr.edu/pxtang/publications.html.www.ualr.edu/pxtang/ is my page. (I will find the file of the second paper and email it to you).

I am sorry that ANU did not treat you well and fairly and you ended up in labor work. (but as you said, the labor work makes you fit). Everything has two sides.

400

I am also sorry about your accident. It is a blessing that you still alive and the damages is minimal.

I am so glad that you have run for the mayor of Brisbane. Your website (http://candobetter.net)is impressive.

If you want to continue work on distributed applications on top multipi, we can think about something. Now, I am mainly interested in mainstrain parallel programming with Intel TBB libary.

Keep in touch and I am glad you found me.

Peiyi"

James wrote back to Prof Tang on Saturday, October 29, 2011:

"Dear Peiyi,

(I trust that you will approve of my cc'ing this to my partner/carer/editor/etc, Sheila. Of course, Sheila would never allow any e-mail address given to her to fall into the hands of others)

Great to hear from you, again!

I tried to talk to you through Skype [1, 2] a few days ago, but couldn't reach you.

Another reason, why I haven't been able to be in touch as much as I had intended to is that my time has been largely consumed trying to [b]uild a new Linux Desktop Computer.[3] I really need a computer that works properly in many more ways for me to be able to work effectively with you and others on the 'Net.

Thank you so much for he copy of that paper. I probably have an earlier version somewhere (and the Java source code) but it's great to have a more current and more succinct version.

I would be most interested in helping you to work on Multpi as you mentioned to me in the e-mail of 13 October. Could you come up with a task that would not be too challenging for me to begin with. What I aim to is to try to get multipi running again, but on my newer 64 bit Intel based PC [4, 5].

Thank you so much.

warmest regards,

James"

65   Obviously it was unlikely at this stage that he would be able to do this work, but James was not sure and wanted to try to be ready.

66   I think it was on our second visit to Brisbane after James came down to Victoria for rehab, perhaps in Christmas 2010. I remember that our mutual friend Greg Wood who was visiting, and I were fascinated to observe James's preoccupation with the progress of his bids on E-Bay during the evening meal and after. It left him no mental energy to pay attention to his visitor. He kept leaving the room for sustained periods of time to check and barely ate. He concluded a very successful deal.

67   In August 2012 we were again in Brisbane but James could not use this computer because he did not remember the password and did not know if or where he might have recorded it. He had to go on the internet and find out how to reset it. Note that he had by this time also become more at ease with his laptop and no longer felt such a need to have a personal computer in Brisbane.

68   At time of writing, I realize that I still don't know if that computer is now in operation or not. James thinks that it isn't. The fact that he has probably not tried it with another motherboard is likely related to time, interruptions, and the perceived need for the new high-end computer for possible professional-level work with his old professor.

69   Students did get fee-free access to Visual Studio through their course, but would need to pay to subscribe after they graduated.

70   "Triage" is a term used by military and civil hospitals for the process of deciding who to save first among the wounded. This process of course inevitably still means that some will die.

71   21 May 2012 Letter to Disability Unit asking for another extension

"Thomas P.

Monash Disability Liaison Unit

Dear Thomas,

I am James Sinnamon's carer and partner. We spoke once and you were very helpful.   I know that he has an extension on Assignment No.2 which is due tomorrow.   I have been aware of his difficulties doing the Components of

Internet Technology Course.   I am writing to say that today he made substantial progress after weeks of disappointing effort.

This is due largely to his being given extra time. For this reason I would like to ask you to support his being given more time, as much as it is possible to give him, so as he can keep on trying to complete as much of the course as he can.

This is what I observed him do today.

After again running into problems with his home computer, he went into the university laboratories. I accompanied him because this increases his ability to self-monitor and I felt that he might benefit from that extra support. He managed to complete an exercise which he had attempted unsuccessfully on several occasions. He then managed, from home, to make significant progress on Assignment No.2. This involved him incrementally adapting his Week 4 assignment which processes a "Books" database so that it could process his xml file of intended micropayments for websites. He is not yet ready to present Assignment 2, but he seems to be getting there.

With regard to his difficulties with programming, I confirmed to him that he needs to very carefully examine all windows and read all instructions. He also can use assistance to spot typos. It seems important to always work as 'Administrator' in Visual Studio. It is very helpful to have one Windows computer with Visual Studio to read instructions from whilst working on a second one. The light is very good in the laboratories and this may also have been important. The lack of distractions in the laboratories may also have helped.

Once James became so tired that he lost his way temporarily and might have given up. It was important that I was there to remember what he had been doing. He then, however, picked up where he had left off. In fact, his ability to work, concentrate and remember has improved since the beginning of the course, so I have some optimism that it may continue to do so over the next month or so as well."

72   Although kindness and the wish to save James the pain of total failure were obviously real motivations for Matthew, it did cross my mind that the pressure for teaching staff at universities to pass students in Australia may also have been a factor in the dedication that Matthew put into helping James. Matthew did actually do some of James's work for him, but he probably did quite a bit for numerous other students. Apparently the C# software carried many problems which required the lecturers' sophisticated skills and time to fix. Allowances were then made for the impact of problematic software on the students' work. Matthew sent James and me a draft report on his performance for James's insurers to use in their assessment of James's capacity. He intended

403

to follow this up with a formal report, but was prevented from doing this by the university administration pending the resolution of a mix-up where James had responded to academic sales pitches by signing up for a second semester without realising that he did not have the insurer's authority to do so and without understanding that he was signing some kind of financial commitment. The administrative mix-up took a couple of months to resolve, by which time the formal task of a report was out of time. Fortunately the draft report was useable and we were very thankful that Matthew had so promptly provided it.

[73]   See http://en.wikipedia.org/wiki/Windows_8 (Accessed 12 December 2014): "In February 2014, Bloomberg reported that Microsoft would be lowering the price of Windows 8 licenses by 70% for devices that retail under US$250; alongside the announcement that an update to the operating system would allow OEMs to produce devices with as little as 1 GB of RAM and 16 GB of storage, critics felt that these changes would help Windows compete against Linux-based devices in the low-end market, particularly those running Chrome OS. Microsoft had similarly cut the price of Windows XP licenses to compete against the early waves of Linux-based netbooks. Reports also indicated that Microsoft was planning to offer cheaper Windows 8 licenses to OEMs in exchange for setting Internet Explorer's default search engine to Bing. Some media outlets falsely reported that the SKU associated with this plan, "Windows 8.1 with Bing", was a variant which would be a free or low-cost version of Windows 8 for consumers using older versions of Windows. On April 2, 2014, Microsoft ultimately announced that it would be removing license fees entirely for devices with screens smaller than 9 inches, and officially confirmed the rumored "Windows 8.1 with Bing" OEM SKU on May 23, 2014.

On the information gathered by Net Applications, adoption rate in June 2014 for Windows 8.1 was at 6.61%, while the original Windows 8 was at 5.93%.

Chinese government ban

In May 2014, the Government of China banned the internal purchase of Windows 8-based products under government contracts requiring "energy-efficient" devices. The Xinhua News Agency claimed that Windows 8 was being banned in protest of Microsoft's support lifecycle policy and the end of support for Windows XP (which, as of January 2014, had a market share of 49% in China), as the government "obviously cannot ignore the risks of running OS [*sic*] without guaranteed technical support." However, Ni Guangnan of the Chinese Academy of Sciences had also previously warned that Windows 8 could allegedly expose users to surveillance by the United States government due to its heavy use of internet-based services.

In June 2014, state broadcaster China Central Television (CCTV) broadcast a news story further characterizing Windows 8 as a threat to national security. The story featured an interview with Ni Guangnan, who stated that operating systems could aggregate "sensitive user information" that could be used to "understand the conditions and activities of our national economy and society", and alleged that per documents leaked by Edward Snowden, the U.S. government had worked with Microsoft to retrieve encrypted information. Yang Min, a computer scientist at Fudan University, also stated that "the security features of Windows 8 are basically to the benefit of Microsoft, allowing them control of the users' data, and that poses a big challenge to the national strategy for information security." Microsoft denied the claims in a number of posts on the Chinese social network Sina Weibo, which stated that the company had never "assisted any government in an attack of another government or clients" or provided client data to the U.S. government, never "provided any government the authority to directly visit" or placed any backdoors in its products and services, and that it had never concealed government requests for client data."

[74] This very popular diet was the child of BBC's Dr Michael Mosely and Mimi Spencer. The concept of modified fasting can be useful but the promises that one can eat anything and as much as one likes on the non-fast days seems to be nonsense, despite the popular science film on the BBC http://www.disclose.tv/action/viewvideo/110651/BBC_Horizon_2012_Eat_Fast_and_Live_Longer/ that helped to launch the book. We have kept up our modified fasts, however, because of the principle of allowing the body to use its energy to repair its systems rather than continuously digest and, of course, because it gives you time and a different perspective. http://thefastdiet.co.uk/

[75] If you are interested in this theory there are several films with Dr Lustig (University of California Children's Hospital endocrinologist) explaining it, for instance, https://www.youtube.com/watch?v=LHlEuDHpE2k . The idea is that the only time most animals consume large quantities of fruit is in autumn and this is also when they acquire fat for winter. Autumn is classically the time when many fruits come into season at once. Dr Lustig does not go into how fruit bats or hummingbrids manage on mostly fruit diets. I am sure that he does not have the entire field covered but James and I have found his theory a very useful approach.

[76] Sheila Newman (Ed.) *The Final Energy Crisis*, 2nd edition, Pluto Press, 2008 consisted of chapters by ten scientists and one economist. I found the authors, edited their work and also wrote chapters myself and presented the work of the others. My 2002 research thesis in environmental sociology compared different historical approaches to economic policies on petroleum by Australia and France as part of its evaluation of different politico-economic systems. So I had a view from two different political systems of long-term interaction with the oil-producing countries of the world from national, international, imperial and corporate entities.

[77] James Q Wilson Ed., *The Politics of Regulation*, Harper, New York, 1980 devised a method to explain why unpopular policies and practises persisted in democracies. He classified four types of politics depending on whether the benefits and costs of policies were concentrated or diffuse: client (cb,dc), interest

group (cb,cc), majoritarian (db,dc), and entrepreneurial (db,cc). Gary Freeman described a situation he perceived in countries where high immigration has become institutionalised despite evidence that most people in these countries want less immigration. Using the Wilson method, he sought reasons for differences in immigration policies between France and Britain in *Immigrant Labor and Racial Conflict in Industrial Societies* (1979) and for those between European and other Western countries, including Australia and France in his much later article, "Modes of Immigration Politics in Liberal Democratic States" (1995). I tested this method and theory in *The Growth Lobby in Australia and its Absence in France*, Swinburn University, 2002, an MA by Research thesis in Environmental Sociology of over 100,000 words., where I identified the growth lobby in Australia and then sought to explain why there was no growth lobby in France. This work later led me to identify different land-tenure and political systems in the world, which ultimately led me back to the natural phenomenon of incest avoidance and the Westermarck Effect (causing kinship rules) in many species as well as humans. To do post graduate work in universities requires the applicant to find at least two academics who have an interest in one's work, but there was no-one in Australia with any interest in my areas of research. My theory and research were too multi-disciplinary in scope and the challenge they posed to established theory and the growth lobby made them too revolutionary to find support in conventional academia, even if someone had already been asking the same questions.

[78] The article was republished on Eye on Immigration and on Online Opinion.

[79] I'm referring to the adage, often quoted in medicine, that when you hear hoof beats you should assume they are horses, not zebras

[80] I know a lot of human beings also go for specials and overspend, but James never did this before his accident. He was unusually able to resist advertising and spending sprees. The exception was chocolate, which he often bought more of if it was on special.

[81] This note may interest people who struggle with knee instability. After tennis I noticed that my knee had swollen somewhat. It did not get much better over the next day or two, so I consulted a doctor. X-rays revealed a worn down cartilage, but the doctor and I supposed that this was an old injury from jogging when I was forty, since I had not had any symptoms (like pain on kneeling) after I had stopped jogging. I wondered about a recent visit to Brisbane where James and I had gone bushwalking with a Vietnamese friend, Ha, who is like a mountain goat but who always starts a hike off by telling us that he is feeling very old and slow. We had followed him down and then up 456 steps of an extremely steep path at Springbrook National Park. I remembered at the time feeling very physically stressed and perhaps I did my calf muscles significant harm, leaving my knee ligaments vulnerable. After some expensive physiotherapy for a torn medial ligament and a couple of months bandaging and relative inactivity, my knee

406

seemed to improve. The treatment was to strengthen the calf and thigh muscles through exercises with the aim of taking over movement and support work from the knee ligaments. I also began taking magnesium supplements for muscles. Although the ligament healed, during the next couple of months I sprained it twice through a particular movement, which was to twist my body and leg slightly whilst maintaining my foot fixed on the ground.

After the first sprain, two months post the initial torsion on the court, and about two and a half months into our low simple carbohydrate diet, I played tennis once, with a bandage on the knee to keep it from moving sideways, and had no problems. More surprising, I had far more energy than I had had on the first tennis occasion, and could run with ease. I was unpuffed after an hour on the court and it was James who called for a halt to the session. Although another slight sprain interrupted my tennis (this one caused simply by keeping my foot on the ground whilst transferring in sitting position from one chair to another at the dining room table), three weeks later I was once again able to play tennis (wearing the knee immobilizing bandage.)

Then we were in Queensland again, in mid-summer, just after Christmas. Within a day of arrival, **both** my knees started to ache! The left one – which was not the injured one – even swelled visibly. I wondered if I was reacting to local mosquito bites, to which I seem to be allergic. The heat can make your muscles ache and certainly seemed to make my knees worse. The strangest thing was that I had done nothing to strain my knees. I particularly noticed my knee pain if I sat on any of the chairs in the Brisbane house because they were all too tall for me. It was very difficult to find anywhere to work. I finished up using my laptop whilst seated between two fans on a low easy chair, which did not exacerbate the pain. I anticipated that my knees would improve once I returned to Victoria, where the heat and humidity in January and February are not quite as continuous as in Brisbane.

To my alarm, when I did return to Victoria and resumed work at my desktop computer, my knees continued to get worse. The physiotherapy exercises seemed unhelpful. Another visit to the doctor got me a diagnosis of osteoarthritis and the suggestion that I begin to take Panadol Osteo to control the pain. I gritted my teeth, bandaged both legs and played a couple of games of tennis with James. We had both improved – which is to say we occasionally managed to hit and return the ball without losing if a few times in succession – and I really enjoyed myself. However now I was thinking on a week to week basis, researching knee replacements, stem-cell cartilage replacements and knee braces, wondering how long I would be able to walk, let alone play tennis. I became very irritable and it didn't help that my knees now ached during the night and that sitting up at my pc now seemed to make my knees ache more, especially when I stood up, so that I began to avoid writing.

One day I realised that the chair I was using was too large for my frame and the long seat pushed my knees into a position that strained their sockets. I changed to a small folding chair and my knee pain on standing reduced considerably. Shortly after this I began taking soluble aspirin 100mg once daily, since theory suggests that reduced blood flow contributed to osteoarthritis by affecting the supply of blood to cartilage (which is relatively poor even in best health.)

My knee problems seemed to improve but my legs felt weak and I became gingerly about descending steps, getting up off couches, or sitting on my haunches. After James and both my parents were diagnosed as B12 deficient, I had myself tested. I tested at 450pg/mL. By this time I knew quite a lot about the measures. Whilst my doctor congratulated me on my 'superb' result, I knew that, if I had been in Japan or Germany, I would have been under the therapeutic level. So I began to take methyl B12 5000mcg sublingual lozenges (which I had to order from overseas). After a month and a half persistent problems with my balance had resolved and I could stand on one leg with my eyes closed for over 30 seconds. Prior to B12 supplements, my maximum had been around 13 seconds. After two months on supplements I realised that I had not had a knee incident for a month and that two mild ones before that had resolved almost immediately. Some mild but persistent word-finding problems and mild depression had also largely resolved. Best of all I felt so solidly in touch with the ground, so sure-footed.

But that was not the end of it. A few months later the front part of my legs swelled up in what is sometimes called 'pretibial swelling'. I had 'pitting odema' right up my shin, although I knew I had no heart problems. I experienced severe aching in both legs and the swelling increased, reaching right round my knee. Both knees ached and became unstable. Walking was problematic. I felt very tired, quite depressed, found I could not initiate or follow through on anything much. I just wanted to look out the window. Something about the mental symptoms and a kind of overall subtle swelling of the face and neck reminded me of how my mother had been for years prior to succumbing to severe, irreversible myxodema, despite her 'thyroid stimulating hormone' (TSH) results in the 'normal' range. Using forums on the internet, I found a GP who would test the level of actual thyroid hormones in patients who presented with symptoms like mine and I told him about my mother's tragic history. Sure enough, I tested borderline low in T4 and T3. He prescribed me dessicated pig thyroid, to augment until I felt well. Within a month I lost 3kg – presumably of fluids. My legs went back to their normal size and I experienced no more knee pain. Gradually, over months, I stopped fearing a knee incident and began to walk less gingerly. I played tennis with James without any knee events. I now associate my 'arthritis' with my thyroid problems. I am also very sensitive to dosage – up or down. I'm happy on one grain for the moment. My doctor suggested that I try a quarter grain more, but my leg pain, weakness, and swelling returned with a vengeance, even affecting my pelvic ligaments. I stopped any dose for a day then began my lower original dose (1 grain). Within two days all my symptoms had gone. I seem to belong to a subset of people who experience these sorts of leg symptoms in association with thyroid problems. I worked this out by asking questions on healthunlocked.com. As for my initial response to reducing simple carbohydrates and increasing B12, I assume that I must share the same syndrome as my mother and James. It is months since I took any painkillers for my knees. I keep up my B12 because my balance problems return when I stop it for a week or two.

[82] My own basic psychiatric nursing education – a three year in-hospital course - had familiarised me with the life or death, sanity or dementia, features of some other B Vitamin deficiencies with similar effects – for instance thiamine (B1) deficiency in Wernikes encephalopathy and Korsakov's psychosis (both associated with alcohol toxicity; Pellagra (Niacin B3 deficiency), which I had learned was a malnutritive condition that elderly people got from poor eating, and Beri-Beri, which was also B1 deficiency, without the alcohol component. Somehow I was never taught about B12 deficiency as a cause of neurological problems. The more I read lately, however, the more blurred the delineations between these separately discovered vitamin-deficiency brain injury and dementias seem; if you've got one, you frequently have more than one.

[83] Pachlok and Stuart's book, *Could it be B12: An epidemic of misdiagnoses*, 2nd Edition, Quill Driver Books, California, 2011, p.204, suggests that insurance streaming of diagnostic categories in the 1980s caused pernicious anaemia to be narrowed down to just a blood disorder involving enlarged red blood cells and for vitamin B_{12} injections to be associated with quackery and fraud.

> *"Many physicians remember the days when thousands of patients got B12 shots whether they needed them or not. Their justified scorn for this practice leads them to overreact now by making the opposite mistake: failing to realize that B12 deficiency is, indeed, a real, common, and serious medical disorder.*
>
> *When diagnosis-related groups (DRGs) came about in the early 1980s, physicians who billed for B_{12} shots had to prove that their patients had 'pernicious anemia' or another medical reason for their malabsorption syndrome, or the doctors could be accused of fraud. Because many physicians weren't testing their patients for B12 deficiency to begin with, they discontinued using B12 injections for fear of legal consequences.*
>
> *Some of these doctors had created more revenue for their practice by administering cheap B_{12} shots and then billing patients' insurance. Many of them stopped B_{12} treatment because they equated B_{12} therapy with placebo, having only given the shots to make money. Unknowingly, these physicians had been doing many patients some good (those with a true B_{12} deficiency), while the others were not harmed. But the practice has unfortunately given vitamin B_{12} a bad name within medical circles."*

[84] In fact, I have since learned that a host of substances can interfere with the uptake of Vitamin B12, as well as a variety of conditions. There are commonly used drugs, including some used for diabetics – for instance metformin - that

greatly increase the risk of B12 deficiency and consequent demyelination of the nervous system, with all the features that go with it, loss of use of limbs, insanity and death. Antacids and nitrous oxide (laughing gas) are two other dangerous drugs where B12 is concerned. The risk rises with contraceptive pills and becomes a sure thing where parts of the intestine have been removed in weight-loss surgery.

[85] As a psychiatric nurse, after dealing with brain injury, I can see how mis-used and abused psychiatric concepts are. It seems that individuals, GPs, the man in the street, are all very much inclined to suppose that much of what they experience in terms of pain and dysfunction is most probably due to overactive imaginations or personality problems.

[86] In Australia the reference range is 150picomols per millilitre and 750 picomols per millilitre. In Japan and Germany the reference range starts at 500 picomols per millilitre. Despite Australia's ridiculously low reference ranges, the literature recommends treatment where there are neuro-symptoms, but a majority of doctors seem to ignore or not notice this protocol. Many believe that pernicious anaemia is not present where there is no enlargement of the red blood cells and ignore the fact that in many cases the red blood cells never do enlarge, and, even if they eventually do, irreversible brain injury can occur well before that happens. If B12 deficiency occurs in the presence of iron deficiency, the enlargement of red blood cells will be compensated by the effect of shrinkage caused by the iron deficiency.

[87] He continued to attend Artwell, but noted that attendance had dwindled to almost nothing, but for his own and one other person's. Recently a session was cancelled with little warning, due to the space having been allocated to some other event.